A. G. Ramm

Theory and Applications of Some New Classes of Integral Equations

Springer-Verlag
New York Heidelberg Berlin

A. G. Ramm

Theory and Applications of Some New Classes of Integral Equations

Springer-Verlag
New York Heidelberg Berlin

A. G. Ramm
Department of Mathematics
University of Michigan
Ann Arbor, Michigan 48109 / USA

AMS Subject Classifications: 45GXX, 93C22, 93E99, 94A99, 94C99, 78A40, 78A45
70C05, 81F99

Library of Congress Cataloging in Publication Data

Ramm, Alexander.
Theory and applications of some new classes of
integral equations.

Bibliography: p.
Includes index.
1. Integral equations. I. Title.
QA431.R35 515.4'5 80-25387

Printed in the United States of America.

9 8 7 6 5 4 3 2 1

ISBN 0-387-90540-5 Springer-Verlag New York Heidelberg Berlin
ISBN 3-540-90540-5 Springer-Verlag Berlin Heidelberg New York

TO MY PARENTS

PREFACE

This book is intended for students, research engineers, and mathematicians interested in applications or numerical analysis. Pure analysts will also find some new problems to tackle. Most of the material can be understood by a reader with a relatively modest knowledge of differential and integral equations and functional analysis. Readers interested in stochastic optimization will find a new theory of practical importance. Readers interested in problems of static and quasi-static electrodynamics, wave scattering by small bodies of arbitrary shape, and corresponding applications in geophysics, optics, and radiophysics will find explicit analytical formulas for the scattering matrix, polarizability tensor, electrical capacitance of bodies of an arbitrary shape; numerical examples showing the practical utility of these formulas; two-sided variational estimates for the polarizability tensor; and some open problems such as working out a standard program for calculating the capacitance and polarizability of bodies of arbitrary shape and numerical calculation of multiple integrals with weak singularities. Readers interested in nonlinear vibration theory will find a new method for qualitative study of stationary regimes in the general one-loop passive nonlinear network, including stability in the large, convergence, and an iterative process for calculation the stationary regime. No assumptions concerning the smallness of the nonlinearity or the filter property of the linear one-port are made. New results in the theory of nonlinear operator equations form the basis for the study.

Readers interested in the theory of open systems will find a general numerical method for calculating losses in quantum mechanics and diffraction theory. Nonselfadjoint integral equations are studied. Readers interested in stable solutions of equations of the first kind will find a method with effective error estimates with applications to the antenna synthesis problem, and an explicit formula for approximation of a given function by entire functions of exponential type. There is a list of unsolved problems in pure and applied mathematics.

ACKNOWLEDGEMENTS

I am thankful to Janet Vaughn who typed the manuscript; to Kate MacDougall for her expert typing of the final manuscript for Springer-Verlag; to Walter Kaufmann-Buhler, mathematics editor of Springer-Verlag, for his help; to Air Force Office of Scientific Research (AFOSR) for its financial support. Special thanks are due to Dr. R. Buchal, Director of the Physical Mathematics Division of AFOSR, for his interest and support. Also I thank Walter Thirring for his hospitality during the Spring 1979 when I was at the University of Vienna. Last but not least, thanks are due to the editor of Springer-Verlag who made many linguistic corrections in the manuscript.

A. G. Ramm

Department of Mathematics
University of Michigan
Ann Arbor, MI 48109

July 4, 1980

TABLE OF CONTENTS

INTRODUCTION

There are many books and papers on integral equations. So the author should first explain why he has written a new book on the subject. Briefly, the explanation is as follows. Almost all the results presented in this book are new. Some new classes of integral equations are defined and investigated in this book. All the equations are closely connected with problems of physics and technology of great interest in applications. Some of the problems which have remained unsolved for years are solved in this book for the first time. Here we mention only three of them (Chapters 1-3);

(1) The basic integral equation of estimation theory for random fields and vector processes is investigated and, in a way, solved explicitly for the general case of an arbitrary smooth bounded domain of signal processing;

(2) Some explicit formulas for the scattering matrix in the problem of wave scattering by a small body of arbitrary shape are obtained; so the classical work initiated by Lord Rayleigh in 1871 (Rayleigh [2]) is in a way completed;

(3) Periodic and almost periodic stationary regimes in a general passive one-loop network are investigated without

any assumptions about the smallness of the nonlinearity or the filter property of the linear one-port. An iterative process for calculating the stationary regime in such a network is given.

It must be emphasized that all results of importance for physics or technology are obtained by means of some new mathematical theory, result, or idea. The table of contents gives an idea of the questions considered. The main results of each of the first three chapters are summarized in the first sections of those chapters. The remaining sections contain proofs, examples, and applications.

Let us now give a brief account of the main results presented in the book.

In Chapter 1, class $\mathscr{R}$ of integral equations is defined and investigated. These equations are of the type

$$Rh = \int_D R(x,y)h(y)dy = f(x), \quad x \in \overline{D} \subset \mathbb{R}^n. \tag{0.1}$$

The kernel $R(x,y) \in \mathscr{R}$ is, roughly speaking, the kernel of a positive rational function of an elliptic selfadjoint operator L on $L^2(\mathbb{R}^n)$. Equation (0.1) is the basic equation of stochastic optimization theory. It seems that even for translation invariant kernels $R(x,y) = R(x-y)$, equation (0.1) has not been investigated for an arbitrary domain D in $\mathbb{R}^n$, $n > 1$. Some generalization of the above description of the kernels of class $\mathscr{R}$ can be obtained if, instead of a single selfadjoint operator L, one takes a family of m commuting selfadjoint operators $L_1, L_2, \ldots, L_m$, where R is, roughly speaking, a positive rational function of the operators $L_1, \ldots, L_m$. For example, if $L_j = -iD_j$, $1 \leq j \leq n$,

$$D_j = \partial/\partial x_j, \quad \tilde{R}(\lambda) = P(\lambda)Q^{-1}(\lambda), \quad \lambda \in \mathbb{R}^n,$$

where $\tilde{R}(\lambda)$ is the Fourier transform of $R(x)$, then $R(x)$ can be considered as a rational function of the selfadjoint operators L_j, $1 \le j \le m$. We study also the case of a matrix kernel $R(x,y)$ without using the cumbersome factorization theory of matrix functions. In Section 5, a new method of approximate analytic solution of some integral equations of the first kind is given. In Section 6, the asymptotic distribution of the eigenvalues of integral operators of class $\mathcal{R}$ is studied. This study is partly based on the general results obtained in Section 7, where we solve the following problem. Suppose A is a closed linear operator with discrete spectrum $\lambda_n(A)$, $B = A+T$. Under what assumptions concerning T does $\lambda_n(B)\lambda_n^{-1}(A) \to 1$ as $n \to \infty$? In Section 8, the results of Section 4 are applied to nonlinear signal estimation.

In Chapter 2, a class of linear operator equations is selected. Bounded operators in this class have a semisimple characteristic value as the lowest point of their spectrum. There are many examples of operator equations with such operators, including equations arising in potential theory. Some stable iterative processes to calculate the solutions of such equations are given. Some approximate analytic formulas for linear functionals of the solutions are also obtained. These formulas allow one to calculate, for example, the electrical capacitance and the polarizability tensor of a single body or a set of bodies of arbitrary shape with a prescribed accuracy. These results permit us to obtain an approximation analytic formula for the scattering matrix in the problem of wave scattering by small bodies of arbitrary shape. This result is of importance in optics, radiophysics, and in other fields. It makes it possible to solve practically some inverse problems, for example, an inverse problem of radiomeasurement theory. The theory developed allows one to consider wave scattering in a medium, consisting of many ($\sim 10^{23}$) small bodies. By a small body we always mean a body with characteristic dimension considerably smaller than the wavelength of the initial wave. In Section 4 a variational method for obtaining two-sided estimates for capacitance and polarizability is given. In particular, necessary and sufficient conditions are given for the Schwinger stationary principle to be extremal. A stable iterative process to solve integral equations of the first kind with positive kernels is also given with its applications to electrostatics and other static fields.

In Chapter 3 we investigate a class of nonlinear equations of the type

$$Tu = Au + Fu = J, \tag{0.2}$$

where A is a closed linear operator on a Hilbert (or reflexive Banach) space and F is a bounded nonlinear operator. Some new conditions on A are imposed which permit us to give a very simple method of investigating the properties of solutions of equation (0.2) if the operator T is monotone. We also give an iterative process to calculate the solution of equation (0.2). Our assumptions are very natural from the phy cal point of view. In a way the results obtained are final. Under these assumptions equation (0.2) describes stationary regimes in a general nonlinear passive one-loop network, consisting of a linear one-port L with admittance operator A, a nonlinear one-port N with voltage-current characteristic $i = Fu$ and external source of current J. The passivity of the network implies the monotonicity of T in equation (0.2). No assumptions concerning the smallness of the nonlinearity or the filter property of L are made. Although the literature on nonlinear vibrations is very extensive, most of it assumes either the smallness of the nonlinearity of the filter property of L. We give a new general method of investigation and obtain new results. Roughly speaking, it is proved that in network described above there exists a single stationary periodic or almost periodic regime depending on the external source J, and this regime can be calculated by means of the iterative process. The absolute stability of this regime and existence of convergence in the network are also studied. In Section 4 the case of discontinuous nonlinearity, which is important from the practical point of view, is treated. In Section 5 we study the stationary regime, taking into account the nonlinearity of the final cascade of an amplifier with a feedback. This problem has not been discussed in the literature as far as the author knows.

In Chapter 4 we study some mathematical and physical questions of the theory of open systems. In Section 1 a general numerical method to calculate the complex poles of Green's function is given. The method is applied to nonrelativistic potential scattering and to scalar diffraction. In Section 2 an integral equation with the nonselfadjoint kernel $\exp\{ib(x,y)\}$ is studied. Some two-sided estimates for diffraction losses in open confocal resonators with arbitrary mirror shape are obtained. It is shown at the physical level of strictness that of all centrally symmetric mirrors with fixed area, the circle mirror has minimal diffraction losses for the zero mode. In Section 3 we give some facts concernin the equation

$$B\psi = \int_S \frac{\exp(ik\,|x-y|)}{4\pi|x-y|}\,\psi\,dy = \lambda\psi(x), \quad x \in S, \tag{0.3}$$

where S is a smooth closed surface in $\mathbb{R}^3$. This equation arises in diffraction theory. The kernel in (0.3) is non-selfadjoint. It is interesting to know whether the root functions of the operator B form a fundamental set in $L^2(S$ and whether they form a basis in $L^2(S)$ in some sense. Thes questions are answered in Section 3.

In Chapter 5 some integral equations arising in the inverse radiation problem (antenna synthesis theory) are studie In Section 1 the equation

$$Au = f \tag{0.4}$$

is studied, where A is a closed linear operator with the location of its spectrum known. We give a stable method to calculate an approximate solution u_δ to equation (0.4) if f_δ, $\|f-f_\delta\| \leq \delta$ is known. Moreover, the error $\|u-u_\delta\|$ of the approximation is effectively estimated. In Section 2 the uniqueness and existence of solutions to the antenna synthesi problem are studied. Roughly speaking, the problem is as follows. Given a vector field $f(n,\nu,k)$, when is it possible to find a current distribution so that f is the radiation pattern for this current distribution with the prescribed accuracy, where n, ν are unit vectors and k is the wave number. In Section 3 an analytic formula to solve the follow ing approximation problem is given. Let a function $f(t)$ be given in some domain $\Delta \subset \mathbb{R}^N$. Denote by W_D the class of entire functions $\psi(t)$, $t \in C^N$, such that $\psi(t) \in L^2(\mathbb{R}^N)$ and $\tilde{\psi}(x) = 0$ if $x \in \mathbb{R}^N \setminus D$, where $\tilde{\psi}(x)$ is the Fourier transform of $\psi(t)$. Let $\varepsilon > 0$ be a given number. It is necessary to find $f_\varepsilon \in W_D$ so that $\|f-f_\varepsilon\| < \varepsilon$, where $\|\cdot\|$ denotes the norm in $L^2(\Delta)$ or $C(\Delta)$. This problem is of interest not only from the mathematical point of view but also for applications, including apodization theory and antenna synthesis. We give the following explicit formula to solve this problem

$$f_\varepsilon = \int_\Delta g_n(t-y)f(y)\,dy, \tag{0.5}$$

where

$$g_n(y) = \left[\frac{1}{|D|}\int_D \exp\left\{-\frac{i(t,y)}{2n+N}\right\}dt\right]^{2n+N}\left(1 - \frac{|y|^2}{R^2}\right)^n\left(\frac{n}{\pi R^2}\right)^{\frac{1}{2}N} \tag{0.6}$$

$|D|$ = meas D, and $R > 0$ is chosen such that the disk $|y| \leq R$ contains all the vectors $t-y$, $t,y \in \Delta$, $n = n(\varepsilon)$. A constructive error estimate is given. In Section 4 nonlinear problems of synthesis theory are considered. A typical example leads to the equation

$$F^2(k+k_o) - F^2(k-k_o) = g(k), \tag{0.7}$$

where $F(k) = \int_{-\pi}^{\pi} j(x)\exp(ikx)\,dx$, k_o is some given small number, $g(k)$ is a given function which is called the pelengation characteristic in applications, $j(x)$, the current distribution in applications, is to be found.

The nonlinear integral equation (0.7) should be investigated. Given a function $g(k)$ one must ask whether equation (0.7) has a solution, whether it has more than one solution, how to calculate the solution, and how to find $j_\varepsilon(x)$ so that the corresponding $g_\varepsilon(k)$ will not differ much from g, $\|g-g_\varepsilon\| < \varepsilon$ in some appropriate norm. In Section 5 the problem of finding the scattering potential and the shape of the reflecting boundary from the scattering data is considered. In Section 6 an optimal solution to some antenna synthesis problems is given.

The concluding section of each chapter contains several unsolved problems.

In Appendix 1 the integral equation

$$\int_{-1}^{1} \frac{\psi(t)}{|t-x|}\, dt = f(x), \quad x > 1 \tag{0.8}$$

is stably solved by means of an iterative process. The method is also valid for the multidimensional equation of the same type:

$$\int_D \frac{\psi(t)\,dt}{|t-x|} = f(x), \quad x \in \Delta, \tag{0.9}$$

where $D, \Delta \subset \mathbb{R}^N$ are given bounded domains, Δ is sufficiently large, and $D \cap \Delta = \emptyset$. In Appendix 2 iterative processes for solving interior and exterior boundary-value problems are given. In Appendices 3-11 some questions connected with the material of Chapters 2 and 4 are discussed.

Each chapter contains a brief bibliographical note. Most of the results were obtained by the author and are presented in book form for the first time. The author hopes that the book will be interesting to a wide circle of readers, including mathematicians interested in analytical methods for estimation problems, signal detection, etc., in numerical solutions to the problems of static and quasistatic field theory (electrostatics and magnetostatics, heat and mass transfer, hydrodynamics, optics of muddy mediums, radiophysics and geophysics, a subject in which the scattering by small bodies is a phenomenon of prime physical importance), in solving the problems of nonlinear vibrations in real network systems; and specialists in computational physics interested in some questions of open resonator theory, diffraction theory, etc. Each chapter can be studied separately and the author hopes that only the first chapter requires a little more prior knowledge of the spectral theory of differential operators than it is reasonable to expect from the reader. In all cases the necessary references are given. The author will be more than satisfied if the theory presented in this book will be applied by some research engineers to their problems. He believes that there are many possibilities for such applications of the developed methods.

A few words about the organization of the material follow. The preface presents a brief account of the questions discussed in the book. The first three chapters each deal with one question of interest for both pure mathematics and applications. The results are presented in the first section of each chapter, and proofs are given in other sections. A summary of the results presented in Chapters 1-3 is given in Ramm [101]. In Chapters 4 and 5 various types of integral equations arising in applications are studied. Near the end of each chapter, a list of research problems is given, many of which are of immediate interest in applications. A brief bibliographical note concludes each chapter.

The bibliography is incomplete because we treat many problems of differing natures. Nevertheless the author hopes that the presentation of the material is self-contained, at least in the first three chapters. In the bibliography, MR 34 #1112, for example, stands for a review number in *Math. Rev.*, PA 1123y (1973) stands for a review number in *Phys. Abstracts* appearing in 1973, and EEA 4750 (1972) stands for a review number in *Electric Electronics Abstracts* appearing in 1972.

We do not give a subject index because of the variety of problems treated. §4.3 means section 3 of Chapter IV.

We denote the end of a proof by □.

CHAPTER I

INVESTIGATION OF A NEW CLASS OF INTEGRAL EQUATIONS AND APPLICATIONS TO ESTIMATION PROBLEMS (FILTERING, PREDICTION, SYSTEM IDENTIFICATION)

1. Statement of the Problems and Main Results

Kolmogorov [1] initiated the study of filtering and extrapolation of stationary time series. These and other related problems were studied by N. Wiener in 1942 for stationary random processes and his results were published later in Wiener [1]. The basic integral equation of the theory of stochastic optimization for random processes is

$$Rh = \int_{t-T}^{t} R(x,y)h(y)dy = f(x), \quad t-T \leq x \leq t, \tag{1.1}$$

where $R(x,y)$ is a nonnegative definite kernel, a correlation function, $f(x)$ is a given function, and $T > 0$ is a given number. In Wiener [1] equation (1.1) was studied under the assumptions that $R(x,y) = R(x-y)$ and $T = +\infty$. We note that in applications T is the time of signal processing and the assumption about the kernel means that only stationary random processes were studied in Wiener [1]. Under these and some additional assumptions concerning the kernel $R(x)$ a theory of the integral equation (1.1), now widely known as the Wiener-Hopf method, was given in Wiener-Hopf [1].

Their results were developed later in Krein [1], Gohberg-Krein [1], and Gohberg-Feldman [1]. Equation (1.1) for $T < \infty$ and $R(x,y) = R(x-y)$ was studied in Yaglom [1], Wiener-Masani [1], Pisarenko-Rosanov [1], Zadeh-Ragazzini [1], Slepian [1], Youla [1], and the results are partly reported in Van Trees [1]. In these papers it was assumed that $\hat{R}(\lambda) = (2\pi)^{-1}\int_{-\infty}^{\infty} R(x)\exp(-i\lambda x)dx = P(\lambda)Q^{-1}(\lambda)$, where $P(\lambda), Q(\lambda) > 0$ are polynomials. Under these assumptions, some analytic formulas to solve equation (1.1) were obtained. Equation (1.1) with matrix kernel $R(x-y)$ and $T = +\infty$ was studied in Gohberg-Krein [1]. The main tool in this paper is the factorization theory for matrix functions. Even for rational matrix functions this method leads to difficult calculations. In applications the cases of nonstationary processes and of random field estimation problems are of great importance. In the latter case the basic equation is

$$Rh = \int_D R(x,y)h(y)dy = f(x), \quad x \in \bar{D} \subset \mathbb{R}^r, \quad r > 1. \quad (1.2)$$

This equation has not been studied in the mathematical and physical literature, as far as the author knows, even in the case $R(x,y) = R(x-y)$. Nevertheless such a study is of great interest to the people working with optimal filters in optics, geophysics, etc.

We define and investigate a new class $\mathscr{R}$ of kernels for which equation (1.2) can be solved analytically and investigated theoretically, and also give an approximate analytical solution to this equation. The kernels $R(x,y) = R(x-y)$ with rational Fourier transform are very special representatives of the class $\mathscr{R}$. In applications the kernel in equation (1.2) is such that the corresponding operator R is compact in

$L^2(D)$. So equation (1.2) is of the first kind. In the literature there are many papers in which the following problem was studied: Given f_δ, $\|f_\delta - f\| < \delta$, and assuming that $\text{Ker } R = \{0\}$, $f \in \text{im } R$, one must find $h_\delta \in L^2(D)$ such that $\|h - h_\delta\| \leq \varepsilon(\delta) \to 0$. Some results in this direction are summarized in Tihonov-Arsenin [1]. In the stochastic optimization and signal estimation problems the situation is entirely different. Equation (1.2) has no solution in $L^2(D)$. Instead, the solution is to be found in the appropriate space of distributions. The solution is usually a sum of a smooth function and a singular function, a distribution with support included in $\partial D = \Gamma$. There are many solutions to equation (1.2) in the space of distributions. The problem is how to reasonably select a single one. The anser we give to this question is very natural from the physical point of view. We describe all solutions to equation (1.2) in the space of distributions, we prove that there exists a unique solution of equation (1.2) of minimal order of singularity, and we give formulas to calculate this unique solution and its order of singularity. Similar results are also obtained for a matrix kernel. We now briefly remind the reader how equation (1.2) arises in estimation problems. Let us consider this simple situation: A signal $u = s + n$, where n is noise and s is the useful signal, should be processed so that $\varepsilon = \mathscr{D}[Lu - s] = \min$, $\mathscr{D}$ is the symbol of variance, L is a linear operator to be found, given by $Lu = \int_D h(x,y)u(y)dy$. Here D is the domain of signal processing, and h is the so-called weight function. This function is to be found so that $\varepsilon = \min$. It is not difficult to show that equation (1.2) is a necessary condition on h. The unimportant dependence of $h(x,y)$ on the second argument is omitted in equations (1.1) and (1.2). But it is important to emphasize that $\varepsilon < \infty$ only if h is the solution of minimal order of singularity to equation (1.2). This solution is stable in some sense. More precisely, we prove that the operator R in (1.2) is a homomorphism of $H_{-\alpha}$ onto H_{α}, where $\alpha \geq 0$ is some integer and $H_{\alpha} = H_{\alpha}(D)$ is the Sobolev space. In this section we formulate the main results, which will be proved in other sections. Some of these results were obtained in Ramm [1]-[17].

1. Consider a bounded domain $D \subset \mathbb{R}^r$, $r \geq 1$, with sufficiently smooth boundary $\partial D = \Gamma$, $\bar{D} = D \cup \Gamma$. The smoothness conditions are such that the theorems of imbedding and continuation are valid for Sobolev spaces of functions with domain D. Let Ω be the domain $\mathbb{R}^r \setminus \bar{D}$. If $r = 1$ the domain D is the segment $(t-T,t)$. Let L be a formally self-adjoint elliptic differential operator of order s defined on the set $C_0^\infty(\mathbb{R}^r)$ and let L be the corresponding self-adjoint elliptic operator on $H = L^2(\mathbb{R}^r)$. It is known that s is even if $r \geq 3$ (see, for example, Beresanskij [1] or Schechter [1]). The coefficients of the operator L are assumed to be sufficiently smooth. The usual requirements concerning the smoothness of the coefficients of L can be found in Beresanskij [1]. We do not want to specify these requirements because they will not be used explicitly in what follows. Let $\Phi(x,y,\lambda)$ and $d\rho(\lambda)$ be the spectral kernel and spectral measure of the operator L and let Λ be the spectrum of L. We say that the kernel $R(x,y) \in \mathscr{R}$ if

$$R(x,y) = \int_\Lambda P(\lambda)Q^{-1}(\lambda)\Phi(x,y,\lambda)d\rho(\lambda) \tag{1.3}$$

where $P(\lambda)$, $Q(\lambda) > 0$ for $\lambda \in J$ are polynomials, $J = (-\infty,\infty)$, $\deg P(\lambda) = p$, and $\deg Q(\lambda) = q$. The convergence of the integral in (1.3) is understood as is customary in distribution theory. If $p = q$ the kernel (1.3) is the kernel of an operator $cI + A$, where $c = \text{const} > 0$, I is the identity operator in H, and A is a compact operator in H with the kernel also given by (1.3) but with $\deg P(\lambda) < \deg Q(\lambda)$. The following generalization is useful from a practical point of view and requires no new ideas. Let $L_1,\ldots,L_m$ be a system of commuting selfadjoint differential operators in H. Then there exists a spectral measure $d\rho(\lambda)$ and spectral kernel $\Phi(x,y,\lambda)$, $\lambda = (\lambda_1,\ldots,\lambda_m)$, such that a function of the operators $L_1,\ldots,L_m$ is given by

$$F(L_1,\ldots,L_m) = \int_\Lambda F(\lambda)\psi(\lambda)d\rho(\lambda),$$

for any $F(\lambda) \in L^2(\Lambda,d\rho(\lambda))$, where Λ is the support of the measure $d\rho(\lambda)$ and ψ is the operator-valued function with kernel $\Phi(x,y,\lambda)$.

Convolution kernels with rational Fourier transforms are a very special case of the kernels of class $\mathscr{R}$.

For example, let $r = 1$, $L = -i\partial$, $\partial = d/dx$. Then $\Phi(x,y,\lambda)d\rho(\lambda) = (2\pi)^{-1}\exp\{i\lambda(x-y)\}d\lambda$. Let us note that

$$\int_\Lambda \Phi(x,y,\lambda)d\rho(\lambda) = \delta(x-y),$$

where $\delta(x)$ is the delta function and the operator $F(L)$ has the kernel

$$F(L) \sim \int_\Lambda F(\lambda)\Phi(x,y,\lambda)d\rho(\lambda),$$

where $F(\lambda) \in L^2(\Lambda,d\rho)$ and the symbol $\sim$ here denotes the words "has the kernel". If $L_j = -i\partial_j$, $\partial_j = \partial/\partial x_j$, $1 \leq j \leq r$, then $\Phi(x,y,\lambda)d\rho(\lambda) = (2\pi)^{-r}\exp\{\lambda\cdot(x-y)\}d\lambda$, $\lambda \in \mathbb{R}^r$, where $\lambda\cdot x$ is the inner product in $\mathbb{R}^r$. Given a selfadjoint ellipt operator L one can construct the kernels belonging to $\mathcal{R}$. For example, if $L = -\partial^2$, $\partial = d/dx$, $H = L^2(0,\infty)$ and boundary condition is $u'(0) = 0$, then

$$R(x,y) = R_1(x,y) = \frac{1}{2}[A_1(|x+y|) + A_1(|x-y|)],$$

$$A_1(t) = 2\pi^{-1}\int_0^\infty P(\lambda)Q^{-1}(\lambda)\cos(t\sqrt{\lambda})(2\sqrt{\lambda})^{-1}d\lambda.$$

For the boundary condition $u(0) = 0$ and the same differential expression in H we have

$$R(x,y) = R_2(x,y) = \frac{1}{2}[A_2(|x+y|) - A_2(|x-y|)],$$

$$A_2(t) = 2\pi^{-1}\int_0^\infty 2\,\sin^2(t\sqrt{\lambda}/2)P(\lambda)Q^{-1}(\lambda)^{-1}d\lambda.$$

In both examples $P(\lambda),Q(\lambda) > 0$ are polynomials, $d\rho(\lambda) = (2\sqrt{\lambda})^{-1}d\lambda$, $\Phi_1(s,y,\lambda) = 2\pi^{-1}\cos(x\sqrt{\lambda})\cos(y\sqrt{\lambda})$, $\Phi_2(x,y,\lambda) = 2\pi^{-1}\sin(x\sqrt{\lambda})\sin(y\sqrt{\lambda})$. If $L = -\partial^2 + (\nu^2 - \frac{1}{4})x^{-2}$, $\nu \geq 0$, $d\rho(\lambda) = \lambda d\lambda$, $\Phi(x,y,\lambda) = \sqrt{xy}\, J_\nu(\lambda x)J_\nu(\lambda y)$, then $R(x,y) = R_3(x,y) = \int_0^\infty P(\lambda)Q^{-1}(\lambda)\sqrt{xy}\, J_\nu(x\lambda)J_\nu(y\lambda)\lambda d\lambda$ is the kernel of class $\mathcal{R}$. Consider the equation $\int_a^t R(xy)h(y)dy = f(x)$, $a \leq x \leq t$. If we put $y = \exp(\eta)$, $x = \exp(-\xi)$, $q(\eta-\xi) \equiv R\{\exp(\eta-\xi)\}$, $\phi(\eta) = \eta\,\exp(\eta)$, $F(\xi) = f(\exp(-\xi))$, $c = \ell n\, t$, $b = \ell n\, a$ and assume $\tilde{g}(\lambda) > 0$ to be rational then we come t the equation of class $\mathcal{R}$:

$$\int_c^b g(\eta-\xi)\phi(\eta)d\eta = F(\xi).$$

If $R(x,y) = (\exp(-a|x-y|)/4\pi|x-y|)$, $a > 0$, $x,y \in \mathbb{R}^3$, $P(\lambda) = Q(\lambda) = \lambda^2 + a^2$, $\Phi(x,y,\lambda)d\rho(\lambda) = (2\pi)^{-3}\exp\{i\lambda\cdot(x-y)\}d\lambda$, $\lambda \in \mathbb{R}^3$, then $R(x,y) \in \mathcal{R}$, $L_j = -i\partial_j$, $1 \leq j \leq 3$. Let us note that if $\lambda \in \mathbb{R}^1$, $P(\lambda),Q(\lambda) > 0$ for $\lambda \in J$, then there are factorizations $P(\lambda) = a_+(\lambda)a_-(\lambda)$, $Q(\lambda) = b_+(\lambda)b_-(\lambda)$, where $\deg a_\pm(\lambda) = \frac{1}{2}p$, $\deg b_\pm(\lambda) = \frac{1}{2}q$, the roots of $a_+(\lambda)$, $b_+(\lambda)$ are in the upper half-plane $\mathrm{Jm}\,\lambda > 0$, the roots of $a_-(\lambda)$, $b_-(\lambda)$ are in the lower half-plane $\mathrm{Jm}\,\lambda < 0$. One says that a distribution h has order of singularity $\sigma = \sigma(h)$ in domain D if $h = D^\sigma h_\sigma$, $h_\sigma \not\equiv 0$ is some locally integrable function in D, $D^j = \partial^{|j|}/\partial x_1^{j_1} \ldots \partial x_r^{j_r}$, $|j| = j_1 + \ldots + j_r$. Here and below differentiation is to be understood in the sense of distribution theory. Actually, it will be convenient to use the space $H_{-\sigma}$ as the space of distributions. In this case σ is the order of singularity of $h \in H_{-\sigma}$. The definition of the scale of spaces H_t can be found in many books, for example in Hörmander [1]. We recall some basic properties of these spaces: $H_o = L^2(D)$, $H_{t'} \supset H_t$ if $t > t'$, and the imbedding operator $i\colon H_t \to H_{t'}$ is bounded; if the domain D is bounded this operator is compact for $t > t'$; the operator D^α is bounded as an operator from $H_{t+|\alpha|}$ into H_t: $H_t(D)$ consists of the restrictions to D of the functions in $H_t(\mathbb{R}^r)$; if a smooth closed surface Γ divides the whole space $\mathbb{R}^r$ into inner part D and outer part $\mathbb{R}^r \setminus D = \Omega$, N is the outward normal pointing to Γ, $u^+ \in H_t(D)$, $u^- \in H_t(\Omega)$, then

$$u = \begin{cases} u^+ & \text{in } D \\ u^- & \text{in } \Omega \end{cases}$$

belongs to $H_t(\mathbb{R}^r)$ iff $\partial_N^j u^+|_\Gamma = \partial_N^j u^-|_\Gamma$, $0 \le j \le t-1$, where $\partial_N^j = \partial^j/\partial N^j$; the spaces H_{-t} and H_t are dual relative to H_o, $H_t \subset H_o \subset H_{-t}$. (This means that the space H_{-t}, $t \ge 0$, can be described as the completion of H_o with respect to the norm $|f|_{-t} = \sup\limits_{g \in H_t} \frac{(f,g)_o}{|g|_t}$.)

Theorem 1. Let the kernel of equation (1.2) be of the form (1.3) with $P(\lambda) = 1$. Let $f(x)$ be an integrable function in $\overline{D}$ and let

$$F(x) = \begin{cases} f(x) & \text{in } D, \\ u(x) & \text{in } \Omega = \mathbb{R}^r \setminus D, \end{cases} \tag{1.4}$$

where $u(x)$ is a solution to the boundary value problem

$$\begin{cases} Q(L)u = 0 \quad \text{in } \Omega, \\ \partial_N^j u|_\Gamma = \phi_j(t), \quad 0 \le j < \frac{1}{2}sq - 1; \quad u(\infty) = 0, \end{cases} \tag{1.5}$$

$s = \text{ord } L$, $q = \deg Q(\lambda)$, and $\phi_j(t)$ is some arbitrary sufficiently smooth functions. Then the set of solutions to equation (1.2) in the space H_{-sq} with sing supp $h = \Gamma$ can be described by the formula

$$h(x) = Q(L)F(x), \quad \sigma(h) \le sq. \tag{1.6}$$

There exists a unique solution to equation (1.2) with minimal order of singularity. This minimal order of singularity is at most $\frac{1}{2}sq$ and the solution is given by formula (1.6) if the function $F(x)$ is defined by formula (1.4), $u(x)$ is the solution to the boundary value problem (1.5) where $\phi_j(t) = \partial_N^j f|_\Gamma$, $0 \le j \le \frac{1}{2}sq - 1$, and f is assumed smooth enough so that the problem (1.5) has solution $u \in H_{\frac{1}{2}sq}(\Omega)$. The map $R^{-1}: f \to h$ is a homomorphism of $H_{\frac{1}{2}sq}$ onto $H_{-\frac{1}{2}sq}$.

Consider the equation

$$\int_D S(x,y)h(y)dy = g(x), \quad x \in \overline{D}, \tag{1.7}$$

where $S(x,y) = \int_\Lambda Q^{-1}(\lambda)\Phi(x,y,\lambda)d\rho(\lambda)$. Let $p = \deg P(\lambda)$ be positive.

Theorem 2. *Let the kernel of equation* (1.2) *be of form* (1.3), g *be a solution to the equation* $P(L)g = f$ *in* D, $f \in H_{\frac{1}{2}s(q-p)}$. *Then there exists a solution to equation* (1.2) *with minimal order of singularity* σ, $\sigma \le \frac{1}{2}s(q-p)$. *This solution is given by the formula*

$$h(x) = Q(L)G, \quad G = \begin{cases} g_o(x) + v(x), & x \in \overline{D}, \\ u(x), & x \in \Omega, \end{cases} \tag{1.8}$$

where $g_o \in H_{\frac{1}{2}s(p+q)}$ *is the solution to the equation* $P(L)g = f$ *in* D *and functions* $u(x)$, $v(x)$ *are the solution to the problem*

$$\begin{cases} Q(L)u = 0 \ \text{ in } \ \Omega, \quad P(L)v = 0 \ \text{ in } \ D, \\ \partial_N^j u|_\Gamma = \partial_N^j(g_o+v)|_\Gamma, \quad 0 \le j \le \frac{1}{2}s(p+q)-1; \quad u(\infty) = 0. \end{cases} \tag{1.9}$$

The map $R^{-1}: f \to h$, *where* h *is given by formula* (1.8) *is a homomorphism of the space* $H_{\frac{1}{2}s(q-p)}$ *onto* $H_{-\frac{1}{2}s(q-p)}$.

Remark 1. The last statement of Theorem 2 means that the solution to equation (1.2) with minimal order of singularity is stable in $H_{-\frac{1}{2}s(q-p)}$ under arbitrary small perturbations of $f(x)$ in $H_{\frac{1}{2}s(q-p)}$. So for equation (1.2) with the kernel of class $\mathscr{R}$ the problem of finding the solution with minimal order of singularity is properly posed.

Let us formulate some results for equation (1.1) analogous to the results given in Theorem 1 and 2. For differential operators in $L^2(\mathbb{R}^1)$, ellipticity means that the coefficient of the highest derivative does not vanish in $\overline{D}$. We omit the description of the set of all solutions of equation (1.1) in the space of distributions. The description is similar to the one given in Theorem 1. Our main interest is the solution of minimal order of singularity.

Theorem 3. Suppose the kernel of equation (1.1) has the form of (1.3), $P(\lambda) = 1$, $0 < Q(\lambda) = a_+(\lambda)a_-(\lambda)$, $\deg a_\pm(\lambda) = \frac{1}{2}q$, $f \in H_{\frac{1}{2}sq}$, $D = (t-T,t)$, and the zeros of polynomial $a_+(\lambda)(a_-(\lambda))$ lie in the half-plane $\mathrm{Im}\,\lambda > 0$ $(\mathrm{Im}\,\lambda < 0)$. Then the solution to equation (1.1) with minimal order of singularity does exist, is unique, has order of singularity at most $\frac{1}{2}sq$, and this solution can be found by the formula

$$h(x) = a_+(L)[1(x-t+T)a_-(L)f(x)] - a_-(L)[1(x-t)a_+(L)f(x)], \tag{1.10}$$

where

$$1(x) = \begin{cases} 1, & x \geq 0, \\ 0, & x < 0. \end{cases}$$

Theorem 4. Let the kernel of equation (1.1) be of the form (1.3), $D = (t-T,t)$ and $f \in H_{\frac{1}{2}s(q-p)}$. Then the solution to equation (1.1) with minimal order of singularity does exist, is unique, has order of singularity at most $\frac{1}{2}s(q-p)$, and the solution itself can be found from the formula

$$h(x) = Q(L)G, \tag{1.11}$$

where

$$G(x) = \begin{cases} \sum_{j=1}^{\frac{1}{2}sq} b_j^- \psi_j^-(x), & x < t - T \\ g(x) & , \quad t - T \leq x \leq t, \\ \sum_{j=1}^{\frac{1}{2}sq} b_j^+ \psi_j^+(x), & t < x, \end{cases}$$

the functions $\psi_j^+(x)(\psi_j^-(x))$ $1 \leq j \leq \frac{1}{2}sq$ form a fundamental system of solutions to the equation $a_+(L)\psi = 0$, $\psi(+\infty) = 0$, $(a_-(L)\psi = 0,\ \psi(-\infty) = 0)$; the function $g(x)$ is defined by the equality

$$g(x) = g_o(x) + \sum_{j=1}^{sp} c_j\phi_j(x),$$

$g_o(x)$ <u>is an arbitrary fixed solution to the equation</u> $P(L)g = f$, <u>while the functions</u> $\phi_j(x)$, $1 \leq j \leq sp$, <u>form a fundamental system of solutions to the equation</u> $P(L)\phi = 0$; <u>and the constants</u> $b_j^{\pm}$, $1 \leq j \leq \frac{1}{2}sq$, <u>and</u> c_j, $1 \leq j \leq sp$, <u>are uniquely determined from the linear system</u> $(D = d/dx)$

$$\begin{aligned} D^k\left\{\sum_{j=1}^{\frac{1}{2}sq} b_j^-\psi_j^-(x)\right\}\bigg|_{x=t-T} &= D^k\left\{g_o(x) + \sum_{j=1}^{sp} c_j\phi_j(x)\right\}\bigg|_{x=t-T}, \\ &\quad 0 \leq k \leq \tfrac{1}{2}sq(p+q)-1, \\ D^k\left\{\sum_{j=1}^{\frac{1}{2}sq} b_j^-\psi_j^-(x)\right\}\bigg|_{x=t} &= D^k\left\{g_o(x) + \sum_{j=1}^{sp} c_j\phi_j(x)\right\}\bigg|_{x=t}, \\ &\quad 0 \leq k \leq \tfrac{1}{2}sq(p+q)-1. \end{aligned} \tag{1.12}$$

<u>The map</u> $R^{-1}: f \to h$, <u>where</u> h <u>is given by formula</u> (1.11), <u>is a homomorphism of the space</u> $H_{\frac{1}{2}s(q-p)}$ <u>onto the space</u> $H_{-\frac{1}{2}s(q-p)}$.

<u>Remark 2</u>. Let the conditions of Theorem 3 be satisfied and equation (1.1) have the form

$$\int_0^\infty R(x-y)h(y)dy = f(x), \quad x \geq 0,$$

so that $t = +\infty$, $t - T = 0$, $P(\lambda) = 1$, $Q(\lambda) = a_+(\lambda)a_-(\lambda)$, $L = -iD$. Then formula (1.10) gives the solution in the form $h(x) = a_+(-iD)[1(x)a_-(-iD)f(x)]$. This result can be obtained also by the traditional factorization method.

Let us consider now systems of integral equations, i.e., equation (1.1) with a matrix kernel. Let L be a selfadjoint elliptic differential operator on $L^2(\mathbb{R}^1)$. Consider the matrix kernel of the type

$$R(x,y) = \int_\Lambda \tilde{R}(\lambda)\Phi(x,y,\lambda)d\rho(\lambda),$$
$$\tilde{R} = (\tilde{R}_{ij}(\lambda)) = (P_{ij}(\lambda)Q_{ij}^{-1}(\lambda)), \quad 1 \le i,j \le d, \tag{1.13}$$

where the polynomials $P_{ij}(\lambda)$, $Q_{ij}(\lambda)$ are relatively prime for $1 \le i,j \le d$, $P_{ij}(\lambda) > 0$, $Q_{ij}(\lambda) > 0$, and the matrix $\tilde{R}(\lambda)$ is positive definite for ρ-almost all $\lambda \in \Lambda$. Let the least common multiple of the polynomials $Q_{ij}(\lambda)$, $1 \le i,j \le d$, be $Q(\lambda) > 0$, $q = \deg Q(\lambda)$, $A_{ij} \equiv P_{ij}(\lambda)Q_{ij}^{-1}(\lambda)Q(\lambda)$, $\det A_{ij}(\lambda) > 0$ for $\Lambda \in J$, and $J = (-\infty,\infty)$. We denote by $A(L)$ the matrix differential operator with the elements $A_{ij}(L)$, and by E the $d \times d$ unit matrix. We note that $\tilde{R}_{ij}(\lambda) = A_{ij}(\lambda)Q^{-1}(\lambda)$. Equation (1.1) with kernel (1.13) can be rewritten in the form

$$A(L)\int_D S(x,y)h(y)dy = f(x), \quad x \in \overline{D} = [t-T,t] \tag{1.14}$$

where the diagonal kernel $S(x,y)$ is

$$S(x,y) = \int_\Lambda Q^{-1}(\lambda)\Phi(x,y,\lambda)d\rho(\lambda)E. \tag{1.15}$$

Let us write equation (1.14) as

$$A(L)v = f \quad \text{in} \quad \overset{\circ}{D}; \quad \int_D S(x,y)h(y)dy = v(x),$$
$$A(L)v = \sum_{j=0}^{m} B_j(x)v^{(j)}, \tag{1.16}$$

where $\det B_m(x) \ne 0$, $x \in J$, $m = sa$, $a = \max_{1\le i,j\le d} \deg A_{ij}(\lambda)$, $s = \text{ord } L$, and $v^{(j)} = D^j v$. The operator $Q(L)E$ has the form of (1.16) and the number $n = sq$ plays the role of m. The number m is assumed to be even and n is even because q is even. Let $\phi_j(x)$, $1 \le j \le m$, be a fundamental system of matrix solutions to the equation $A(L)\phi_j = 0$ and let $\psi_j^{\pm}(x)$, $1 \le j \le \frac{1}{2}n$ be a fundamental system of matrix solution to the equation $Q(L)E\psi_j = 0$, $\psi_j^+(+\infty) = 0$, $\psi_j^-(-\infty) = 0$. If $Q(\lambda) > 0$ and $L = -iD$ then such a partition of the set $\{\psi_j\}$ into two subsets $\{\psi_j^+\}$, $\{\psi_j^-\}$ is evidently possible. For an operator L of general form, $Lu = \sum_{j=0}^{s} P_j(x)D^j u$, $P_s(x) \neq 0$, $x \in \overline{D}$, such a partition is also possible, see Naimark [1, p. 118]. We rewrite equation (1.16) as

$$\int_D S(x,y)h(y)dy = g_0(x) + \sum_{j=1}^{m} \phi_j(x)c_j, \quad x \in \overline{D}, \qquad (1.17)$$

where c_j are arbitrary constant vectors and $g_0(x)$ is an arbitrary fixed solution of the equation $A(L)g = f$. The idea is similar to the idea used in studying the scalar equation. That is why we use similar notations. Comparing Theorem 5 below with Theorem 4, we note that the roles of $\psi_j^\pm$, ϕ_j, sp, sq are now played by the matrix functions $\psi_j^\pm$, ϕ_j and numbers m,n. The spaces H_t of scalar functions should be replaced by the similar spaces of vector functions, which we denote by $\mathscr{H}_t$.

Theorem 5. *Let the kernel of equation* (1.1) *be of the form* (1.13) *and let* $f(x) \in \mathscr{H}_{\frac{1}{2}(n-m)}$. *Then there exists a unique solution to equation* (1.1) *with minimal order of singularity, having order of singularity at most* $\frac{1}{2}(n-m)$, *and the solution itself can be found from the formula*

$$h(x) = Q(L)G,$$

where

$$G(x) = \begin{cases} \sum_{j=1}^{\frac{1}{2}n} \psi_j^+(x) b_j^+, & x > t \\ g_o(x) + \sum_{j=1}^{m} \phi_j(x) c_j, & t - T \le x \le t \\ \sum_{j=1}^{\frac{1}{2}n} \psi_j^-(x) b_j^-, & x > t - T. \end{cases} \tag{1.18}$$

Here the functions $\psi_j^{\pm}$, $g_o(x)$, ϕ_j were defined above, the vectors $b_j^{\pm}$, c_j are uniquely determined by the system

$$D^k\left\{\sum_{j=1}^{\frac{1}{2}n} \psi_j^-(x) b_j^-\right\}\Bigg|_{x=t-T} = D^k\left\{g_o(x) + \sum_{j=1}^{m} \phi_j(x) c_j\right\}\Bigg|_{x=t-T}$$

$$D^k\left\{\sum_{j=1}^{\frac{1}{2}n} \psi_j^+(x) b_j^+\right\}\Bigg|_{x=t} = D^k\left\{g_o(x) + \sum_{j=1}^{m} \phi_j(x) c_j\right\}\Bigg|_{x=t},$$

$$0 \le k \le \frac{1}{2}(n+m)-1. \tag{1.19}$$

The map $R^{-1}: f \to h$, where h is defined by formula (1.8), is a homomorphism of the space $\mathscr{H}_{\frac{1}{2}(n-m)}$ of vector functions onto the space $\mathscr{H}_{-\frac{1}{2}(n-m)}$ of vector functions.

Now we pass to the study of the asymptotic distribution and some properties of the eigenvalues of the kernel

$$R(x,y) = \int_{\Lambda} \omega(\lambda) \Phi(x,y,\lambda) d\rho(\lambda), \tag{1.20}$$

where Φ and $d\rho$ are, as above, the spectral kernel and spectral measure of an elliptic selfadjoint operator L in $H = L^2(\mathbb{R}^r)$, ord $L = s$, $\omega(\lambda)$ is some continuous function on Λ, $\omega = \sup_{\lambda \in \Lambda} |\omega(\lambda)|$, and $\omega(\infty) = 0$. The eigenvalues are enumerated below according to their multiplicity.

Theorem 6. Let $\omega(\lambda) \ge 0$ and let $\lambda_j = \lambda_j(D)$, $\lambda_1 \ge \lambda_2 \ge \ldots \ge 0$, be the eigenvalues of the operator $R: L^2(D) \to L^2(D)$ with kernel (1.20). If $D' \supset D$ then $\lambda_j' \ge \lambda_j$, $\lambda_j' = \lambda_j(D')$.

If $\sup_{x \in \mathbb{R}^r} \int |R(x,y)|dy < \infty$, *then* $\lim_{D \to \mathbb{R}^r} \lambda_1(D) = \lambda_{1\infty} = \omega$. *Here* $\int = \int_{\mathbb{R}^r}$, $D \to \mathbb{R}^r$ *means that* D *uniformly expands to the whole space*.

Remark 3. Similar results are valid for equation (1.1) with a matrix kernel.

Theorem 7. *Let the kernel* $R(x,y)$ *be of the form* (1.20), $\omega(\lambda) = (1+\lambda^2)^{-\frac{1}{2}a} (1 + g(\lambda))$, $a > 0$, *where* a *is an even integer*, $g(\lambda)$ *is a continuous function on* J, $g(\infty) = 0$, *and* $1 + g(\lambda) > 0$. *Let* λ_j, $\lambda_1 \geq \lambda_2 \geq \ldots > 0$, *be the eigenvalues of the operator* $R: L^2(D) \to L^2(D)$ *with kernel* $R(x,y)$. *Then*

$$\lambda_n \sim (\gamma n^{-1})^{as/r} \quad \text{as} \quad n \to \infty, \tag{1.21}$$

where

$$\gamma = (2\pi)^{-r} \int_D \eta(x)dx$$

$$\eta(x) = \text{meas}\{t: t \in \mathbb{R}^r, \sum_{|\alpha|=|\beta|=\frac{1}{2}s} a_{\alpha\beta}(x) t^{\alpha+\beta} \leq 1\}. \tag{1.22}$$

Here α, β *are multi-indices and the form* $a_{\alpha\beta}(x)$ *generates the principal part of the selfadjoint elliptic operator* $Lu = \sum_{|\alpha|=|\beta|=\frac{1}{2}s} D^\alpha(a_{\alpha\beta}(x)D^\beta u)$. *If* $\omega(\lambda)$ *in formula* (1.20) *is equal to* $P(\lambda)Q^{-1}(\lambda)$ *then* $a = q - p$, *where* a *is the number in formula* (1.21), $q = \deg Q(\lambda)$, $p = \deg P(\lambda)$.

Theorem 8. *Let the kernel be of the form*

$$R(x,y) = \int_\Omega \omega(\lambda_1,\ldots,\lambda_m)\Phi(x,y,\lambda_1,\ldots,\lambda_m)d\rho(\lambda_1,\ldots,\lambda_m), \tag{1.20'}$$

and let $\omega(\lambda_1,\ldots,\lambda_m) = (1 + |\lambda|^2)^{-a/2} (1 + g(\lambda))$, $\lambda = (\lambda_1,\ldots,\lambda_m)$, $|\lambda|^2 = \lambda_1^2 + \ldots + \lambda_m^2$, a *is an even integer*, $1 + g(\lambda) \neq 0$, $g(\lambda) \to 0$ *as* $\lambda \to \infty$, $L = L_1^2 + \ldots + L_m^2$,

ord $L_j = s$, where the L_j are commuting selfadjoint elliptic operators in $L^2(\mathbb{R}^r)$. Let the principal part of the operator $L^{a/2}$ be generated by the form $b_{\alpha\beta}(x)$,
$L^{a/2}u = \sum_{|\alpha|=|\beta|=\frac{1}{2}sa} D^\alpha(b_{\alpha\beta}D^\beta u) + \dots$. Let λ_j,
$\lambda_1 \geq \lambda_2 \geq \dots$, be the eigenvalues of the operator in $L^2(D)$ generated by the kernel (1.20'),

$$\gamma_m = (2\pi)^{-r}\int_D \eta_m(x)dx, \qquad (1.23)$$

$$\eta_m(x) = \text{meas}\{t:\ t \in \mathbb{R}^r,\ \sum_{|\alpha|=|\beta|=\frac{1}{2}aq} b_{\alpha\beta}(x)t^{\alpha+\beta} \leq 1\}.$$

Then

$$\lambda_n \sim (\gamma_m n^{-1})^{as/r}, \quad n \to \infty.$$

In particular, if $\omega(\lambda_1,\dots,\lambda_m) = P(\lambda_1,\dots,\lambda_m)Q^{-1}(\lambda_1,\dots,\lambda_m)$, where $P(\lambda_1,\dots,\lambda_m)$, $Q(\lambda_1,\dots,\lambda_m)$ are polynomials of the principal type, i.e., there exist positive constants C_j, $1 \leq j \leq 4$, such that $C_1(1+\lambda^2)^{p/2} \leq P(\lambda) \leq C_2(1+\lambda^2)^{p/2}$, $C_3(1+\lambda^2)^{q/2} \leq Q(\lambda) \leq C_4(1+\lambda^2)^{q/2}$, $\lambda = (\lambda_1,\dots,\lambda_m)$, $\lambda^2 = \lambda_1^2 + \dots + \lambda_m^2$, then the number a in Theorem 8 is equal to $\frac{1}{2}(q-p)$.

In the proofs of Theorems 7 and 8 some general abstract results of the theory of linear operators are used. These results are formulated below.

2. Let H be the Hilbert space, A be a closed linear densely defined operator on H, $D(A)$, $R(A)$, and $N(A)$ denote th domain, range, and kernel of A. We say that the spectrum $\sigma(A)$ of the operator A is discrete if the spectrum consists of isolated eigenvalues of finite algebraic multiplicity. If $A = A^* \geq m > 0$ then we denote by H_A the Hilbert space which is the completion of $D(A)$ in the norm

$\|u\| = (Au,u)^{\frac{1}{2}}$. Let $\{0\}$ be the set consisting only of zero. If A is compact we call $s_n(A) = \lambda_n\{(A^*A)^{\frac{1}{2}}\}$ a singular number (s-number) of A. If A^{-1} is compact we call $s_n(A) \equiv s_n^{-1}(A^{-1})$ a singular number of A. If A^{-1} does not exist, but the resolvent $(A - \lambda_o I)^{-1}$ is compact for some λ_o, we can decompose H into the direct sum $M_o \dotplus M_1$ of subspaces invariant for A. Here M_o is the root subspace which corresponds to the eigenvalue $\lambda = 0$, and M_1 is a subspace in which operator A is invertible. Denote by A_1 the restriction of A to M_1. The compactness of A_1^{-1} follows from the compactness of $(A - \lambda_o I)^{-1}$. Let us define $s_n(A) \equiv s_n^{-1}(A_1^{-1})$. Then the case $0 \in \sigma(A)$ is reduced to the case in which A^{-1} exists.

Theorem 9. *Let* A *be a closed densely defined linear operator on the Hilbert space* H *with discrete spectrum* $\sigma(A)$ *such that* A^{-1} *exists and is defined everywhere on* H. *Let* T *be a linear operator on* H *such that* $D(A) \subset D(T)$, $B = A + T$, *and* $D(B) = D(A)$. *If the operator* TA^{-1} *is compact, then* B *is closed. If, in addition,* A^{-1} *is compact and* $N(B+kI) = \{0\}$ *for some number* $k \notin \sigma(A)$, *then* $\sigma(B)$ *is discrete and* $s_n(B)s_n^{-1}(A) \to 1$ *as* $n \to \infty$. *If the operators* $A^{-1}T$ *and* TA^{-1} *are compact,* $A = A^*$ *is semibounded from below, and* B *is normal, then* $\lambda_n(B)\lambda_n^{-1}(A) \to 1$ *as* $n \to \infty$.

Theorem 10. *Let* A *be a closed densely defined linear operator in* H *with discrete spectrum such that* A^{-1} *exists and is defined on all* H, *and let* T *be a linear operator in* H *such that* $D(T) \supset D(A)$, $B = A + T$, *and* $D(B) = D(A)$. *If* $A = A^* \geq m > 0$, *the operator* $A^{-1}T$ *is compact in* H_A,

$D(T) \supset H_A$, *and* $B = B^*$, *then* $\lambda_n(B)\lambda_n^{-1}(A) \to 1$ *as* $n \to \infty$.

Theorem 11. Let Q, S *be compact linear operators on* H, $\dim R(Q) = \infty$, *and* $N(I + S) = \{0\}$. *Then* $s_n(Q + QS)s_n^{-1}(Q) \to 1$ *and* $s_n(Q + SQ)s_n^{-1}(Q) \to 1$, *as* $n \to \infty$.

In the following theorem we consider perturbations of quadratic forms preserving the asymptotics of the spectrum. The spectrum of a closed sectorial quadratic form is the spectrum of the operator generated by the form (see Kato [1]). We let $D[A]$ denote the domain of the definition of the quadratic form. A quadratic form $T[f,f]$ is called compact relative to a positive definite quadratic form $A[f,f]$ if from any sequence $\{f_n\}$ such that $A[f_n,f_n] \leq 1$ one can select a subsequence $\{f_{n_k}\}$ such that $T[f_{n_{k+p}} - f_{n_k}, f_{n_{k+p}} - f_{n_k}] \to 0$ as $k \to \infty$ uniformly with respect to $p = 1,2,3,\ldots$.

Theorem 12. Let $A[f,f]$ *be a positive definite quadratic form on* H *with discrete spectrum* $\lambda_n(A)$ *and* $T[f,f]$ *be a real-valued closed quadratic form such that* $D[T] \supset D[A]$, $B[f,f] = A[f,f] + T[f,f]$, *and* $D[B] = D[A]$. *If* $T[f,f]$ *is compact relative to* $A[f,f]$ *then the spectrum of* $B[f,f]$ *is discrete and* $\lambda_n(B)\lambda_n^{-1}(A) \to 1$ *as* $n \to \infty$.

3. Consider an approximate analytical solution of equations (1.1) and (1.2). For the sake of simplicity we discuss in detail only the scalar equation

$$Rh = f, \quad Rh = \int_D R(x,y)h(y)dy, \qquad R(x,y) = \int_\Lambda \tilde{R}(\lambda)\Phi(x,y,\lambda)d\rho(\lambda). \tag{1.24}$$

We assume that $0 < \tilde{R}(\lambda) \sim (1 + \lambda^2)^{-\beta_A}$ as $|\lambda| \to \infty$,

$A = \text{const} > 0$, $\lambda \in \mathbb{R}^1$, and $\beta > 0$ is integer. Then for any $\varepsilon > 0$ polynomials $P_\varepsilon(\lambda)$, $Q_\varepsilon(\lambda)$ can be found such that $\deg Q_\varepsilon - \deg P_\varepsilon = 2\beta$, $\|\tilde{R} - \tilde{R}_\varepsilon\| < \varepsilon$, where $\tilde{R}_\varepsilon = \tilde{R}_\varepsilon(\lambda) \equiv P_\varepsilon(\lambda)Q_\varepsilon^{-1}(\lambda)$ and $\|\tilde{R}\| \equiv \sup_{\lambda \in J} \{(1 + \lambda^2)^\beta |\tilde{R}(\lambda)|\}$. Our idea of an approximate analytical solution of equation (1.24) in the space $H_{-\beta s}$ under the above assumptions concerning $R(x,y)$ can be described as follows. Let $f \in H_{\beta s}$. Consider the equation $R_\varepsilon h_\varepsilon = f$, where R_ε is the operator of the same type as R generated by the function $\tilde{R}_\varepsilon(\lambda) = P_\varepsilon(\lambda)Q_\varepsilon^{-1}(\lambda)$, $\deg Q_\varepsilon - \deg P_\varepsilon = 2\beta$ according to formula (1.24), and $\|R(\lambda) - \tilde{R}_\varepsilon(\lambda)\| < \varepsilon$. The equation $R_\varepsilon h_\varepsilon = f$ can be solved by the formulas given in Theorems 2 and 4. As a result we find $h_\varepsilon \in H_{-\beta s}$. It can be proved that $|h - h_\varepsilon|_{-\beta s} \le C\varepsilon$, where $C = \text{const}$, $|\cdot|_s$ is the norm in H_s. An explicit value for C will be given below. So the idea consists of approximating the kernel $R(x,y)$ by the kernel $R_\varepsilon(x,y)$ and applying Theorem 2 or 4. In the literature such an idea was neglected in the theory of integral equations of the first kind because the solutions of such equations were sought in $L^2(D)$ and such solutions are unstable under small perturbations of the kernel. The stability in $H_{-\beta s}$ of the solution h_ε constructed above under small perturbations of f in $H_{\beta s}$ and small (in the norm $\|\cdot\|$) perturbations of the kernel $R(x,y)$ is stated in the following theorem. Formula (1.24) can be considered as the Fourier transform of $R(x,y)$ with respect to the eigenfunctions of the operator L, $\text{ord}\, L = s$.

Theorem 13. *Suppose that* $\tilde{R} \sim A(1 + \lambda^2)^{-\beta}$ *as* $\lambda \to \infty$ *and* $f \in H_{s\beta}$. *Let* $R_\varepsilon h_\varepsilon = f$, *where* R_ε *was defined above. Then there is a unique solution* h *in* $H_{-\beta s}$ *of equation* $Rh = f$

and the map $R^{-1}: f \to h$ is a homomorphism of $H_{\beta s}$ onto $H_{-\beta s}$. Moreover the following estimate holds: $|R_\varepsilon^{-1} - R^{-1}|_{\beta s \to -\beta s} \le 2M^2 \varepsilon (1 - 2\varepsilon M)^{-1}$ if $2\varepsilon M < 1$, where $M = \{\inf_{\varepsilon > 0} |||\tilde{R}_\varepsilon(\lambda)|||\}^{-1}$, $|||\tilde{R}_\varepsilon(\lambda)||| \equiv \inf_{\lambda \in J}\{(1 + \lambda^2)^\beta |\tilde{R}_\varepsilon(\lambda)|\}$, $|\cdot|_{t \to -t}$ is the usual norm of a linear operator mapping H_t into H_{-t}.

Remark 4. For the approximation of a function $\tilde{R}(\lambda)$ by rational functions see Ahieser [1], Remes [1]. Similar results can be stated for equation (1.1) with a matrix kernel.

4. Let us discuss some applications of the developed theory. We restrict ourselves to problems of stochastic optimization theory, which includes estimation problems, filtering, signal detection and discrimination, etc. Let a random field $u = s + n$ be the input to a linear instrument or filter, where $s(x)$ is the useful signal and $n(x)$ is noise. The covariance function of $u(x)$ and the mutual correlation function of $u(x)$ and $s(x)$ are assumed to be known. A typical problem of the theory consists of finding a linear operator A of the type $Au = \int_D h(x,y)u(y)dy$ such that $\mathscr{D}[Au - s] = \min$, where $\mathscr{D}$ is the symbol of variance and D is the domain of signal processing. It is not difficult to show that the optimal function $h(x,y)$ satisfies the equation

$$\int_D R(x,y)h(y,z)dy = f(x,z), \quad x \in \bar{D}, \tag{1.25}$$

where $R(x,y)$ is the covariance function of the field $u(x)$ and $f(x,z)$ is the correlation function of the fields $s(x)$ and $u(x)$. If the field $u(x)$ is homogeneous then $R(x,y) = R(x-y)$. The argument z in (1.25) is a parameter, so equation (1.25) is of the form (1.2). If the kernel $R(x,y)$ in equation (1.25) belongs to class $\mathscr{R}$ then the results of

Theorems 1-8 are fully applicable. All the other problems mentioned above can be reduced to equation (1.25). Consider, for example, the problem of the discrimination of signals against a background of noise. Let signal $u(x)$ be observed in domain D, $u(x) = n(x) + s(x)$, where $n(x)$ is noise, $s(x)$ is either $s_o(x)$ or $s_1(x)$, $s_j(x)$, $j = 0,1$, are deterministic signals, $n(x)$ is a Gaussian random field with mean value zero. The problem is to discriminate between two hypotheses H_o: $u = n + s_o$ and H_1: $u = n + s_1$, using the values of $u(x)$ observed in the domain D. The statistical test to solve this problem by the maximal likelihood method consists of the following: if

$$\operatorname{Re} \int_D u(y)V^*(y)\,dy \geq \frac{1}{2} \int_D [s_o(y)V^*(y) + s_1^*(y)V(y)]\,dy, \quad (1.26)$$

then hypothesis H_1 is accepted, otherwise hypothesis H_o is accepted. Here $V(x)$ is the solution to the integral equation

$$RV = s_1(x) - s_o(x), \quad x \in \overline{D}, \qquad RV \equiv \int_D R(x,y)V(y)\,dy. \quad (1.27)$$

If $R(x,y) \in \mathscr{R}$ the developed theory is applicable to equation (1.27). It is not very restrictive in applications to assume that $R(x,y) \in \mathscr{R}$. For example, in many electrical engineering problems the covariance functions $R(x,y)$ are assumed to be translation invariant, i.e., $R(x,y) = R(x-y)$, and the Fourier transform $\tilde{R}(\lambda)$ is rational or can be approximated by rational functions so that Theorem 13 is applicable.

It is interesting to note that the developed theory is also useful in nonlinear estimation theory. We restrict ourselves only to minimization of the variance of the error of estimate. There have been few papers on the subject. The main paper seems to be that of Katznelson-Gould [1]. Unfortunately, there are some errors in their paper as will be shown below.

Let $u = s + n$. We introduce the notations:

$$Au = \sum_{j=0}^{n} H_j u^{[j]}, \tag{1.28}$$

$$H_j u^{[j]} = \int_D \cdots \int_D h_j(\tau_1, \ldots, \tau_j) u(t-\tau_1) \cdots u(t-\tau_j)\, d\tau_1 \cdots d\tau_j, \tag{1.29}$$

$$u^{[0]} = 1, \quad H_o u^{[0]} = h_o(t), \quad u^{[j]} = u(t-\tau_1) \cdots u(t-\tau_j). \tag{1.30}$$

The problem is to find among all estimates (1.28) the one minimizing $\mathscr{D}[Au - f]$ where $\mathscr{D}$ is the variance symbol and f is a given random function. If $f = s(t+\tau)$, for a given number $\tau > 0$, the posed problem is the problem of filtering and extrapolation. The optimal estimate is defined by $n+1$ functions $(h_o, \ldots, h_n)$. It is not difficult to prove that a necessary condition for the functions $(h_o, \ldots, h_n)$ to be optimal is the following system of integral equations

$$\sum_{j'=0}^{n} a_{jj'} H_{j'} = b_j, \quad 0 \le j \le n, \tag{1.31}$$

where

$$b_j = \overline{u^{*[j]} f}, \quad a_{jj'} = \overline{u^{*[j]} u^{[j']}},$$

$$a_{jj'} H_{j'} = \underbrace{\int_D \cdots \int_D}_{j' \text{ times}} h_{j'}(\tau_1', \ldots, \tau_{j'}') a_{jj'}\, d\tau_1' \cdots d\tau_{j'}', \tag{1.32}$$

the line denotes statistical mean, and the star denotes complex conjugation. If the system (1.31) has a unique solution which gives a finite value for the variance of error of the estimate, i.e. $\mathscr{D}[Au - f] \to \infty$, then the solution of the system (1.31) is the solution of our stochastic optimization problem. In Katznelson-Gould [1] the following approach to solving the system (1.31) was proposed. Considering the functions h_j, $0 \le j \le n - 1$, as known, we rewrite equation (1.31) as

$$a_{nn}H_n = b_n - \sum_{j'=0}^{n-1} a_{nj'}H_{j'} \equiv \Phi_n \tag{1.33}$$

where the right-hand side of equation (1.33) is known. The equation $a_{nn}H_n = \Phi_n$ is an equation of type (1.2) for the function h_n. If its kernel $a_{nn} \in \mathscr{R}$ we can use the developed theory to find h_n. So step by step we can find $(h_o, h_1, \ldots, h_n)$. Of course in such manner we can find not the optimal but a quasi-optimal solution of the basic problem. In Katznelson-Gould [1] an iterative process to solve the system (1.31) was proposed. The first step of this process consists in finding step by step functions $h_j^{(1)}(x)$, $0 \le j \le n$, from the system (1.33). After the r-th step is finished we find functions $h_j^{(r+1)}(x)$, $0 \le j \le n$, step by step from the equation

$$a_{jj}H_j^{(r+1)} = b_j - \sum_{j'=0}^{j-1} a_{jj'}H_{j'}^{(r+1)} - \sum_{j'=j+1}^{n} a_{jj'}H_{j'}^{(r)}. \tag{1.34}$$

Putting $j = 0,1,2,\ldots,n$, in (1.34) we find $h_j(x)$, $0 \le j \le n$. This completes the description of the iterative process. Denote by A_r the operator (1.28) where H_j are substituted by $H_j^{(r)}$, and set $\varepsilon_r = \mathscr{D}[A_r u - f]$. It is clear that $\varepsilon_r \ge \varepsilon_{r+1} \ge 0$. Thus the limit $\lim_{r\to\infty} \varepsilon_r \equiv \varepsilon \ge 0$ exists. Katznelson-Gould 1 tried to prove convergence in $L^2(D)$ of the iterative process (1.34) for any $f \in L^2(D)$, but this is impossible. One can consider equation (1.31) as a linear operator equation involving a compact operator in the space of vector functions $(h_o,\ldots,h_n)$, $h_j \in L^2(D)$, $0 \leq j \leq n$. The range of this compact operator is not closed in $L^2(D)$, so equation (1.31) cannot be solved for any right-hand side from $L^2(D)$. Moreover, the iterative process (1.34) cannot converge for any $b_j \in L^2(D)$. But if the kernels in equation (1.31) belong to the class $\mathcal{R}$ and f is smooth enough, then it is possible to calculate the solution of system (1.31) by means of an iterative process. We illustrate this by taking the simpler equation (1.2) as an example. In Theorem 2 it was stated that the map $R^{-1}: H_\alpha \to H_{-\alpha}$, $\alpha = \frac{1}{2}s(q-p)$, is a homomorphism of H_α onto $H_{-\alpha}$. These spaces are Hilbert spaces. The adjoint R^* of R maps H_α onto $H_{-\alpha}$. Let us construct the iterative process mentioned above for the equation $Rh = f$, $f \in H$, $R \in \mathcal{R}$. Consider the equivalent equation $Bh \equiv R^*Rh = R^*f$, where $B: H_{-\alpha} \to H_{-\alpha}$ is a selfadjoint positive operator. If B is a linear operator in the Hilbert space H, $B = B^* \geq 0$, and $g \in R(B)$, then the iterative process $Bh_n + h_n = h_{n-1} + g$, $h_o \in H$, converges to a solution of the equation $Bh = g$. So the iterative process $Bh_n + h_n = h_{n-1} + R^*f$, $h_o \in H_{-\alpha}$, converges in $H_{-\alpha}$ to the (unique) solution of the equation $Bh = R^*f$ and simultaneously to the solution of the equivalent equation $Rh = f$.

2. Investigation of the Scalar Equations

Here we prove Theorems 1-4 of Section 1. Consider the equation

$$\int_{t-T}^{t} R(x,y)h(y)dy = f(x), \quad t - T \leq x \leq t, \tag{2.1}$$

where the kernel $R(x,y)$ is nonnegative definite, $T > 0$. We suppose that $R(x,y) \in \mathscr{R}$ is of the form (1.3) and the conditions of Theorem 3 from Section 1 are fulfilled.

In order to prove this theorem we first prove a lemma.

Lemma 1. The set of solutions of equation (2.1) with the kernel

$$R(x,y) = \int_{\Lambda} Q^{-1}(\lambda)\Phi(x,y,\lambda)d\rho(\lambda), \quad Q(\lambda) = a_{+}(\lambda)a_{-}(\lambda) \tag{2.2}$$

in the space H_{-sq} is in one-to-one correspondence with the set of the solutions of the equation

$$\int_{-\infty}^{\infty} R(x,y)H(y)dy = f(x), \quad x \in J. \tag{2.3}$$

in the space $H_{-sq}(\mathbb{R}^1)$, $\operatorname{supp} H \subset \overline{D} = [t - T, t]$. Here

$$F(x) = \begin{cases} \sum_{j=R}^{\frac{1}{2}qs} b_j^- \psi_j^-(x), & x < t, \\ f(x), & t - T \leq x \leq t, \\ \sum_{j=1}^{\frac{1}{2}qs} b_j^+ \psi_j^+(x), & x > t, \end{cases} \tag{2.4}$$

and the functions $\{\psi_j^{\pm}(x)\}$, $1 \leq j \leq \frac{1}{2}qs$, form a fundamental system of solutions of the equation $Q(L)\psi = 0$, $\psi_j^+(+\infty) = 0$, $\psi_j^-(-\infty) = 0$, $1 \leq j \leq \frac{1}{2}qs$, and $b_j^{\pm}$, $1 \leq j \leq \frac{1}{2}qs$, are arbitrary constants.

<u>Proof of Lemma 1</u>: Let $h \in H_{-qs}$ be a solution of equation (2.1) with $\text{supp } H \subseteq \bar{D}$. Denote $\psi(x) \equiv \int_{t-T}^{t} R(x,y)h(y)dy$ by $\psi^{+}(x)$ for $x > t$, by $\psi^{-}(x)$ for $x < t - T$. As $Q(L)R(x,y) = \delta(x-y)$, it is clear that $Q(L)\psi = 0$ for $x > t$, $x < t - T$. Because $Q(\lambda) > 0$ it is clear that $|R(x,y)| \to 0$ if $|x-y| \to \infty$. Thus $\psi^{+}(+\infty) = 0$, $\psi^{-}(-\infty) = 0$. The function

$$H(x) = \begin{cases} h(x), & x \in \bar{D} \\ 0 \quad , & x \in \Omega, \end{cases}$$

where $D = (t - T, t)$, $\Omega = \mathbb{R}^{1} \setminus \bar{D}$, is a solution to equation (2.3), $\text{supp } H \subseteq \bar{D}$, $H \in H_{-sq}(\mathbb{R}^{1})$.

Conversely, let H be a solution of (2.3), $H \in H_{-sq}(\mathbb{R}^{1})$, $\text{supp } H \subseteq \bar{D}$. Then setting $h(x) = H(x)$, we obtain a solution of equation (2.1), $h \in H_{-sq}$. □

<u>Proof of Theorem 3</u>: By Lemma 1 there is a one-to-one correspondence between the set of solutions of equation (2.1) with kernel (2.2) in the space H_{-sq} and the set of solutions of equation (2.3) with support in $\bar{D}$ in the space $H_{-sq}(\mathbb{R}^{1})$. So we consider equation (2.3) with the right-hand side defined by formula (2.4). This equation has solutions in $H_{-sq}(\mathbb{R}^{1})$ if $f \in L^{2}(\mathbb{R}^{1})$. All of these solutions can be found from the formula

$$H(x) = Q(L)F(x), \tag{2.5}$$

because $Q(L)R(x,y) = \delta(x-y)$. From the definition (2.4) of the function $F(x)$ it follows that $Q(L)F = 0$ in Ω. Thus $\text{supp } H \subseteq \bar{D}$. That is why the function $h(x) = H(x)$ is a solution of equation (2.1), $h(x) \in H_{-sq}$. We see that the set of all solutions of equation (2.1) in H_{-sq} can be described by the formula $h(x) = Q(L)F(x)$. Every solution is uniquely determined if we fix the constants $b^{\pm}$ in formula (2.4). In order that the function $h(x)$ have minimal order of singularity it is necessary and sufficient that $F(x)$ is maximally smooth. If $f \in H_{\frac{1}{2}sq}$ the function $F(x)$ defined by formula (2.4) will be maximally smooth if and only if the following conjugation conditions hold:

$$F^{(k)}(t-0) = F^{(k)}(t+0), \quad F^{(k)}(t-T-0) = F^{(k)}(t-T+0), \qquad 0 \leq k \leq \frac{1}{2}qs-1. \tag{2.6}$$

There are precisely qs conditions according to the number of the constants $b^{\pm}$ in formula (2.4). So it is impossible to impose any more conditions. Conditions (2.6) can be written as

$$\begin{cases} D^k \sum_{j=1}^{\frac{1}{2}sq} b_j^- \psi_j^-(x) \Big|_{x=t-T} = D^k f(x) \Big|_{x=t-T}, \quad 0 \leq k \leq \frac{1}{2}qs-1, \\ D^k \sum_{j=1}^{\frac{1}{2}sq} b_j^+ \psi_j^+(x) \Big|_{x=t} = D^k f(x) \Big|_{x=t}, \quad 0 \leq k \leq \frac{1}{2}qs-1. \end{cases} \tag{2.7}$$

Conditions (2.7) are the linear system from which the unknown coefficients $b_j^{\pm}$ can be uniquely determined. If the $b_j^{\pm}$ are already found and put in formula (2.4), then we can find $F(x)$ and $h(x)$ from the formula

$$h(x) = Q(L)F(x). \tag{2.8}$$

It is obvious that the order of singularity of the obtained solution $\sigma(h)$ is at most $\frac{1}{2}sq$. It remains to prove that the system (2.7) has a unique solution and the obtained solution of minimal order of singularity can be written as in formula (1.10). To prove uniqueness it is sufficient to prove that $f = 0$ implies $b_j^{\pm} = 0$, for all j. Let $f \equiv 0$,

$\psi^- \equiv \sum_{j=1}^{\frac{1}{2}sq} b_j^- \psi_j^-(x)$. As $a_-(L)\psi^- = 0$, $D^k\psi^-\Big|_{x=t-T} = 0$, $0 \leq k \leq \frac{1}{2}sq-1$, and $\operatorname{ord} a_-(L) = \frac{1}{2}sq$ we conclude that $\psi^- \equiv 0$. From this and the linear independence of the set of functions $\{\psi_j^-\}$, it follows that $b_j^- = 0$, for all j. Similarly, one proves that $b_j^+ = 0$, for all j. The fact that the solution of minimal order of singularity can be written in the form (1.10) can be verified by direct calculation. □

Proof of Theorem 4: Let the kernel of equation (2.1) be of the form

$$R(x,y) = \int_\Lambda P(\lambda)Q^{-1}(\lambda)\Phi(x,y,\lambda)d\rho(\lambda). \tag{2.9}$$

Equation (2.1) can be rewritten as:

$$P(L)\int_{t-T}^{t} S(x,y)h(y)dy = f(x), \quad t-T \leq x \leq t, \tag{2.10}$$

where

$$S(x,y) = \int_\Lambda Q^{-1}(\lambda)\Phi(x,y,\lambda)d\rho(\lambda). \tag{2.11}$$

Equation (2.10) is equivalent to the equation

$$\int_{t-T}^{t} S(x,y)h(y)dy = g_o(x) + \sum_{j=1}^{sp} c_j\phi_j(x), \quad t-T \leq x \leq t, \tag{2.12}$$

where c_j are arbitrary constants, $\{\phi_j\}$, $1 \leq j \leq ps$, is a fundamental system of solutions of the equation $P(L) = 0$, and $g_o(x)$ is some arbitrary fixed solution of the equation $P(L)g = f$. We can apply Theorem 3 to equation (2.12). Denote by $g(x)$ the right-hand side of equation (2.12) and define the function $G(x)$ by the formula

$$G(x) = \begin{cases} \psi^-(x), & x < t - T \\ g(x), & t - T \leq x \leq t, \\ \psi^+(x), & t < x, \end{cases} \tag{2.13}$$

where

$$\psi^-(x) = \sum_{j=1}^{\frac{1}{2}qs} b_j^- \psi_j^-(x), \quad \psi^+(x) = \sum_{j=1}^{\frac{1}{2}qs} b_j^+ \psi_j^+(x).$$

As was proved, the solution of equation (2.12) can be found from the formula

$$h(x) = Q(L)G(x). \tag{2.14}$$

The function $G(x)$ depends on $(p + q)s$ arbitrary constants C_j, $1 \leq j \leq ps$, $b_j^{\pm}$, $1 \leq j \leq \frac{1}{2}qs$. In order for the function (2.14) to be of minimal order of singularity it is necessary and sufficient that the function $G(x)$ be maximally smooth. It will be so if and only if the conjugation conditions

$$\begin{aligned} G^{(k)}(t-T-0) &= G^{(k)}(t-T+0), \\ G^{(k)}(t+0) &= G^{(k)}(t-0), \quad 0 \leq k \leq \tfrac{1}{2}(p+q)s-1 \end{aligned} \tag{2.15}$$

hold. These conditions are similar to conditions (2.6). They can be written in the form given in formula (1.12). To prove the existence and uniqueness of the solution of equation (2.1) with minimal order of singularity, we must prove that conditions (2.15), which can be considered as a linear system for the unknowns $b_j^{\pm}$, c_j, allow us to find $b_j^{\pm}$, c_j uniquely. As above, it is sufficient to prove the uniqueness of the solution to the homogeneous system. To prove this, let us suppose that $f = 0$ and h is given by formula (2.14), where $G(x)$ is defined by formula (2.13) and satisfies conditions (2.15). Then $h(x)$ is a solution of equation (2.1) for $f = 0$, $h(x) \in H_{-\frac{1}{2}(q-p)s}$. Actually, finding the function $G(x)$ for which $Q(L)G(x)$ is the solution with minimal order of singularity of equation (2.1) with kernel (2.9) is equivalent to solving the boundary-value problem

$$\begin{cases} P(L) = f \quad \text{in} \quad D, \quad Q(L)\psi = 0 \quad \text{in} \quad \Omega, \\ \psi(\pm\infty) = 0, \quad D^k g = D^k \psi \quad \text{at} \quad x = t-T, \ x = t, \\ \qquad\qquad 0 \le k \le \frac{1}{2}(p+q)s-1. \end{cases} \tag{2.16}$$

The homogeneous problem corresponds to $f = 0$. The function G in formulas (2.13)-(2.15) can be expressed in terms of g, by formula (2.13) in which one should put $\psi^- = \psi$ for $x < t-T$ and $\psi^+ = \psi$ for $x > t$.

So it remains to prove that there is no nontrivial solution to the homogeneous equation (2.1) in the space $H_{-\frac{1}{2}(q-p)s}$. Let $Rh = 0$, $h \in H_{-\frac{1}{2}(q-p)s}$. Using Parseval's equality for eigenfunction expansions for the operator L we get

$$0 = \int_D Rh \cdot h^* dx = \int_\Lambda Rh \cdot h^* d(\lambda) = \int_\Lambda \tilde{R}(\lambda)|\tilde{h}|^2 d\rho(\lambda), \tag{2.17}$$

where $\tilde{R}(\lambda) = P(\lambda)Q^{-1}(\lambda)$, $\tilde{h}(\lambda)$ is the Fourier transform of $h(x)$ according to the eigenfunctions of the operator L. For details concerning generalized eigenfunction expansions see Beresanskij [1]. As $\tilde{R}(\lambda) > 0$ we conclude that $\tilde{h} = 0$, $h(x) = 0$. It is interesting to note that the condition $h \in H_{-\frac{1}{2}(q-p)s}$ implies the convergence of the integral in the right-hand side of formula (2.17). Indeed $|\tilde{R}| = O(|\lambda|^{p-q}$ as $|\lambda| \to \infty$ while the membership $h \in H_{\alpha s}$, in terms of the eigenfunction expansions for the differential elliptic self-adjoint operator L, ord $L = s$, implies that $\int_\Lambda |\tilde{h}(\lambda)|^2 (1+\lambda^2)^\alpha d\rho(\lambda) < \infty$.

In our case $\alpha = -\frac{1}{2}(q-p)$ so that $\int_\Lambda |\tilde{h}|^2 (1+\lambda^2)^{-\frac{1}{2}(q-p)} d\rho(\lambda) < \infty$. It means that the integral in the right-hand side of formula (2.17) converges. To end the proof we must prove that the map R^{-1} is a homomorphism of the space $H_{\frac{1}{2}(q-p)s}$ onto $H_{-\frac{1}{2}(q-p)s}$. We have already proved that the linear map $R: H_{-\frac{1}{2}(q-p)s} \to H_{\frac{1}{2}(q-p)s}$ is injective and surjective. So the inverse map is also continuous. □

Remark 1. In applications to stochastic optimization theory it is interesting to find solutions of equation (2.1) for which the integral

$$\int_D Rh \cdot h^* dx = \int_\Lambda \tilde{R}(\lambda)|\tilde{h}|^2 d\rho(\lambda) \tag{2.18}$$

is finite, because only in this case is the variance of the error of estimate finite. As was proved above, the integral (2.18) is finite only for the solution of minimal order of singularity of equation (2.1). So the requirement of minimal order of singularity of the solution of equation (2.1) is equivalent to the requirement of finiteness of the variance of the error of estimate. This explains the important role of the solution of equation (2.1) with minimal order of singularity. Let us demonstrate the above statement in detail. We have the expression $\varepsilon = \mathscr{D}[Au - f]$, for the dispersion of the estimate, where $Au = \int_D h(x,y)u(y)dy$, $u = s + n$, u,s,n are random functions, and $h(x,y)$ is the impulse response of the filter corresponding to the operator A. It is easy to obtain

$$\begin{aligned} \varepsilon = & \int_D\int_D h^*(x,y)h(x,y')R(y,y')dydy' \\ & - 2\text{Re}\int_D h^*(x,y)R_1(y,x)dy + R_2(x,x), \end{aligned} \tag{2.19}$$

where $R(y,y') = \overline{u^*(y)u(y')}$, $R_1(y,x) = \overline{u^*(y)f(x)}$, $R_2(x,x) = \overline{|f(x)|^2}$, and the line denotes the mean value.

A necessary condition for the functional (2.19) to be

minimal is

$$\int_D R(y,y')h(x,y')dy' = R_1(y,x), \quad y \in \bar{D}. \tag{2.20}$$

If $h_o(x,y)$ satisfies equation (2.20), then ε from (2.19) can be written as

$$\varepsilon = R_2(x,x) - \int_D h_o(x,y')R_1^*(y',x)dy'. \tag{2.21}$$

Let us note that

$$\int_D h_o(x,y')R_1^*(y',x)dy' = \int_D\int_D h_o(x,y')h_o^*(x,y)R(y',y)dydy'$$
$$= \int_\Lambda \tilde{R}(\lambda)|\tilde{h}(\lambda)|^2 d\rho(\lambda). \tag{2.22}$$

The dependence of $\tilde{h}$ on x is omitted. From formulas (2.21 and (2.22) it follows that $\varepsilon < \infty$ if and only if the integral (2.18) converges.

Remark 2. From the mathematical point of view, the assumptions $P(\lambda) > 0$, $Q(\lambda) > 0$ are not necessary. Suppose that $P(\lambda) = 1$, $Q(\lambda) = q_+(\lambda)q_-(\lambda)$, the polynomial $q_+(\lambda)$ has q_+ roots in the upper half-planes, the polynomial $q_-(\lambda)$ has q_- roots in the lower half-plane, $q_+ + q_- = q$, and $q_+ \neq q_-$. Solutions to equation $q_+(L)\psi^+ = 0$ tend to zero as $x \to +\infty$, while solutions of equation $q_-(L)\psi^- = 0$ tend to zero as $x \to -\infty$. We can find some solutions of equation (2.1) in the space H_{-qs} using the method given in the proof of Theorem 3. But it is not clear how to select a unique solution in this situation. There is no unique solution of minimal order of singularity under the above assumptions. In stochastic optimization theory equation (2.1) with kernel (2.2) and $Q(\lambda) = q_+(\lambda)q_-(\lambda)$, $\deg q_+ \neq \deg q_-$ seems to be of no interest.

3. Investigation of the Vector Equations

We prove here Theorem 5. The idea of the proof is similar to that of the proof of Theorem 4. We consider equation (1.1) with matrix kernel (1.13) and write it in the form (1.16). The equivalent form is (1.17).

Lemma 1. The set of solutions to equation (1.17) in the space $\mathcal{H}_{-n}$, $n = sq$, $s = \text{ord } L$, $q = \deg Q(\lambda)$ is in one-to-one correspondence with the set of solutions of the equation

$$\int_J S(x,y)H(y)dy = G(x), \quad x \in J, \quad \text{supp } H \subseteq \bar{D} \tag{3.1}$$

in $\mathcal{H}_{-n}(\mathbb{R}^1)$, where the vector function $G(x)$ is defined by formula (1.18), $\bar{D} = [t-T,t]$.

Proof of Lemma 1: Let $h(x) \in \mathcal{H}_{-n}$ be a solution of equation (1.17), $H(x) = h(x)$ in $\bar{D}$, $H(x) = 0$ in Ω. Then $\text{supp } H \subseteq \bar{D}$, $H \in \mathcal{H}_{-n}(\mathbb{R}^1)$. Denoting by $\psi(x)$ the left-hand side of equation (3.1), we see that $Q(L)\psi = 0$ for $x \in \Omega$ because $Q(L)S = \delta(x-y)E$, where E is the unit matrix in $\mathbb{R}^d$. Because of the condition $Q(\lambda) > 0$, the kernel $S(x,y) \to 0$ as $|x-y| \to \infty$. So $\psi(\infty) = 0$. There exists a fundamental system of matrix solutions of the matrix equation $Q(L)\psi_j = 0$ such that $\frac{1}{2}n$ of the solutions ψ_j^+, $1 \leq j \leq \frac{1}{2}n$, satisfy the condition $\psi_j^+(+\infty) = 0$, while $\frac{1}{2}n$ of the solutions ψ_j^-, $1 \leq j \leq \frac{1}{2}n$, satisfy the condition $\psi_j^-(-\infty) = 0$. So the vector function $\psi(x)$ is a linear combination of ψ_j^+, $1 \leq j \leq \frac{1}{2}n$, for $x > t$ and a linear combination of ψ_j^-, $1 \leq j \leq \frac{1}{2}n$, for $x < t-T$, where the coefficients of the linear combinations are arbitrary constant vectors in $\mathbb{R}^d$. So we have proved that the function $H(x)$ satisfies equation (3.1) with

$$G(x) = \begin{cases} \sum_{j=1}^{\frac{1}{2}n} \psi_j^+(x) b_j^+, & x > t, \\ g_0(x) + \sum_{j=1}^{m} \phi_j(x) c_j, & t-T \le x \le t, \\ \sum_{j=1}^{\frac{1}{2}n} \psi_j^-(x) b_j^-, & x < t-T, \end{cases} \tag{3.2}$$

where vectors $b_j^\pm$, C_j are arbitrary. Conversely, $H(x) \in \mathcal{H}_{-n}(\mathbb{R}^1)$ is a solution of equation (3.1), if $\operatorname{supp} H(x) \subseteq \bar{D}$, $G(x)$ is defined by formula (3.2), and $h(x) = H(x)$ in $\bar{D}$, then $h \in \mathcal{H}_{-n}$ and

$$\int_D S(x,y)h(y)dy = g_0(x) + \sum_{j=1}^{m} \phi_j(x) c_j, \quad x \in \bar{D}. \tag{3.3}$$

So $h(x)$ is a solution of equation (1.17), $h \in \mathcal{H}_{-n}$. □

Proof of Theorem 5: To prove Theorem 5, we note that

$$Q(L)S(x,y) = \delta(x-y)E. \tag{3.4}$$

So any solution $H(x) \in \mathcal{H}_{-n}$ of equation (3.1) can be written in the form

$$H(x) = Q(L)G(x). \tag{3.5}$$

As $Q(L)\psi_j^\pm = 0$, we conclude from formulas (3.5) that $\operatorname{supp} H(x) \subseteq \bar{D}$.

According to Lemma 1, the function

$$h(x) = Q(L)G(x) \tag{3.6}$$

is a solution of equation (3.3) and is therefore a solution of vector equation (1.1) with the kernel (1.13). From formula (3.6) with various $b_j^\pm$, c_j we obtain various solutions of the vector equation (1.1) in $\mathcal{H}_{-n}$. Now we prove that the solution with minimal order of singularity of the vector equation (1.1) does exist, is unique, has order of singularit at most $\frac{1}{2}(n-m)$, and can be calculated from formula (3.6), where $G(x)$ is defined by formulas (3.2) and the vectors $b_j^{\pm}$, $1 \leq j \leq \frac{1}{2}n$, and c_j, $1 \leq j \leq m$, satisfy the linear system (1.19). To prove this we note that the vector function $h(x)$ in formula (3.6) has minimal order of singularity if and only if $G(x)$ has minimal order of singularity, i.e., $G(x)$ is maximally smooth. If $f \in \mathscr{H}_{\frac{1}{2}(n-m)}$ then $g_o(x)$ in formula (3.2) belongs to $\mathscr{H}_{\frac{1}{2}(n+m)}$ because it is a solution of equation $A(L)g = f$, where $A(L) = (A_{ij}(L))$, $1 \leq i,j \leq d$, is a matrix differential operator of order at most m,

$$A(L)v = \sum_{j=0}^{m} B_j(x)v^{(j)}, \quad \det B_m(x) \neq 0, \quad x \in J, \tag{3.7}$$

and $B_j(x)$ are matrix coefficients of size $d \times d$. So for the function $G(x)$ to be maximally smooth it is necessary and sufficient that the conjugation conditions (1.19) are satisfied. It remains to prove that these conditions define the vectors $b_j^{\pm}$, c_j uniquely. This is true if the homogeneous linear system (1.19) has no nontrivial solutions. Every solution to system (1.19) generates by formula (3.6) with $G(x)$ defined by formula (3.2) and $b_j^{\pm}$, c_j the solution to system (1.19), a solution to the equation

$$Rh = f, \quad x \in \overline{D}, \quad h \in \mathscr{H}_{-\frac{1}{2}(n-m)}. \tag{3.8}$$

If $f = 0$, this equation becomes the homogeneous system

$$Rh = 0, \quad x \in \overline{D}, \quad h \in \mathscr{H}_{-\frac{1}{2}(n-m)}. \tag{3.9}$$

Hence

$$0 = \int_D Rh \cdot h^* dx = \int_\Lambda \tilde{R}(\lambda)\tilde{h}(\lambda) \cdot \tilde{h}^* d\rho(\lambda), \tag{3.10}$$

where $a \cdot b$ is the inner product in $\mathbb{R}^d$. By our assumption the matrix $\tilde{R}(\lambda)$ is positive definite, so from (3.10) it follows that $\tilde{h}(\lambda) = 0$ and hence $h(x) = 0$. That is why the system (1.19) has a unique solution. This means that there exists a unique solution with minimal order of singularity, which is at most $\frac{1}{2}(n-m)$, of the vector equation (1.1 with kernel (1.13). It remains to prove that the map $R^{-1}: \mathscr{H}_{\frac{1}{2}(n-m)} \to \mathscr{H}_{-\frac{1}{2}(n-m)}$ is continuous. We have already proved that the map $R: \mathscr{H}_{-\frac{1}{2}(n-m)} \to \mathscr{H}_{\frac{1}{2}(n-m)}$ is injective and surjective. As R is a linear map, the inverse map R^{-1} is continuous. □

We now discuss some problems of stochastic optimization theory which can be reduced to the vector equation (1.1). Let some vector random process $u = s(t) + n(t)$ be the input of a filter with impulse response $h(x,y)$, where $s(t)$ is the useful signal, $n(t)$ is noise, and $s(t)$ and $n(t)$ have mean values zero. Let $R(x,y)$ denote $(R_{ij}(x,y)) \equiv \overline{u_i^*(x)u_j(}$ $1 \le i,j \le d$, where the line denotes statistical average and the star denotes complex conjugation. The output signal is $Au = \int_D h(x,y)u(y)dy$, where h is a $d \times d$ matrix. Let f be a given random function. The problem consists of finding $h(x,y)$ (or A) such that

$$\varepsilon \equiv \mathscr{D}[Au - f] = \min, \tag{3.11}$$

where $\mathscr{D}$ is the variance. If $f = s(t)$ we have a filtering problem, if $f = s(t+t_0)$, $t_0 > 0$, we have a filtering and extrapolation problem. It is not difficult to prove that the necessary condition for $h(x,y)$ to be optimal is the equation

$$\int_D R(x,y')h^T(y,y')dy' = R_1(x,y), \quad x \in \overline{D}, \tag{3.12}$$

where $(R_1(x,y))_{ij} = \overline{u_i^*(x)f_j(y)}$, h^T is the matrix transpose of h, and y is a parameter. The matrix equation (3.12) is equivalent to d vector equations with the same kernel in which the right-hand side is one of the columns of the matrix R_1, $h^T = h(y')$ is the corresponding column in the matrix $h^T(y,y')$, and the dependence on y is omitted. We assume that $R(x,s) \in \mathcal{R}$, i.e., has the form of (1.13) so that Theorem 5 is applicable. Denoting by h_o the solution of matrix equation (3.12) with minimal order of singularity, we can write the following formula for the dispersion of error of an optimal filter:

$$\begin{aligned}\varepsilon_{min} &= R_2(x) - 2\mathrm{Re}\int_D \mathrm{tr}\{h_o^*(x,y)R_1(y,x)\}dy \\ &+ \int_D\int_D \mathrm{tr}\{h_o^*(x,y)R(y,y')h_o^T(x,y')\}dydy',\end{aligned} \tag{3.13}$$

where $R_2(x) = \mathrm{tr}\ \overline{\{f_i^*(x)f(x)\}} = \overline{|f(x)|^2}$, tr denotes the trace of a matrix, $h_o(x,s)$ is the solution with minimal order of singularity of equation (3.12), and the star denotes here complex (not hermitian) conjugation (so that $(h^*)_{ij} = h_{ij}^*$). Using equation (3.12), one can rewrite formula (3.13) as

$$\varepsilon_{min} = R_2(x) - \int_D \mathrm{tr}\{h_o(x,y)R_1^*(y,x)\}dy. \tag{3.14}$$

This formula is similar to formula (2.21). As in Section 2, it can be shown that $\varepsilon < \infty$ if and only if the order of singularity of solution of equation (3.12) is minimal.

4. Investigation of the Multidimensional Equations

Here proofs of Theorems 1 and 2 will be given. We start with a lemma.

Lemma 1. The set of solutions of equation (1.2) *with kernel* (1.3) *with* $P(\lambda) = 1$ *in the space* $H_{-\frac{1}{2}sq}$ *is in one-to-one correspondence with the set of solutions of the equation*

$$\int_{\mathbb{R}^r} R(x,y)H(y)dy = F(x), \qquad x \in \mathbb{R}^r. \tag{4.1}$$

belonging to $H_{-\frac{1}{2}sq}(\mathbb{R}^r)$ *with* $\operatorname{supp} H(x) \subseteq \overline{D}$, *where*

$$F(x) = \begin{cases} f(x), & x \in \overline{D} \\ u(x), & x \in \Omega = \mathbb{R}^r D, \end{cases} \tag{4.2}$$

and $u(x)$ *is the solution to the boundary-value problem*

$$\begin{cases} Q(L)u = 0 \quad \text{in} \\ \partial_N^j u\big|_\Gamma = \psi_j(s), \quad 0 \le j \le \frac{1}{2}sq-1, \quad u(\infty) = 0, \end{cases} \tag{4.3}$$

where ∂_N^j *denotes the derivative along the outward pointing normal to the boundary* $\partial D = \Gamma$ *and the* $\psi_j(s)$ *are arbitrary sufficiently smooth functions.*

Proof of Lemma 1: Let $h \in H_{\frac{1}{2}sq}$ be a solution of equation (1.2). Then the function $u(x) = \int_D R(x,y)h(y)dy$ satisfies equation (4.3) in Ω because $Q(L)R(x,y) = \delta(x-y)$. As $Q(\lambda) > 0$, the selfadjoint elliptic operator $Q(L)$ is positive definite in $L^2(\mathbb{R}^r)$. Hence $R(x,y) \to 0$ as $|x-y| \to \infty$.

Therefore the above defined function $u(x) = \int_D R(x,y)h(y)dy$ satisfies the condition $u(\infty) = 0$ and also the equation $Q(L)u = 0$ in Ω. Let us put

$$H(x) = \begin{cases} h(x), & x \in \bar{D}, \\ 0, & x \in \Omega. \end{cases} \tag{4.4}$$

Then $H(x)$ satisfies equation (4.1) with $F(x)$ defined by formula (4.2) in which $u(x)$ is the solution to problem (4.3), and $H \in H_{-\frac{1}{2}sq}(\mathbb{R}^r)$. If $h(x) \in H_t$, $t \geq -\frac{1}{2}sq$, then the function

$$u(x) = \int_D R(x,s)h(y)dy = \int_{\mathbb{R}^r} R(x,s)H(y)ds \in H_{t+sq}(\mathbb{R}^r)$$

and $t+sq \geq \frac{1}{2}sq$, so that $\partial_N^j u\big|_\Gamma$, $0 \leq j \leq \frac{1}{2}sq - 1$, belongs to $L^2(\Gamma) = H_o(\Gamma)$. Actually, it is known (see, for example, Beresanskij [1]) that for $u \in H_t(\mathbb{R}^r)$, $t > \frac{1}{2}$, and sufficiently smooth Γ the restriction of u to Γ is a bounded operator from $H_t(\mathbb{R}^r)$ onto $H_{t-\frac{1}{2}}(\Gamma)$. (The spaces $H_{-(t-\frac{1}{2})}(\Gamma)$ are considered to be the negative spaces constructed by $H_o(\Gamma) = L^2(\Gamma)$ and the positive space $H_{t-\frac{1}{2}}(\Gamma)$, $t \geq \frac{1}{2}$.)

Conversely, if $H \in H_{-\frac{1}{2}sq}(\mathbb{R}^r)$, supp $H \subseteq \bar{D}$, is a solution of equation (4.1), where $F(x)$ is given by formula (4.2) in which u is the solution to problem (4.3), then

$$H(x) = \begin{cases} H(x), & x \in \bar{D} \\ 0, & x \in \Omega \end{cases}$$

is the solution of equation (1.2). □

Remark 1. Problem (4.3) is the well-known Dirichlet problem for the positive definite selfadjoint elliptic operator $Q(L)$. It is well-known (Beresanskij [1]) that this problem has a unique solution if the boundary Γ, the functions $\psi_j(s)$, and the coefficients of the operator L are sufficiently smooth. Precise restrictions concerning the smoothness can be found in Beresanskij [1]. It is more than enough to assume that the boundary and the coefficients are qs-smooth, $\psi_j \in H_{\frac{1}{2}qs-j-\frac{1}{2}}(\Gamma)$. If $H(y) \in H_{-sq}$ in formula (4.1) but $f(x)$ is smooth in D, and $u(x)$ is smooth in Ω, then sing supp $H = \Gamma$. So the set of solutions H in $H_{-sq}(\mathbb{R}^r)$ of equation (4.1) with sing supp $H = \Gamma$ is in one-to-one correspondence with the set of solutions h in H_{-sq} of equation (1.2) with kernel (1.3), $P(\lambda) = 1$, and sing supp $h = \Gamma$.

Proof of Theorem 1: Taking into account that $Q(L)R(x,y) = \delta(x-y)$, we conclude that the set of solutions of equation (4.1) can be found from the formula

$$H(x) = Q(L)F(x), \tag{4.5}$$

where $F(x)$ is given by formula (4.2) in which $u(x)$ is the solution to problem (4.3). The latter problem is the Dirichlet problem which, as it is well-known, has a unique solution under the above assumptions concerning smoothness of the boundary Γ, the coefficients of L, and the functions ϕ_j. From equalities (4.3) and (4.5) we see that $H(x) = 0$ in Ω. So

$$h(x) = \begin{cases} H(x), & x \in \overline{D} \\ 0 \quad , & x \in \Omega \end{cases}$$

is a solution to equation (1.2). If $H \in H_{-\frac{1}{2}sq}(\mathbb{R}^r)$ then $h \in H_{-\frac{1}{2}sq}$. According to Remark 1 if $H \in H_{-sq}(\mathbb{R}^r)$ and sing supp $H = \Gamma$, then $h \in H_{-sq}$ and sing supp $h = \Gamma$. This is so when $f(x)$ and $u(x)$ are sufficiently smooth in D and Ω, respectively. In order to find the solution of equation (1.2) with minimal order of singularity, we must take $F(x)$ in (4.5) maximally smooth. That, in turn, is equivalent to finding $u(x)$ such that the conjugation boundary conditions hold:

$$[\partial_N^j F]\Big|_\Gamma = 0, \quad 0 \leq j \leq \tfrac{1}{2}sq - 1, \tag{4.6}$$

where $[\partial_N^j F]_\Gamma$ denotes the jump of $\partial_N^j F$ on Γ and we assume F is smooth in D and Ω so that sing supp $H = \Gamma$. The conditions (4.6) are equivalent to the conditions

$$\partial_N^j f\Big|_\Gamma = \partial_N^j u\Big|_\Gamma, \quad 0 \leq j \leq \tfrac{1}{2}sq - 1. \tag{4.7}$$

So the solution of equation (1.2) with minimal order of singularity exists and can be found from

$$h(x) = Q(L)F(x), \tag{4.8}$$

where $F(x)$ is given by formula (4.2) in which $u(x)$ is the solution of the Dirichlet problem (4.3) with $\phi_j = \partial_N^j f\Big|_\Gamma$. This solution $h \in H_{-\frac{1}{2}sq}$. It is obvious that in general the order of singularity of $h(x)$ cannot be made less than $\frac{1}{2}sq$ because one cannot impose more than $\frac{1}{2}sq$ boundary conditions (4.7) in the Dirichlet problem (4.3). Nevertheless, the solution $h(x)$ actually can be smooth under some appropriate assumptions concerning $f(x)$. For example, if $f(x) \in C_0^\infty(D)$ and the coefficients of L are infinitely smooth then, by formula (4.8), $h(x) = Q(L)F(x) = Q(L)f(x) \in C_0^\infty(D)$. We end the proof by remarking that the map $R^{-1}: f \to h$ is a homomorphism of $H_{\frac{1}{2}sq}$ onto $H_{-\frac{1}{2}sq}$.

Indeed it was already proved that R is injective and surjective. As R is linear, the inverse R^{-1} is continuous.

<u>Proof of Theorem 2</u>: Let us rewrite equation (1.2) with kernel (1.3) as

$$P(L)\int_D S(x,y)h(y)dy = f(x), \quad x \in \overline{D}, \tag{4.9}$$

where

$$S(x,y) = \int_\Lambda Q^{-1}(\lambda)\Phi(x,y,\lambda)d\rho(\lambda). \tag{4.10}$$

Equation (4.9) is equivalent to the equation

$$\int_D S(x,y)h(y)dy = g_o(x) + v(x), \quad x \in \overline{D}, \tag{4.11}$$

where $g_o(x)$ is an arbitrary fixed solution to the equation

$$P(L)g = f, \quad x \in D \tag{4.12}$$

and v is an arbitrary solution to the homogeneous equation

$$P(L)v = 0, \quad x \in D. \tag{4.13}$$

The kernel of equation (4.11) satisfies the hypotheses of Theorem 1, so we can find the set of solutions of equation (4.11) with sing supp $h = \Gamma$ from the formula

$$h(x) = Q(L)G(x), \tag{4.14}$$

where

$$G(x) = \begin{cases} g_o(x) + v & \text{in } D, \\ u(x) & \text{in } \Omega, \end{cases} \tag{4.15}$$

and $u(x)$ is a solution to the equation $Q(L)u = 0$ in Ω. The solution $h(x)$ in formula (4.14) will have minimal order of singularity if and only if $G(x)$ is maximally smooth. This will be the case if and only if the conjugation condition

$$[\partial_N^j G]\Big|_\Gamma = 0, \quad 0 \le j \le \tfrac{1}{2}s(p+q) - 1, \tag{4.16}$$

hold. Roughly speaking, $\frac{1}{2}sq$ conditions determine $u(x)$ and $\frac{1}{2}ps$ conditions determine $v(x)$. The conditions (4.16) lead us to a nonlocal boundary-value problem from which $u(x)$, $v(x)$ are to be found:

$$\begin{cases} P(L)G = f \quad \text{in} \quad D, \quad Q(L)G = 0 \quad \text{in} \quad \Omega, \quad u(\infty) = 0, \\ [\partial_N^j G]\Big|_\Gamma = 0, \quad 0 \le j \le \frac{1}{2}s(p+q) - 1. \end{cases} \tag{4.17}$$

We note that $G \in H_{\frac{1}{2}(p+q)s}(\mathbb{R}^r)$. The peculiarity of this problem is that the orders of the elliptic operators $P(L)$ and $Q(L)$ are different. So we have an elliptic boundary-value problem with the order of the operator depending on x; the order is ps in D and qs in Ω. We assume for a moment that problem (4.17) has a unique solution and proceed with the proof of Theorem 2. If $f \in H_{\frac{1}{2}s(q-p)}$ then the right-hand side of equality (4.11) belongs to $H_{\frac{1}{2}s(q+p)}$. Hence the solution to equation (1.2) given by formula (4.14) belongs to $H_{-\frac{1}{2}s(q-p)}$. The continuous linear map $R:h \to f$ defined by $Rh = f$ maps $H_{-\frac{1}{2}(q-p)s}$ onto $H_{\frac{1}{2}s(q-p)}$. So the inverse map R^{-1} is continuous and is a homomorphism of $H_{\frac{1}{2}(q-p)s}$ onto $H_{-\frac{1}{2}s(q-p)}$.

It remains to study the problem (4.17). First we prove that the solution to this problem is unique. Assume that $G(x)$ is a solution to the problem (4.17) with $f = 0$. By formula (4.19) we find the solution h of the homogeneous equation (1.2), $h \in H_{-\frac{1}{2}s(q-p)}$. Using Parseval's inequality and the equation $Rh = 0$, we obtain

$$0 = \int_D Rh \cdot h^* dy = \int_{\mathbb{R}^r} Rh \cdot h^* dy = \int_\Lambda \tilde{R}(\lambda)|\tilde{h}|^2 d\rho(\lambda), \tag{4.18}$$

where $\tilde{h}(\lambda)$ is the Fourier transform of $h(x)$ with respect to eigenfunction expansions of the operator L. As $\tilde{R}(\lambda) \equiv P(\lambda)Q^{-1}(\lambda) > 0$, we have $h = 0$. So the solutions of equation (1.2) and problem (4.17) in $H_{-\frac{1}{2}s(q-p)}$ are unique. We note that the condition $h \in H_{-\frac{1}{2}s(q-p)}$ implies the convergence of the integrals in (4.18). To prove the existence of the solution to problem (4.17) we consider the bilinear form

$$[\phi,\psi] \equiv \int_\Lambda P(\lambda)Q(\lambda)\tilde{\phi}(\lambda)\tilde{\psi}^*(\lambda)d\rho(\lambda), \tag{4.19}$$

defined on the set $H_{\frac{1}{2}(q+p)s}(\mathbb{R}^r) \cap H_{sq}(\Omega)$ of functions satisfying the equation $Q(L)G = 0$ in Ω. We denote this set by V. As $P(\lambda)Q(\lambda) > 0$, $\lambda \in \Lambda$, we conclude that $[\phi,\phi] \geq C\,\|\phi\|^2$, where $C = \text{const} > 0$ and $\|\cdot\|$ is the norm in $L^2(\mathbb{R}^r)$. The integral in (4.19) converges because $\phi,\psi \in H_{\frac{1}{2}s(q+p)}(\mathbb{R}^r)$. Indeed, $|\phi|(1+\lambda^2)^{\frac{1}{4}(p+q)} \in L^2(\Lambda,d\rho)$, $|\psi|(1+\lambda^2)^{\frac{1}{4}(p+q)} \in L^2(\Lambda,d\rho)$, so that $\phi\psi(1+\lambda^2)^{\frac{1}{2}(p+q)} \in L^1(\Lambda,d\rho)$. But $P(\lambda)Q(\lambda) \sim C(1+\lambda^2)^{\frac{1}{2}(p+q)}$ as $|\lambda| \to \infty$. From here the convergence of integral (4.19) follows. According to Parseval's equality we have

$$[G,\psi] = \int_{\mathbb{R}^r} P(L)G\{Q(L)\psi\}^*dx = \int_D P(L)G\{Q(L)\psi\}^*dx.$$

That is why the boundary-value problem (4.17) is equivalent to the equality

$$[G,\psi] = \int_D f\{Q(L)\psi\}^*dx, \quad \psi \in V. \tag{4.20}$$

For $f \in H_{\frac{1}{2}(q-p)s}$, the left-hand side of equality (4.20) is a linear functional on the Hilbert space W, which is the completion of the set V with respect to the norm generated by the bilinear form (4.19). Indeed, extending the function $f(x) \in H_{\frac{1}{2}s(q-p)}$ throughout $\mathbb{R}^r$ so that the extension operator $j: H_{\frac{1}{2}s(q-p)} \to H_{\frac{1}{2}s(q-p)}(\mathbb{R}^r)$ is bounded and using Parseval's equality we obtain

$$\left|\int_D f\{Q(L)\psi\}^* dx\right| = \left|\int_\Lambda \tilde{f}(\lambda)Q(\lambda)\tilde{\psi}(\lambda)d\rho(\lambda)\right|$$
$$\leq \left\{\int_\Lambda P(\lambda)Q(\lambda)|\tilde{\psi}|^2 d\rho\right\}^{\frac{1}{2}}\left\{\int_\Lambda |f|^2 \frac{Q(\lambda)}{P(\lambda)}d\rho(\lambda)\right\}^{\frac{1}{2}}$$
$$\leq C(f)[\psi,\psi]^{\frac{1}{2}}. \tag{4.21}$$

Here $f \in H_{\frac{1}{2}s(q-p)}(\mathbb{R}^r)$, $C^2(f) = \int_\Lambda |\tilde{f}|^2 Q(\lambda)P^{-1}(\lambda)d\rho < \infty$. According to Riesz's theorem about linear functionals on the Hilbert space, equality (4.20) can be written in the form

$$[G,\psi] = [Tf,\psi], \quad Tf \in W, \quad \psi \in V. \tag{4.22}$$

Hence $G = Tf$ and the solvability of problem (4.17) in W for any $f \in H_{\frac{1}{2}s(q-p)}$ is proved. It is quite clear from the given proof that the constructed solution is unique in W. □

5. Approximate Solution of the Integral Equations in the Space of Distributions

Here we prove Theorem 13. For the sake of brevity, we do not repeat the notations used in the formulation of Theorem 13 in Section 1. We consider equation (1.2) with the kernel

$$R(x,y) = \int_\Lambda \tilde{R}(\lambda)\Phi(x,y,\lambda)d\rho(\lambda). \tag{5.1}$$

Lemma 1. Let $H_+ \subset H_0 \subset H_-$ be a triple of Hilbert spaces, where H_- is the negative space of H_+ relative to H_0. Let $R: H_- \to H_+$ be a linear map satisfying the inequalities

$$C_1|h|_-^2 \leq (Rh,h) \leq C|h|_-^2, \quad C_1, C > 0, \quad h \in H_-. \tag{5.2}$$

Then

$$|R| \leq 2C, \quad |R^{-1}| \leq C_1^{-1}. \tag{5.3}$$

Here $|R|$ is the norm of the operator $R: H_- \to H_+$, $|R^{-1}|$ is

<u>the norm of the operator</u> $R^{-1}: H_+ \to H_-$, (f,h) <u>is the value of the functional</u> $h \in H_-$ <u>on the element</u> $f \in H_+$, <u>and</u> $|h|_-$ <u>is the norm of</u> h <u>in</u> H_-.

<u>Proof of Lemma 1</u>: By assumption, the linear operator R is defined everywhere on H_-. From (5.2) it follows that $C_1|h|_-^2 \le |Rh|_+|h|_-$. Hence

$$|Rh|_+ \ge C_1|h|_-. \tag{5.4}$$

Therefore R^{-1} is defined on $\operatorname{im} R$ and the inequality $|R^{-1}| \le C_1^{-1}$ holds. To prove the first inequality in (5.3) we note that

$$|R| = \sup_{|h|_-\le 1, |g|_-\le 1} |(Rg,h)| = \sup_{|h|_-\le 1, |g|_-\le 1} |\tfrac{1}{4}\{(R(h+g),h+g)$$

$$- (R(h-g),h-g) - i[R(h+ig),h+ig) - (R(h-ig),h-ig)]\}|$$

$$\le \sup \tfrac{1}{4}\{C|h+g|_-^2 + C|h-g|_-^2 + C|h+ig|_-^2 + C|h-ig|_-^2\}$$

$$\le C \sup\{|h|_-^2 + |g|_-^2\} \le 2C.$$

<u>Remark 1</u>. Under the assumptions of Lemma 1, the map R is surjective and R is a homomorphism of H_- onto H_+.

<u>Proof of Remark 1</u>: Indeed R is monotone, continuous, and coercive. Hence R is surjective (see Lions [1]). From (5.3) it follows that R and R^{-1} are continuous linear operators. Therefore R is a homomorphism of H_- onto H_+. □

<u>Lemma 2</u>. <u>Let</u> $R_\varepsilon: H_- \to H_+$ <u>be a linear bijective map of</u> H_- <u>onto</u> H_+ <u>for any</u> ε, $0 < \varepsilon < \varepsilon_0$, <u>where</u> $\varepsilon_0 > 0$ <u>is fixed</u>. <u>Let</u> $R: H_- \to H_+$ <u>be a linear map such that</u> $|R_\varepsilon^{-1}| \le M$ <u>and</u> $|R-R_\varepsilon| < \varepsilon$, <u>where</u> $M = \text{const} > 0$ <u>does not depend on</u> ε,

$0 < \varepsilon < \varepsilon_0$. Then $R: H_- \to H_+$ is a bijection and $|R^{-1}| \leq M(1 - \varepsilon M)^{-1}$ for $\varepsilon M < 1$.

Proof of Lemma 2: We have $R = R_\varepsilon + R - R_\varepsilon = R_\varepsilon[I + R_\varepsilon^{-1}(R-R_\varepsilon)]$, where I is the identify operator on H_-. The operator $R_\varepsilon^{-1}(R - R_\varepsilon)$ acts on H_- and its norm does not exceed εM. So if $\varepsilon M < 1$ the operator $[I + R_\varepsilon^{-1}(R - R_\varepsilon)]^{-1}$ exists and is a bounded operator on H_- with norm at most $(1 - \varepsilon M)^{-1}$. Hence the operator $R^{-1} = [I + R_\varepsilon^{-1}(R - R_\varepsilon)]^{-1}R_\varepsilon^{-1}$ exists and is defined everywhere on H_+, $\operatorname{im} R^{-1} = H_-$, and $|R^{-1}| \leq M(1 - \varepsilon M)^{-1}$. □

Remark 2. In what follows the roles of H_-, H_0, and H_+ are played by $H_{-s\beta}$, $L^2(D)$, and $H_{s\beta}$.

Proof of Theorem 13: Under the assumptions of Theorem 13 using Parseval's equality, we obtain

$$\begin{aligned}(Rh,h) &= \int_\Lambda \tilde{R}(\lambda)|\tilde{h}(\lambda)|^2 d\rho \\ &\leq C\int_\Lambda (1+\lambda^2)^{-\beta}|\tilde{h}|^2 d\rho = C|h|^2_{-s\beta},\end{aligned} \tag{5.6}$$

where $C = \sup_{\lambda\in J}\{(1+\lambda^2)^\beta\tilde{R}(\lambda)\}$. Letting C_1 denote $\inf_{\lambda\in J}\{(1+\lambda^2)^\beta\tilde{R}\}$, we obtain the inequality

$$C_1|h|^2_{-s\beta} \leq (Rh,h) \leq C|h|^2_{-s\beta}. \tag{5.7}$$

From here and Lemma 1, inequalities (5.3) follow. Hence, according to Remark 1, the map R is surjective and R is a homomorphism of $H_{-s\beta}$ onto $H_{s\beta}$. Furthermore, we have

$$\begin{aligned}|R^{-1} - R_\varepsilon^{-1}| &= |R_\varepsilon^{-1}(R_\varepsilon - R)R^{-1}| \leq |R_\varepsilon^{-1}||R - R_\varepsilon||R^{-1}| \\ &\leq |R_\varepsilon^{-1}||R - R_\varepsilon||R^{-1}| \leq 2\varepsilon M^2(1 - 2\varepsilon M)^{-1} \quad \text{if} \quad 2\varepsilon M < 1.\end{aligned} \tag{5.8}$$

Here we took into consideration that the estimate

$\|R - R_\varepsilon\| < \varepsilon$ and the first estimate in (5.3) imply the inequality $|R - R_\varepsilon| \leq 2\varepsilon$, M is the constant from the estimate $|R_\varepsilon^{-1}| \leq M$, and we make use of Lemma 2. According to the second estimate in (5.3), we note that

$$M = \inf_{0<\varepsilon<\varepsilon_0} \inf_{\lambda \in J} [(1 + \lambda^2)^\beta \tilde{R}_\varepsilon(\lambda)]^{-1}$$

Remark 3. We can give an estimate of $|R_\varepsilon^{-1}|$ using only values connected with $\tilde{R}(\lambda)$. Indeed, arguments similar to those used in the proof of Lemma 2 lead to the estimate

$$|R_\varepsilon^{-1}| \leq (1 - \varepsilon C_1^{-1})^{-1} C_1^{-1}, \tag{5.9}$$

where $C_1 = \inf_{\lambda \in J}\{(1 + \lambda^2)^\beta \tilde{R}(\lambda)\}$.

We can omit all direct assumptions concerning $\tilde{R}(\lambda)$ and introduce the following assumptions: For any $\varepsilon > 0$ there exists $\tilde{R}_\varepsilon(\lambda) = P_\varepsilon(\lambda) Q_\varepsilon^{-1}(\lambda), P_\varepsilon(\lambda) > 0$, $Q_\varepsilon(\lambda) > 0$ for $\lambda \in J$, $P_\varepsilon(\lambda)$, $Q_\varepsilon(\lambda)$ are polynomials, $\deg Q_\varepsilon - \deg P_\varepsilon = 2\beta > 0$, β does not depend on ε, $0 < \varepsilon < \varepsilon_0$,

$$\|R(\lambda) - R_\varepsilon(\lambda)\| < \varepsilon, \tag{5.10}$$

where $\|\tilde{R}\| \equiv \sup_{\lambda \in J}\{(1+\lambda^2)^\beta \tilde{R}(\lambda)\}$, $\inf_{0<\varepsilon<\varepsilon_0} \inf_{\lambda \in J} [(1+\lambda^2)^\beta \tilde{R}_\varepsilon(\lambda)] > 0$, and $\sup_{0<\varepsilon<\varepsilon_0} \|R_\varepsilon(\lambda)\| < \infty$.

From (5.10) it is easy to deduce that R is a homomorphism of $H_{-s\beta}$ onto $H_{s\beta}$. The reader may do this as an exercise.

Remark 4. Assume that equation (1.2) with kernel (1.3) is being considered in $L^2(D)$ and R is compact in $L^2(D)$. We consider also an approximate equation $R_\varepsilon h_\varepsilon = f_\delta$, where $|f - f_\delta|_{L^2(D)} \leq \delta$, $|R - R_\varepsilon|_{L^2(D)} \leq \varepsilon$. The solution h_ε may exist and differ greatly from the solution h of equation $Rh = f$ in $L^2(D)$, or it may even not exist in $L^2(D)$. But it does exist and is stable under small perturbations of $f \in H_{s\beta}$ as an element of $H_{-s\beta}$. We have the estimate

$$|h_\varepsilon - h|_{-s\beta} = |R_\varepsilon^{-1} f_\delta - R^{-1} f|_{-s\beta} \leq |f - f_\delta|_{s\beta} |R_\varepsilon^{-1}| + |R_\varepsilon^{-1} - R_\varepsilon^{-1}| |f|_{s\beta}$$
$$\leq |f - f_\delta|_{s\beta} M + M^2 \varepsilon (1 - 2\varepsilon M)^{-1} |f|_{s\beta}.$$

The approximate analytical method of solving equation (1.2) given in Theorem 13 is stable under small perturbations of the right-hand side of equation (1.2) in $H_{s\beta}$ and small perturbations of the kernel such that both constants C_1, C remain in some interval $[a,b]$, $a > 0$, $b < \infty$.

6. Asymptotics of the Spectrum of the Investigated Integral Equations

Here we prove Theorems 6, 7, and 8.

Proof of Theorem 6: Using the well-known minimax description of the eigenvalues of nonnegative selfadjoint operators we can write

$$\lambda_j(D') = \min_{\psi_1, \ldots, \psi_{j-1}} \max_{\substack{(\phi,\psi_i)'=0, 1 \leq i \leq j-1 \\ (\phi,\phi)'=1}} (R\phi, \phi)' \equiv \min_{\psi} \mu_j(\psi), \tag{6.1}$$

where

$$(\phi, \psi)' \equiv \int_{D'} \phi(x) \psi^*(x) dx,$$

the star denotes complex conjugation, and $\mu(\psi)$ is the expression to be minimized. We put $\Delta = D' \setminus D$. If we impose an additional restriction on ϕ in formula (6.1) by requiring $\phi = 0$ in Δ; then $\mu_j(\psi)$ cannot increase. We shall write

$$(\phi,\psi) = \int_D \phi(x)\psi^*(x)dx.$$

Then

$$\mu_j(\psi) \geq N_j(\psi) \equiv \max_{\substack{(\phi,\psi_i)'=0,1\leq i\leq j-1,\\ (\phi,\phi)'=1,\phi=0 \text{ in } \Delta}} (R\phi,\phi)'$$

$$= \max_{\substack{(\phi,\psi_i)=0,1\leq i\leq j-1\\ (\phi,\phi)=1}} (R\phi,\phi). \tag{6.2}$$

As $\lambda_j(D) = \min_\psi N_j(\psi)$ we obtain the inequality

$$\lambda_j(D) \leq \lambda_j(D') \quad \text{for all } j \text{ if } D \subset D'. \tag{6.3}$$

This inequality holds also for selfadjoint, not necessarily nonnegative operators if we replace $\lambda_j(D)$ by $\lambda_j^+(D)$, $\lambda_j(D')$ by $\lambda_j^+(D')$, where the $\lambda_j^+(D)$ are the positive eigenvalues of the operator.

Let us suppose now that $D \to \mathbb{R}^r$. This means that D expands to infinity uniformly with respect to direction. We assume that

$$\sup_{x\in\mathbb{R}^r} \int |R(x,y)|dy = A < \infty, \quad \int \equiv \int_{\mathbb{R}^r}, \tag{6.4}$$

and prove that the limit $\lim_{D\to\mathbb{R}^r} \lambda_1(D) \equiv \lambda_{1\infty}$ exists and $\lambda_{1\infty} \leq A$. Indeed, $\lambda_1(D)$ increases monotonically as D expands, so it is sufficient to prove that $\lambda_1(D) \leq A$. But we have

$$\lambda_1(D) \max_x|\phi_1(x)| = \max_x \int_D |R(x,y)||\phi_1(y)|dy$$

$$\leq \max_x|\phi_1(x)| \max_x \int_D |R(x,y)|dy \leq A \max_x|\phi_1(x)|.$$

Therefore $\lambda_1(D) \leq A$. Here $\phi_1(x)$ is the first eigenfunction, which corresponds to $\lambda_1(D)$. It remains to prove that for the kernel (1.20), $\lambda_{1\infty} = \omega$, where $\omega \equiv \sup_{\lambda\in\Lambda} |\omega(\lambda)|$. We hav

$$\lambda_1(D) = \int_D \int_D R(x,y)\phi_1(y)\phi_1^*(x)\,dy\,dx = \int_\Lambda \omega(\lambda)|\tilde{\phi}_1(\lambda)|^2 d\rho(\lambda)$$
$$\leq \omega \int_\Lambda |\tilde{\phi}_1|^2 d\rho = \omega. \qquad (6.5)$$

Here we make use of Parseval's equality and the normalization condition for ϕ_1. From (6.5) it follows that $\lambda_{1\infty} \leq \omega$. Let us prove that the inequality $\lambda_{1\infty} < \omega$ is impossible. Let ϕ be an arbitrary function such that $|\phi|_{L^2(D)} = 1$. Then

$$\lambda_{1\infty} \geq \lambda_1(D) \geq \int_\Lambda \omega(\lambda)|\tilde{\phi}|^2 d\rho. \qquad (6.6)$$

Assuming that $\omega = \omega(\lambda_0) \geq \omega(\lambda)$, we choose the domain D and the function $\phi(x)$ so that the function $|\tilde{\phi}|^2 d\rho$ is close to $\delta(\lambda - \lambda_0)d\lambda$ in the following sense: $\int_\Lambda \omega(\lambda)|\tilde{\phi}|^2 d\rho \geq \omega - \varepsilon$, for a given small $\varepsilon > 0$. Then $\omega \geq \lambda_{1\infty} \geq \omega - \varepsilon$. As ε is arbitrary we conclude that $\omega = \lambda_{1\infty}$. □

Remark 1. If $R(x,y) = R(x-y)$, $R(-x) = R(x)$, the domain $D \subset \mathbb{R}^r$ is centrally symmetric with respect to the origin, r is even, and λ_j is a simple eigenvalue of the operator R, then the corresponding eigenfunction is either even or odd. We call the eigenvalue λ_j simple if its eigenspace is one-dimensional.

To prove the above statement we assume that $R\phi_j = \lambda_j\phi_j$ and have

$$\lambda_j\phi_j(-x) = \int_D R(-x-y)\phi_j(y)\,dy = \int_D R(x+y)\phi_j(y)\,dy$$
$$= (-1)^r \int_D R(x-z)\phi_j(-z)\,dz = \int_D R(x-z)\phi_j(-z)\,dz.$$

As λ_j is simple we conclude that $\phi_j(-x) = k\,\phi_j(x)$, $k =$ const. Hence $k^2 = 1$, $k = \pm 1$, $\phi_j(-x) = \pm\phi_j(x)$. If λ_j is not simple the corresponding eigenfunction need not be even or odd.

Example: $R\phi = \int_{-\pi}^{\pi} \phi(y)dy = 0$ has eigenvalue $\lambda = 0$ which is is not simple. The eigenfunction $\cos x + \sin x$ of this operator is neither odd nor even.

Proof of Theorem 7: First we assume that $\omega(\lambda) = Q^{-1}(\lambda) \sim (1 + \lambda^2)^{-\frac{1}{2}q}$ as $\lambda \to \infty$, and $q = \deg Q(\lambda)$ is even.

Instead of the problem

$$R\phi_n \equiv \int_D R(x,y)\phi_n(y)dy = \lambda_n\phi_n(x), \quad x \in \overline{D}, \tag{6.7}$$

we consider the problem

$$\int_{\mathbb{R}^r} R(x,y)\phi_n(y)dy = \begin{cases} \lambda_n\phi_n(x), & x \in \overline{D}, \\ u_n(x), & x \in \Omega, \end{cases} \tag{6.8}$$

with the additional restrictions

$$\begin{gathered} \phi_n(x) = 0 \quad \text{in} \quad \Omega, \quad Q(L)u_n = 0 \quad \text{in} \quad \Omega; \\ \partial_N^j u_n \Big|_\Gamma = \partial_N^j \lambda_n \phi_n \Big|_\Gamma, \quad 0 \le j \le \frac{1}{2}sq - 1. \end{gathered} \tag{6.9}$$

Every solution $\{\phi_n, u_n\}$ to problem (6.8) and (6.9) is an eigenfunction of equation (6.7) in D. The converse is also valid: every eigenfunction of equation (6.7) which is extended to have value zero in Ω, generates a solution to problem (6.8) and (6.9). The problem (6.8) and (6.9) is equivalent to the problem

$$\lambda_n Q(L)\phi_n = \kappa(x)\phi_n(x) \quad \text{in} \quad \mathbb{R}^r, \tag{6.10}$$

$$\kappa(x) = \begin{cases} 1, & \text{in} \quad D, \\ 0, & \text{in} \quad \Omega. \end{cases} \tag{6.10'}$$

Agmon [1] investigated this problem and proved that

$$\lambda_n \sim (\gamma n^{-1})^{qs/r}, \quad \text{as} \quad n \to \infty, \tag{6.11}$$

where

$$\begin{cases} \gamma = (2\pi)^{-r} \int_D \eta(x)dx \\ \eta(x) = \text{meas}\left\{t: t \in \mathbb{R}^r, \sum_{|\alpha|=|\beta|=\frac{1}{2}s} a_{\alpha\beta}(x)t^{\alpha+\beta} \leq 1\right\}. \end{cases} \tag{6.12}$$

Here α,β are multi-indices, the forms $a_{\alpha\beta}(x)$ generates the principal part of the operator L, $q = \deg Q(\lambda)$, and $s = \text{ord } L$. This completes the proof under the assumption on $\omega(\lambda)$ given in the first sentence.

In the general case, $\omega(\lambda) = (1 + \lambda^2)^{-\frac{1}{2}a}(1 + g(\lambda))$, where $\omega(\lambda) > 0$, $g(\lambda) \to 0$ as $\lambda \to \infty$, $g(\lambda)$ is continuous on J, and a is an even integer. The operator R with kernel (1.2) can be represented in the form $R = B(I+G)$, where the kernels of the operators B and $I + G$ have the form (1.20) with $\omega(\lambda) = \omega_B(\lambda) = (1 + \lambda^2)^{-a/2}$ and $\omega(\lambda) = \omega_{I+G}(\lambda) = 1 + g(\lambda)$, respectively. The operator $(I+G)^{-1}$ is a bounded selfadjoint positive definite operator in $H = L^2(D)$. In Lemma 1 below we prove that B and G are compact in H. From here and Theorem 11, which will be proved in Section 7, we deduce that $\lambda_n(R)\cdot\lambda_n^{-1}(B) \to 1$ as $n \to \infty$. We have already found above the asymptotics of $\lambda_n(B)$ (see formula (6.10)). To complete the proof of Theorem 7 it remains to prove the following lemma.

Lemma 1. Let R be the operator on $H = L^2(D)$ with kernel (1.20), and assume $\omega(\lambda)$ is continuous on Λ with $\lim_{\lambda\to\infty} \omega(\lambda) = 0$. Then $R: H \to H$ is compact.

Proof of Lemma 1: Let $\varepsilon > 0$ be an arbitrary small given number, and assume $\max_{\lambda\in\Lambda} |\omega(\lambda) - \omega_\varepsilon(\lambda)| < \varepsilon$, $\omega_\varepsilon(\lambda) = 0$ for $|\lambda| > A$, where $A > 0$ is some number. We denote by R_ε the operator with kernel (1.20), in which $\omega(\lambda)$ is replaced by

$\omega_\varepsilon(\lambda)$. Letting T denote the operator with kernel (1.20) acting on $L^2(\mathbb{R}^r)$, we can write $\|T\| = \max_{\lambda\in\Lambda} |\omega(\lambda)|$. Let us introduce the orthogonal projection P on $L^2(\mathbb{R}^r)$ onto $L^2(D)$. Then we can write $R = PTP$. Hence $|R| \leq \|T\|$ and $|R - R_\varepsilon| < \varepsilon$, where $|\cdot|$ denotes the operator norm in $L^2(D) = H$. Hence to prove the compactness of R in H it is sufficient to prove the compactness of R_ε in H. Let $\|f\|_{L^2(\mathbb{R}^r)} = 1$. Then

$$R_\varepsilon f = \int_D\left(\int_{-A}^{A} \omega_\varepsilon(\lambda)\Phi(x,y,\lambda)\,d\rho\right)f(y)\,dy,$$
$$\|LR_\varepsilon f\|^2_{L^2(\mathbb{R}^r)} = \int_{-A}^{A} \lambda^2|\omega_\varepsilon(\lambda)|^2 d(E_\lambda f,f) \leq M, \tag{6.13}$$

where $d(E_\lambda f,f) = \int_D\int_D \Phi(x,y,\lambda)f(y)f^*(x)\,dy\,dx\,d\rho(\lambda)$ and $M = A^2 \max_{|\lambda|\leq A} |\omega_\varepsilon(\lambda)|^2$.

The following estimate is well-known (see, for example, Beresanskij [1])

$$\|u\|_{H_s(D_1)} \leq C\left(\|Lu\|_{L^2(D_2)} + \|u\|_{L^2(D_2)}\right), \quad D_1 \subset D_2, \tag{6.14}$$

where $C = C(D_1,D_2)$ and D_1 is a strictly inner subdomain of D_2. From (6.13), (6.14), and the Sobolev imbedding theorem it follows that the set $\{R_\varepsilon f\}$, $\|f\|_{L^2(\mathbb{R}^r)} = 1$, is compact in $H_t(D_1)$ for any $t < s$. Hence $R : H \to H$ is compact. □

The proof of Theorem 8 is left as an exercise for the reader.

7. General Theorems about Perturbations Preserving the Asymptotics of a Spectrum

Here we prove Theorems 9 through 12. Our main concern is the problem of under what conditions linear operators A and $B = A + T$ with discrete spectrum have the same asymptotics. These results can be applied to the spectral theory of integral and differential operators. Gohberg and Krein [2, p. 338] studied similar questions in the framework of perturbation theory for selfadjoint operators. Our results are more general while the method of the proofs is simpler.

1. Let H be a separable Hilbert space with inner product $(\cdot,\cdot)$, and let $\|f\|$ denote $(f,f)^{\frac{1}{2}}$. Assume A is a closed linear operator with discrete spectrum, A^{-1} is a bounded linear operator, and T is a linear operator such that $D(T) \supset D(A)$, $B = A + T$, and $D(B) = D(A)$. If $A = A^* \geq m > 0$ then by $H_A = D[A]$ we denote the Hilbert space which is the completion of $D(A)$ with respect to the norm $\|f\| = [f,f]^{1/2} = (Af,f)^{1/2}$. Clearly, $D[A] = D(A^{1/2})$ and $|A^{1/2}f|^2 = [f,f]$. By $\lambda_n(A) = \lambda_n$ we denote the eigenvalues of operator A, $|\lambda_1| \leq |\lambda_2| \leq \ldots$, by L_n we denote an n-dimensional subspace of H, $L_n \subset L_{n+1}$, and by $L_n(A)$ we denote the n-dimensional linear span of the first n eigenvectors of operator A. We count eigenvalues according to their multiplicities. If A^{-1} is compact we define s-numbers of A by the equality $s_n(A) = s_n^{-1}(A^{-1})$, where $s_n(Q) \equiv \lambda_n\{(Q^*Q)^{1/2}\}$ for a compact linear operator Q in H. If A^{-1} does not exist, but for a certain λ the resolvent $(A-\lambda I)^{-1}$ is compact, then we can decompose H into the direct sum $H = M_0 \dotplus M_1$, where M_0 is the root subspace

of A corresponding to the eigenvalue $\lambda = 0$ and M_1 is an invariant subspace of A in which the restriction A_1 of A to M_1 is invertible. It is possible to define the s-numbers of A by the equality $s_n(A) = s_n^{-1}(A_1^{-1})$, because A_1 is compact whenever $(A-\lambda I)^{-1}$ is compact. We denote by $\emptyset$ the empty set, by $\{0\}$ the set consisting of zero, by $\to$ convergence in H, by $\rightharpoonup$ weak convergence in H, by $\rho(f,L)$ the distance from the element f to the subspace $L \subset H$, by $\sigma(F)$ the spectrum of a closed linear operator F in H. It is known (see Glazman [1]) that $\sigma(F) = \sigma_p(F) \cup \sigma_e(F)$, where $\sigma_p(F)$ is the set of eigenvalues of F, $\sigma_d(F) \subset \sigma_p(F)$ is the set of isolated eigenvalues of finite algebraic multiplicity, and $\sigma_e(F) = \sigma_c(F) \cup \sigma_r(F)$. We say that $\lambda \in \sigma_c(F)$ if there exists a bounded noncompact sequence $f_n \in H$ such that $|Ff_n - \lambda f_n| \to 0$, as $n \to \infty$. The set $\sigma_r(F)$ consists of the points λ for which $R(F-\lambda I) = \mathrm{cl}\{R(F-\lambda I)\} \neq H$ and $\lambda \notin \sigma_p(F)$, where $\mathrm{cl}\{L\}$ denotes the closure of the set L. If $\lambda \notin \sigma_r(F)$ then $\bar{\lambda}$ is an eigenvalue of the adjoint operator F^* of F in H. We note that if $\lambda, \mu \notin \sigma(F)$ and $(F-\lambda I)^{-1}$ is compact, then $(F-\mu I)^{-1}$ is compact. The spectrum of F is called discrete if $\sigma(F) = \sigma_d(F)$. If the operator F is normal then $\sigma_r(F) = \emptyset$. A quadratic form $A[f,f]$ is called sectorial if the set $\{A[f,f]: \|f\| = 1, f \in D[A]\}$ is a subset of a sector $|\arg(z-\gamma)| \leq \theta$, $0 \leq \theta < \pi/2$, for some real γ. A sequence f_n is called A-convergent if $f_n \in D[A]$, $f_n \to f$, and $A[f_n - f_m, f_n - f_m] \to 0$ as $n,m \to \infty$. The form A is called closed if $f_n \underset{A}{\to} f$ implies $f \in D[A]$, $A[f_n - f, f_n - f] \to 0$ as $n \to \infty$, where $\underset{A}{\to}$ means A-convergence. A sectorial closed form defines a unique m-sectorial operator A such that $D(A) \subset D[A]$, $D(A)$ is dense in $D[A]$ with respect to A-convergence, and $A[f,g] = (Af,g)$ for all $f \in D(A)$, $g \in D[A]$. This is a generalization of the theorem of Friedrichs concerning semibounded forms. A proof can be found in Kato [1].

Lemma 1. *Let* A *be a closed densely defined linear operator in* H. *The operator* $(I+A^*A)^{-1}$ *is compact if and only if the operator* $(A-\lambda I)^{-1}$ *is compact for some* $\lambda \notin \sigma(A)$.

Proof of Lemma 1: Suppose $(I+A^*A)^{-1}$ is compact, $\lambda \notin \sigma(A)$, $|g_n| \leq C$, and $(A-\lambda I)^{-1}g_n = f_n$. Then $|f_n| \leq C$ (C denotes various constants) and $|Af_n| \leq C$, hence $|(I+A^*A)^{1/2}f_n|^2 = |f_n|^2 + |Af_n|^2 \leq C$. With $(I+A^*A)^{-1}$, also $(I+A^*A)^{-1/2}$ is compact. Therefore the sequence f_n is compact. This implies that the operator $(A-\lambda I)^{-1}$ is compact. Conversely, suppose that $(A-\lambda I)^{-1}$ is compact and $|f_n| \leq C$. Then the sequence $(A-\lambda I)^{-1}f_n = g_n$ is compact. We must prove that the sequence $(I+A^*A)^{-1}f_n = h_n$ is compact. Since $|(I+A^*A)h_n| \leq C$, $|h_n| \leq C$ we conclude that $((I+A^*A)h_n,h_n) = |h_n|^2 + |Ah_n|^2 \leq C$. Hence $|(A-\lambda I)h_n| \leq C$. Therefore the sequence h_n is compact. □

Remark 1. We can use Lemma 1 in order to give another definition of the s-numbers of unbounded operators with compact resolvent. If $(A-\lambda I)^{-1}$ is compact for some λ, then we set $s_n^2(A) = \lambda_n(A^*A+I) - 1$ and $\lambda_n(A^*A+I) = \lambda_n^{-1}\{(A^*A+I)^{-1}\}$.

2. Theorem 14. *Let* A *be a closed densely defined linear operator in* H *and* T *a linear operator in* H *such that* $D(T) \subset D(A)$, $B = A + T$, $D(A) = D(B)$ *and* A^{-1} *is a bounded operator defined everywhere on* H. *If the operators* TA^{-1}, A^{-1} *are compact and* $N(B+kI) = \{0\}$ *for some number*

$k \notin \sigma(A)$, *then the operator* B *is closed and* $\sigma(B)$ *is discrete*.

The following lemma is of use.

Lemma 2. *From any noncompact bounded sequence* $\{f_n\} \subset H$ *one can construct a noncompact sequence* $\{\psi_m\} = \{f_{n_{m+1}} - f_{n_m}\}$ *so that* $\psi_m \rightharpoonup 0$ *as* $m \to \infty$.

Proof of Lemma 2: As $\{f_n\}$ is bounded, we can assume that $f_n \rightharpoonup f$. As $\{f_n\}$ is noncompact, we can find a subsequence such that $|f_{n_k} - f_{n_m}| \geq \varepsilon > 0$ for all $k, m, k \neq m$. Let us construct $\{\psi_m\}$ as in the statement of Lemma 2. Then $\psi_m \rightharpoonup 0$ as $m \to \infty$ and $\{\psi_m\}$ is noncompact because $|\psi_m| \geq \varepsilon > 0$ and if $\{\psi_m\}$ were compact it would converge to zero. □

Proof of Theorem 14: First we prove that B is closed. Assume $f_n \to f$, $Bf_n = Af_n + Tf_n \to g$, and $f_n \subset D(A) = D(B)$. If $|Af_n| \leq C$ then the sequence $Tf_n = TA^{-1}Af_n$ is compact because TA^{-1} is compact. We also denote the convergent subsequence by Tf_n and note that $Af_n = Bf_n - Tf_n \to h$, where $h = g - \lim Tf_n$. As A is closed, we conclude that $f \in D(A)$ $Af = h$, $Bf = g$. Hence B is closed. It remains to prove that $|Af_n| \leq C$. Suppose that $|Af_n| \to \infty$. Let $g_n \equiv f_n|Af_n|^{-1}$. Then $|Ag_n| = 1$ and $g_n \to 0$. As above, we can assume that $Ag_n \to u$, $|u| = 1$. But A is closed, hence $u = 0$. This contradiction proves that $|Af_n| \leq C$. We already proved that B is closed. Let us prove that $\sigma(B)$ is discrete. We have $(A+T-\lambda I)^{-1} = (A+kI)^{-1}(I+Q-\mu S)^{-1}$, where $S = (A+kI)^{-1}$, $Q = T(A+kI)^{-1}$, $\mu = \lambda+k$, $k \notin \sigma(A)$, S and Q are compact. If $N(B+kI) = \{0\}$, then $N\{(I+Q)(A+kI)\} = \{0\}$ and $N(I+Q) = \{0\}$. Thus $(I+Q)^{-1}$ exists on the whole space H. Therefore by the well-known result (analytic Fredholm theorem) $(I+Q-\mu S)^{-1}$ is a finite-meromorphic operator function. It means that the operator function $(I+Q-\mu S)^{-1}$ is meromorphic in μ and its Laurent coefficients are finite-rank operators. This completes the proof. □

Let us show another way to prove that $\sigma_C(B) = \emptyset$ and $\sigma_r(B) = \emptyset$. Suppose that $\lambda \in \sigma_C(B)$, i.e., there exists a noncompact bounded sequence $\{f_n\}$, such that $Bf_n - \lambda f_n \to 0$. By Lemma 2 we construct the noncompact sequence $\psi_m \rightharpoonup 0$, $\psi_m = f_{n_{m+1}} - f_{n_m}$. We have $A\psi_m + T\psi_m - \lambda\psi_m \to 0$. Therefore $\psi_m + A^{-1}T\psi_m - \lambda A^{-1}\psi_m \to 0$. If $\lambda \neq 0$ then $\lambda^{-1} \in \sigma_C(A^{-1}) = \emptyset$. But this is impossible. If $\lambda = 0$ then $\psi_m \to 0$. It is impossible because $\{\psi_m\}$ is noncompact; so $\sigma_C(B) = \emptyset$. In both cases we took into consideration that the compactness of $A^{-1}T$ and the relation $\psi_m \rightharpoonup 0$ imply that $A^{-1}T\psi_m \to 0$. No use of the compactness of A^{-1} was made. It remains to prove that $\sigma_r(B) = \emptyset$. Let $\lambda \in \sigma_r(B)$, so that $B^*f = \bar{\lambda}f$, $\lambda \notin \sigma_p(B)$. Here the line denotes complex conjugation. First we assume that $\bar{\lambda} \notin \sigma(A^*)$. Then $f + (A^*-\bar{\lambda}I)^{-1}T^*f = 0$. The operator $(A^*-\bar{\lambda}I)^{-1}T^*$ is compact whenever TA^{-1} is compact. Hence the equation $g + T(A-\lambda I)^{-1}g = 0$ has a nontrivial solution by the Fredholm alternative. Therefore $(A+T-\lambda I)h = 0$, where $h = (A-\lambda I)^{-1}g$. Hence $\lambda \in \sigma_p(B)$. This is impossible. Let us assume now that $\bar{\lambda} \in \sigma(A^*)$. As the spectrum of A is discrete, we can find $\varepsilon > 0$ such that $\bar{\lambda} + \varepsilon \notin \sigma(A^*)$. Then $f + [A^* - (\bar{\lambda}+\varepsilon)I]^{-1}(T^*+\varepsilon I)f = 0$. Using the compactness of TA^{-1} and A^{-1} (only here we use the compactness of A^{-1}) we conclude that the operator $[A^* - (\bar{\lambda}+\varepsilon)I]^{-1}(T^*+\varepsilon I)$ is compact. As above, we deduce from this that $\lambda \in \sigma_p(B)$. This is impossible. Therefore $\sigma_r(B) = \emptyset$. For normal operators

$\sigma_r = \emptyset$, $\sigma \setminus \sigma_c = \sigma_d$.

Remark 2. The operator T is called A-compact if for any sequence $\{f_n\}$ with $|f_n| + |Af_n| \leq C$ the sequence Tf_n is compact, i.e., contains a convergent subsequence. <u>If the operator A^{-1} is bounded, then the operator TA^{-1} is compact if and only if T is A-compact.</u> Indeed, if the operator T is A-compact, $|f_n| \leq C$, and $g_n = A^{-1}f_n$, then $|g_n| + |Ag_n| \leq C_1$. Hence the sequence $Tg_n = TA^{-1}Ag_n = TA^{-1}f_n$ is compact. Conversely, if the operator TA^{-1} is compact and $|f_n| + |Af_n| \leq C$, then $Tf_n = TA^{-1}Af_n$ is compact.

Lemma 3. <u>The operator T is compact in H if and only if one of the following conditions holds:</u>

(1) $\sup_{h \perp L_n} \frac{|Th|}{|h|} \to 0$ as $n \to \infty$, <u>where the sequence of subspaces L_n, $L_n \subset L_{n+1}$, $\dim L_n = n$ is limit-dense in H, i.e., $\rho(f, L_n) \to 0$ as $n \to \infty$, for all $f \in H$.</u>

(2) $(Tg_n, g_n) \to 0$ <u>as $n \to \infty$ for any sequence g_n such that $g_n \rightharpoonup 0$.</u>

Proof of Lemma 3: Let T be compact, $\gamma_n \equiv \sup_{h \perp L_n} \frac{|Th|}{|h|}$, and let $h_1, \ldots, h_n$ be an orthonormal basis of L_n. It is clear that $0 \leq \gamma_{n+1} \leq \gamma_n$. So $\lim \gamma_n = \gamma \geq 0$ exists. If $\gamma > 0$ there exists a sequence f_n, $|f_n| = 1$, $f_n \perp L_n$, $|Tf_n| \geq \gamma > 0$ (*) Since the sequence of subspaces L_n is limit-dense in H, we can choose from the sequence f_n a subsequence also denoted by f_n such that $f_n \rightharpoonup 0$. As T is compact, $Tf_n \to 0$. This contradicts (*). Hence $\gamma = 0$. Conversely, assuming condition (1) holds we set $g_n = h - \psi_n$, $\psi_n = \sum_{j=1}^{n} (h, h_j)h_j$, $\psi_n \in L_n$, $g_n \perp L_n$ and define the operator

T_n by equality $T_n h \equiv T\psi_n$. Then

$$|T-T_n| = \sup_{|h|=1} |(T-T_n)h| = \sup_{\substack{g_n \perp L_n \\ |g_n| = (1-|\psi_n|^2)^{1/2}}} |Tg_n|$$

$$\leq \sup_{\substack{g_n \perp L_n \\ |g_n| \leq 1}} |Tg_n| = \gamma_n \to 0$$

as $n \to \infty$. Hence T is a uniform limit of finite-dimensional operators and hence T is compact. The necessity of condition (2) for compactness of operator T is obvious. To prove sufficiency we assume that condition (2) holds and using the polarization identity (see formula (5.5)) we deduce that $(Tf_n, g_n) \to 0$ whenever $f_n \rightharpoonup 0$, $g_n \rightharpoonup 0$. This means that T is compact. □

Proof of Theorem 10: If the operator $A^{-1}T$ is compact in H_A, then $H_B = H_A$ and the spectrum of the operator B is discrete. We prove this statement below in Lemma 4. Let us denote by $\perp_A$ orthogonality in H_A, by $\perp$ orthogonality in H, and by $L_n(A)$ the linear span of the first n eigenvectors of the operator A. We note that $f \perp L_n(A)$ is equivalent to $f \perp_A L_n(A)$ and that $\inf \alpha(1+\beta) \geq \inf \alpha(1-\sup \beta)$ for $\alpha \geq 0$, $-1 < \beta < 1$. We also note that

$$\gamma_n \equiv \sup_{f \perp L_n(A)} \frac{(Tf,f)}{A[f,f]} = \sup_{f \perp_A L_n(A)} \frac{[A^{-1}Tf,f]}{[f,f]} \to 0, \text{ as } n \to \infty$$

according to Lemma 3, because $A^{-1}T$ is compact in H_A. Here $A[f,f] \equiv [f,f] = \|f\|^2$. Now we have

$$\lambda_{n+1}(B) = \sup_{L_n} \inf_{\substack{f \perp L_n \\ f \in H_A}} \frac{B[f,f]}{(f,f)} \geq \inf_{\substack{f \perp L_n(A) \\ f \in H_A}} \left\{ \frac{A[f,f]}{(f,f)}\left(1 + \frac{(Tf,f)}{A[f,f]}\right) \right\}$$

$$\geq \lambda_{n+1}(A)(1-\gamma_n), \quad \gamma_n \to 0 \text{ as } n \to \infty.$$

By symmetry $\lambda_{n+1}(A) \geq \lambda_{n+1}(B)(1-\delta_n)$, where $\delta_n \to 0$ as

$n \to \infty$. Hence $\lambda_{n+1}(B)\lambda_n^{-1}(A) \to 1$ as $n \to \infty$. This shows that only a finite number of eigenvalues $\lambda_j(B)$ can be negative. ∎

Remark 3. We give a direct proof of the last statement in the proof of Theorem 10. If $(A+T)\phi_j = \lambda_j\phi_j$ and $||\phi|| = 1$, then $1 + (S\phi_j,\phi_j) = \lambda_j(\phi_j,\phi_j)$, $S = A^{-1}T$. By Lemma 3, $[S\phi_j,\phi_j] \to 0$ as $j \to \infty$. As the λ_j are real and $(\phi_j,\phi_j) >$ we conclude that $\lambda_j > 0$ for all sufficiently large j.

Lemma 4. *Under the assumptions of Theorem 10,* $H_B = H_A$ *and the spectrum of* B *is discrete.*

Proof of Lemma 4: We note that $B = A(I + S)$, $S = A^{-1}T$. As S is compact in H_A, it can be represented in the form $S = Q + \mathscr{K}$, where $||Q|| < 1$, $\mathscr{K}$ is a finite-dimensional operator and $||Q||$ denotes the operator norm in H_A. Since $[Sf,g] = (Tf,g)$ and $T = B - A$ is symmetric on $D(A)$, we conclude that S is symmetric in H_A. Since S is a densely defined bounded symmetric operator on H_A, it is essentially self-adjoint in H_A. We denote its closure by S. The operators Q, $\mathscr{K}$ can also be assumed to be selfadjoint in H_A. Hence $I+Q$ is positive definite in H_A, while $[\mathscr{K}f,f] = \sum_{j=1}^{N} a_j|[f,\psi_j]|^2$ with $\psi_j \in H_A$ and $a_j = \text{const}$. We can find $w_j \in D(A)$, $||w_j - \psi_j|| < \varepsilon$, so that

$$|[\mathscr{K}f,f]| \le C_1\varepsilon||f||^2 + \sum_{j=1}^{N}|a_j||(f,Aw_j)|^2$$

$$\le C_1\varepsilon||f||^2 + C_2|f|^2, \quad C_1,C_2 = \text{const}, \quad ||\cdot|| = ||\cdot||_{H_A}.$$

Therefore

$$(Bf,f) = (A(I+Q)f,f) + (A\mathscr{K}f,f) = [(I+Q)f,f] + [\mathscr{K}f,f]$$

$$\ge C||f||^2 - C_1\varepsilon||f||^2 - C_2|f|^2.$$

As $\varepsilon > 0$ can be chosen arbitrary small, we see that (Bf,f)

is semibounded from below in H. It is obvious that $(Bf,f) \leq C_3\|f\|^2$. Hence the metrics generated by the forms (Af,f) and (Bf,f) are equivalent and $H_A = H_B$. To prove that $\sigma(B)$ is discrete, we show that $\sigma_r(B) = \emptyset$ and $\sigma_c(B) = \emptyset$. The first statement is true since $B = B^*$. Assume that $\lambda \in \sigma_C(B)$, i.e., there exists a noncompact sequence f_n bounded in H such that $Bf_n - \lambda f_n \to 0$. By Lemma 2, we construct a noncompact sequence $\psi_m \rightharpoonup 0$, $A\psi_m + T\psi_m - \lambda\psi_m \to 0$. Hence $\psi_m - \lambda A^{-1}\psi_m + A^{-1}T\psi_m \to 0$. Since $A^{-1}T$ is compact in H_A, we have $|(Tf,f)| \leq \varepsilon\|f\|^2 + C(\varepsilon)|f|^2$, $\varepsilon > 0$, $f \in D(A)$. From here and the relation $(A\psi_m,\psi_m) + (T\psi_m,\psi_m) - \lambda|\psi_m|^2 \to 0$ we conclude that $\|\psi_m\| \leq C$. Hence a subsequence $A^{-1}T\psi_{m_j}$ converges in H_A and therefore in H. Since $\psi_m \rightharpoonup 0$ we conclude that $AT\psi_{m_j} \to 0$ and $\psi_{m_j} - \lambda A^{-1}\psi_{m_j} \to 0$. This means that $\lambda \in \sigma_C(A^{-1})$. But this is impossible if $\lambda \neq 0$ since the spectrum of A is discrete. If $\lambda = 0$ we see that $\psi_{m_j} \to 0$. This is also impossible since the sequence ψ_m is noncompact. Hence $\sigma_C(B) = \emptyset$ and $\sigma(B)$ is discrete. □

<u>Theorem 15</u>. <u>Assume</u> $A = A^* \geq m > 0$, $0 \notin \sigma(B)$, $\sigma(A)$ <u>is discrete</u>, <u>the operators</u> $A^{-1}T$ <u>and</u> TA^{-1} <u>are compact in</u> H, <u>and</u> B <u>is normal</u>. <u>Then</u> $\lambda_n(B)\lambda_n^{-1}(A) \to 1$ <u>as</u> $n \to \infty$.

<u>Proof of Theorem 15</u>: Since $A = A^* > 0$, we have $\lambda_j(A) = s_j(A)$. Since B is normal, we have $|\lambda_j(B)| = s_j(B)$. The operator $A \geq m > 0$ is selfadjoint and its spectrum is discrete. This means that A^{-1} is compact, $B^{-1} = A^{-1}(I+C)^{-1}$, $C \equiv TA^{-1}$. It follows from Theorem 14 that $\sigma(B)$ is discrete. It follows from Theorem 11 that $s_n(B)s_n^{-1}(A) \to 1$ as $n \to \infty$. The proof of Theorem 11 is given below. Now we complete the

proof of Theorem 15 by showing that $\lambda_n^{-1}(B)|\lambda_n(B)| \to 1$ as $n \to \infty$. Let $A\phi_j + T\phi_j = \lambda_j\phi_j$, $\lambda_j \equiv \lambda_j(B)$, $Q \equiv A^{-1} > 0$, $S \equiv A^{-1}T$, and $(\phi_j,\phi_i) = \delta_{ji}$. Then $1 + (S\phi_j,\phi_j) = \lambda_j(Q\phi_j,\phi_j)$. Since the operator S is compact, we see that $(S\phi_j,\phi_j) \to 0$ as $j \to \infty$. From this and the inequality $(Q\phi_j,\phi_j) > 0$, it follows that $|\mathrm{Im}\,\lambda_j||\mathrm{Re}\,\lambda_j|^{-1} \to 0$ as $j \to \infty$. Hence $\lambda_j|\lambda_j|^{-1} \to$ as $j \to \infty$. □

Remark 4. If $T \geq 0$ the operator A^{-1} is compact in H_A if and only if the imbedding operator from H_A into H_T is compact. If T is not necessarily nonnegative, but for some $Q \geq 0$ the inequality $|(Tf,f)| \leq (Qf,f)$ holds and the imbedding from H_A into H_Q is compact, then $A^{-1}T$ is compact in H_A.

The reader can prove this statement as an exercise or find its proof in Glasman [1, p. 37].

Proof of Theorem 11: We have $s_n^2(Q + QS) = \lambda_n\{(I+S^*)U(I+S)\}$, $U = Q^*Q \geq 0$, $\lambda_n(U) = s_n^2(Q)$. As S is compact and $N(I+S) = \{0\}$, it is clear that $(I+S)^{-1} = I+\Gamma$, $(I+S^*)^{-1} = I+\Gamma^*$, where Γ is a compact operator. Let $v = (I+S^*)U(I+S)$. If

$$\lambda_n(v) \leq \lambda_n(U)(1+\alpha_n), \quad \alpha_n \to 0 \quad \text{as} \quad n \to \infty, \tag{7.1}$$

then by symmetry

$$\lambda_n(U) \leq \lambda_n(v)(1+\beta_n), \quad \beta_n \to 0 \quad \text{as} \quad n \to \infty. \tag{7.2}$$

From (7.1) and (7.2) it follows that $\lambda_n(v)\lambda_n^{-1}(U) \to 1$ as $n \to \infty$. This is the first statement of Theorem 11. Its second statement can be proved similarly. It remains to prove (7.1). We have

$$\lambda_{n+1}(v) = \inf_{L_n} \sup_{f \perp L_n} \frac{(vf,f)}{|f|^2} \leq \sup_{f \perp M_n} \left\{ \frac{(Ug,g)}{(g,g)} \frac{(g,g)}{(f,f)} \right\}$$

$$\leq \sup_{f \perp M_n} \frac{(Ug,g)}{(g,g)} \sup_{f \perp M_n} \frac{(g,g)}{(f,f)} \leq \lambda_n(U)(1+\alpha_n),$$

$$\alpha_n \to 0 \quad \text{as} \quad n \to \infty. \tag{7.3}$$

Here $g = (I+S)f$ and M_n is an n-dimensional subspace chosen so that the condition $f \perp M_n$ is equivalent to the condition $g \perp L_n(U)$, where $L_n(U)$ is the span of the first n eigenvectors ϕ_j of the operator U. Such a subspace M_n can be chosen. Indeed, the conditions $(g,\phi_j) = 0$, $1 \leq j \leq n$, can be writen in the form $(f,(I+S^*)\phi_j) = 0$, $1 \leq j \leq n$, so that we can take M_n to be the span of the elements $\psi_j = (I+S^*)\phi_j$, $1 \leq j \leq n$. Since the operator $I+S^*$ is invertible, the system $\{\psi_j\}$ is linearly independent and $\dim M_n = n$. Moreover, the sequence of subspaces M_n is limit-dense in H. In proving (7.3) we use the inequality

$$\sup_{f \perp M_n} \frac{(g,g)}{(f,f)} \leq 1 + \sup_{f \perp M_n} \frac{|(Sf,f)| + |(f,Sf)| + |Sf|^2}{(f,f)} = 1 + \alpha_n,$$

$$\alpha_n \to 0 \quad \text{as} \quad n \to \infty, \tag{7.4}$$

which follows from Lemma 3 and the compactness of S. □

Proof of Theorem 9: The first statement of Theorem 9 has been proved above as Theorem 14. The second statement can be proved as follows. $s_n^{-2}(B) = s_n^{-2}(B^{-1}) = s_n^{-2}\{(I+C)^{-1}A^{-1}\} = s_n^{-2}\{(I+S)Q\}$, where $Q = A^{-1}$, $C = A^{-1}T$, $S = (I+C)^{-1} - I$, and S is compact whenever C is compact. The operator $I+C$ is invertible if A and B are invertible. From Theorem 11, it follows that $s_n\{(I+C)Q\}s_n^{-1}(Q) \to 1$ as $n \to \infty$. But $s_n^{-1}(Q) = s_n(A)$. Hence $s_n(B)s_n^{-1}(A) \to 1$ as $n \to \infty$. The third statement of Theorem 9 coincides with Theorem 15. □

Remark 5. For a comparison of the results presented in Theorems 9 - 12 with the known results see the Bibliographical Notes.

3. Here we prove Theorem 12. This theorem is particularly useful when the domains of the initial and perturbed operators are different but the domains of the corresponding quadratic forms are identical. For example, if $A[f,f] = \int_D(|\nabla f|^2 + |f|^2)dx$ and $A_\sigma[f,f] = A[f,f] + \int_\Gamma \sigma(t)|f(t)|^2 dt$, where D is a bounded somain in $\mathbb{R}^r$ with smooth boundary Γ, $\sigma(t) \in C^1(\Gamma)$, then $D[A] = D[A_\sigma] = H_1(D)$, but the corresponding operators A and A_σ have different domains.

The spectrum of a closed sectorial form is the spectrum of the operator generated by the form. A form $T[f,f]$ is said to be compact relative to a positive definite form $A[f,f]$ (A-compact) if any sequence f_n such that $A[f_n,f_n] \leq C$ contains a subsequence f_{n_j} such that $T\left[f_{n_{j+p}} - f_{n_j}, f_{n_{j+p}} - f_{n_j}\right] \to 0$ as $j,p \to \infty$.

Proof of Theorem 12: The proof consists of the following lemmas from which the statement of Theorem 12 follows immediately. We assume that the hypotheses of Theorem 12 hold for all these lemmas.

Lemma 5. *The quadratic form* $T[f,f]$ *can be uniquely represented in the form* $T[f,f] = [tf,f]$, *where* $[u,v]$ *is the inner product in* H_A *and* t *is a selfadjoint and compact operator in* H_A.

Lemma 6. *The quadratic form* $T[f,f]$ *can be represented in the form* $T[f,f] = T_n[f,f] + T_\varepsilon[f,f]$, $\varepsilon > 0$, *where* $|T_\varepsilon[f,f]| \leq \varepsilon\|f\|^2$, $n = n(\varepsilon)$, *and* $T_n[f,f]$ *is a quadratic form in* n-*dimensional space*. *The form* $B[f,f]$ *is semi-bounded from below in* H, *and* $H_B = H_A$.

Lemma 7. *The spectrum of the form* $B[f,f]$ *is discrete and* $\lambda_n(B)\lambda_n^{-1}(A) \to 1$ *as* $n \to \infty$.

Proofs of Lemmas 5 - 7:

(a) The representation formula of Lemma 5 is a well-known fact (see Kato [1]). The operator T is compact since for any sequences f_n, g_n weakly convergent in H_A the sequence $T[f_n,g_n] = [tf_n,g_n]$ is convergent because of the A-compactness of the form $T[f,f]$. Indeed, the sesquilinear form $T[f,g]$ can be expressed by the quadratic form $T[f,f]$ by means of the polarization identity (see formula (5.5)). The operator T is sefladjoint because the form $T[f,f]$ is real-valued.

(b) The proof of Lemma 6 is very similar to the proof of Lemma 4. Details can be reconstructed by the reader.

(c) Proving Lemma 7 we can assume that the form $B[f,f]$ is positive definite in H. Otherwise we can consider the form $B_m[f,f] = B[f,f] + m(f,f)$, where $m > 0$ is large enough so that the form $B_m[f,f]$ is positive definite in H. Passing to $A_m[f,f] = A[f,f] + m(f,f)$, we do not change the form $T[f,f] = B_m[f,f] - A_m[f,f]$. The spectra of the forms $A_m[f,f]$, $B_m[f,f]$ can be obtained from the spectra of the forms $A[f,f]$ and $B[f,f]$ by a shift to the right by m. Since $\lambda_n(A) \to +\infty$ as $n \to \infty$ the relation $\lambda_n(B)\lambda_n^{-1}(A) \to 1$ holds if and only if $\lambda_n(B_m)\lambda_n^{-1}(A_m) \to 1$ as $n \to \infty$. So we may assume the forms $B[f,f]$ and $A[f,f]$ are positive definite in H. Let A and B be the corresponding positive definite selfadjoint operators, $D(A^{1/2}) = H_A$, $H_A = H_B$,

$D(A^{1/2}) = D(B^{1/2})$. For the spectrum of B to be discrete it is necessary and sufficient that the imbedding operator $i: H_B \to H$ is compact. Since the metrics of H_B and H_A are equivalent, $i: H_B \to H$ is compact if and only if $i: H_A \to H$ is compact. Since A has a discrete spectrum, $i: H_A \to H$ is compact, hence $i: H_B \to H$ is compact and B has discrete spectrum. Furthermore, we have

$$\begin{aligned}\lambda_{n+1}(B) &= \sup_{L_n} \inf_{f \perp L_n} \frac{B[f,f]}{(f,f)} \geq \inf_{f \perp L_n(A)} \frac{A[f,f]}{(f,f)}\left(1 + \frac{T[f,f]}{A[f,f]}\right) \\ &\geq \lambda_{n+1}(A)\left(1 - \sup_{f \perp L_n(A)} \frac{|T[f,f]|}{A[f,f]}\right) = \lambda_{n+1}(A)(1-\gamma_n),\end{aligned} \tag{7.5}$$

where, by Lemma 3, $\gamma_n \to 0$ as $n \to \infty$ and $L_n(A)$ is the span of the first n eigenvectors of A. Interchanging A and B we get

$$\lambda_{n+1}(A) \geq \lambda_{n+1}(B)(1 - \delta_n), \text{ where } \delta_n \to 0 \text{ as } n \to \infty. \tag{7.6}$$

From (7.5) and (7.6) it follows that $\lambda_n(B)\lambda_n^{-1}(A) \to 1$ as $n \to \infty$. □

8. Remarks and Examples

1. Let us prove the statement we used near the end of Section 1.

Lemma 1. *Let $B = B^* \geq 0$ be a linear operator in a Hilbert space H, with $g \in R(B)$, $Bh = g$. Then $h = \lim_{n\to\infty} h_n$, where $h_n + Bh_n = h_{n-1} + g$ and $h_o \in H$ is arbitrary.*

Proof of Lemma 1: We have

$$h_n = Ah_{n-1} + f, \quad h_o \in H, \quad f = Ag, \quad A = (I+B)^{-1}. \tag{8.1}$$

It is clear that $A \geq 0$ and $||A|| \leq 1$. Hence $\lim_{n\to\infty} h_n = h$

exists (see Krasnoselskij et. al. [1, p. 71]). Hence $h = Ah + f$. Therefore $Bh = g$. □

2. The following equation is often mentioned in the literature on stochastic optimization:

$$\int_{-b}^{b} \exp(-a|x-y|)h(y)dy = f(x), \quad -b \leq x \leq b. \tag{8.2}$$

Since

$$\exp(-a|x|) = a\pi^{-1} \int_{-\infty}^{\infty} \exp(i\lambda x)(a^2+\lambda^2)^{-1}d\lambda,$$

and

$$L = -iD, \quad \Phi(x,y,\lambda)d\rho = (2\pi)^{-1}\exp\{i\lambda(x-y)\}d\lambda,$$

$$P(\lambda) = 1, \quad Q(\lambda) = \frac{\lambda^2+a^2}{2a},$$

we can use formula (1.10) to find the solution with minimal order of singularity of equation (8.2):

$$h(x) = \frac{1}{2a}\{-f''(x) + a^2f(x)\} + \frac{\delta(x+b)}{2a}\{-f'(-b) + af(-b)\}$$

$$+ \frac{\delta(x-b)}{2a}\{f'(b) + af(b)\}. \tag{8.3}$$

Let us consider the equation

$$\int_D \frac{\exp\{-a|x-y|\}}{4\pi|x-y|} h(y)dy = f(x), \quad x \in \bar{D} \subset \mathbb{R}^3, \quad a > 0, \tag{8.4}$$

where D is the unit ball. Here $L_j = -iD_j$, $\tilde{R}(t) = (t^2+a^2)^{-1}$, $t^2 = t_1^2 + t_2^2 + t_3^2$, $P(t) = 1$, and $Q(t) = t^2 + a^2$. From formula (1.6), we find the solution with minimal order of singularity of equation (8.4),

$$H(x) = (-\Delta+a^2)F(x), \quad F(x) = \begin{cases} f(x) & \text{in } D, \\ u(x) & \text{in } \Omega, \end{cases} \tag{8.5}$$

where Δ is the Laplacian and

$$\begin{cases} -\Delta u + a^2 u = 0 \quad \text{in} \quad \Omega, \\ u\big|_\Gamma = f\big|_\Gamma, \quad u(\infty) = 0. \end{cases} \tag{8.6}$$

Up to now we have not used the special shape of the domain D. If D is the unit ball we can solve problem (8.6) explicitly and find an explicit expression for $h(x)$ from formula (8.5).

3. In Section 1 we used an assumption concerning the self-adjointness of the operator L. This assumption can be dropped. What is essential for the developed theory is the spectral representation

$$F(L) = \int_\Lambda F(\lambda)\Phi(x,y,\lambda)d\rho(\lambda).$$

This representation can be obtained for some nonselfadjoint operators (see, for example, Ramm [16]-[20]).

4. The theory developed was applied to the resolution ability theory in Ramm [6], [21]-[25].

5. Let $A[f,f] = \int_D \{|\nabla f|^2 + |f|^2\}dx$, $A_\sigma[f,f] = A[f,f]$ $T_\sigma[f,f]$, $T_\sigma[f,f] = \int_\Gamma \sigma(t)|f(t)|^2 dt$, where D is a bounded domain with smooth boundary Γ. From Theorem 12 and the A-compactness of T_σ for $\sigma \in C^1(\Gamma)$, it follows that $\lambda_n(\sigma)\lambda_n^{-1} \to 1$ as $n \to \infty$, where $\lambda_n, \lambda_n(\sigma)$, are the eigenvalues of the problems

$$-\Delta f + f = \lambda f \quad \text{in} \quad D, \quad \partial_N f\big|_\Gamma = 0; \quad -\Delta f + f = \lambda_n(\sigma) f \quad \text{in} \quad D,$$
$$\partial_N f + \sigma f\big|_\Gamma = 0.$$

<u>Exercise</u>. Prove that if $\min \sigma(s) \to +\infty$ then $\lambda_n(\sigma) \to \mu_n$, where the μ_n are the eigenvalues of the problem $-\Delta f + f = \mu f$, $f\big|_\Gamma = 0$.

9. Research Problems

1. It would be interesting to develop a theory of equation (1.2) with kernel (1.3) if $P(\lambda)$ or $Q(\lambda)$ can vanish at some points.
2. It would be of special interest for diffraction theory to develop a theory of equation (1.2) with kernel (1.3) for $Q(\lambda) = (\lambda^2 - k^2)^{-1}$, $k^2 > 0$. In particular, it would be interesting to find the asymptotic behavior of solution as $k \to \infty$.
3. It would be interesting to find some problems from other fields, for example, elasticity theory, which can be reduced to equations of class $\mathcal{R}$.
4. It would be interesting to design and build an optimal optical filter using the developed theory.
5. It would be interesting to find theorems similar to Theorems 9 and 10 in order to have some conditions for preserving the asymptotics of a negative spectrum. A detailed statement of the problem follows:

   Let A be a selfadjoint operator with continuous positive spectrum which consists of the positive semiaxis $(0,\infty)$ and discrete negative spectrum, which consists of eigenvalues

$$\lambda_n(A), \quad \lambda_n(A) \leq \lambda_{n+1}(A) \ldots < 0,$$

   such that $\lambda_n(A) \to 0$ as $n \to \infty$. Consider the operator $B = A + T$. Under what assumptions does $\lambda_n(B)\lambda_n^{-1}(A) \to 1$ as $n \to \infty$?

6. How can one approximate with prescribed accuracy an arbitrary nonnegative definite kernel $R(x,y)$ by a kernel of class $\mathscr{R}$? How can one find an operator L so that for given accuracy of approximation the degrees of the polynomials $P(\lambda)$, $Q(\lambda)$ in formula (1.3) will be minimal in some sense? The notion of accuracy should be specified.

   It can be shown that a nonnegative definite kernel $R(x,y)$ can be approximated with the prescribed accuracy by a kernel of class $\mathscr{R}$, e.g., in $L^2(D)$.

7. Find a method (approximate analytical or purely analytical) to solve the equation

$$\int_{-a}^{a} R(x-y)h(y)\,dy = f(x), \quad -a \le x \le a, \tag{9.1}$$

   with

$$R(x) = \int_{-\infty}^{\infty} \exp(i\lambda x)(\lambda^2 + a^2)^{-\beta}\,d\lambda, \quad \beta > 0. \tag{9.2}$$

   Such a method can be used as in Theorem 13.

8. Find an analog to Theorem 12 in the case $D[A] \neq D[B]$.

9. Remove the assumption concerning the normality of operator B in Theorem 9 if possible.

10. Study equation (1.2) in domains with nonsmooth or with infinite boundaries.

11. Remove the condition that the form $T[f,f]$ is real-valued in Theorem 12.

10. Bibliographical Note

Some of the results presented in Section 1 were obtained by the author Ramm [1]-[17]. It seems that equations of class $\mathscr{R}$ have not been studied in the literature. Remarks concerning equation (1.1) with convolution kernel were made in Section 1. Here we mention the book Middleton [1] where many problems of communication theory are reduced to equations (1.1) and (1.2). In the theory of one-dimensional convolution kernels, the theory of Riemann boundary problem is helpful (see Gahov [1], Gahov-Cherski [1], Mushelishvili [1], Zabreiko et al. [1], Noble [1]). The theory of spaces with negative norm is given in many papers (see Beresanskij [1]). In this book the theory of elliptic boundary-value problems is presented. For other presentations of this theory see Schechter [1], Hörmander [1]. All facts from linear functional analysis necessary for Chapter 1 can be found in Kato [1]. The spectral theory of differential operators is discussed in Glazman [1]. In Gohberg-Krein [3] a method of investigation of one-dimensional linear integral equations due to Krein is presented. Let us compare the results of Theorems 9 and 11 with the results due to Gohberg-Krein [3]. In Gohberg-Krein [3, p. 351] the following theorems are proved.

Theorem A. Assume $Q \geq 0$, $\dim R(Q) = \infty$, Q is compact in H, $K = Q(I+S)$, $K = K^*$, S is compact, $I+S$ is invertible or $\lambda_{n+1}(Q)^{-1}\lambda_n(Q) \to 1$ as $n \to \infty$. Then $\lambda_n(K)\lambda_n^{-1}(Q) \to 1$ as $n \to \infty$.

Theorem B. If $Q = Q^*$, $\dim R(Q) = \infty$, and Q and S are compact, $N(Q) = \{0\}$, $I+S$ is invertible or $s_{n+1}(Q)s_n^{-1}(Q) \to 1$ as $n \to \infty$, then $s_n(K)s_n^{-1}(Q) \to 1$ as $n \to \infty$, where $K = Q(I+S)$.

Theorems A and B follow from Theorem 11. Theorem B is a particular case of Theorem 11 with the additional assumption $Q = Q^*$ which can be omitted. The condition $s_{n+1}(Q)s_n^{-1}(Q) \to 1$ as $n \to \infty$ imposed if $I+S$ is not invertible plays the following role. As S is compact, $N(I+S)$

is finite dimensional. It is possible to reduce the problem to the case in which $I+S$ is invertible using the condition $s_{n+1}(Q)s_n^{-1}(Q) \to 1$ as $n \to \infty$. Theorem A follows also from Theorem 11 if we take into consideration that for self-adjoint operator K, $s_n(K) = |\lambda_n(K)|$ and $\lambda_n(K) > 0$ for sufficiently large n under the assumptions of Theorem A. Indeed, if $\lambda_j Q(I+S)\phi_j = \phi_j$ then

$$\lambda_j = \frac{((I+S)\phi_j,\phi_j)}{(Q(I+S)\phi_j,(I+S)\phi_j)} .$$

As $K = K^*$, the eigenvalues λ are real. The denominator of the above ratio is nonnegative, and $(S\phi_j,\phi_j) \to 0$ as $j \to \infty$ because S is compact, $|\phi_j| = 1$, $(\phi_i,\phi_j) = \delta_{ij}$. Hence $\lambda_j > 0$ for sufficiently large j. Lemma 1 from Section 8 can be found in Krjanev [1]. Proofs of Lemmas 4-6 in Section 7 are close to some arguments in Kato [1] and Glazman [1]. Some theorems concerning asymptotics of spectrums of linear nonselfadjoint operators can be found in the appendix to Voitovich et al. [1] written by M. Agranovich, and in Appendix 10 [Ramm 110].

CHAPTER II

INVESTIGATION OF INTEGRAL EQUATIONS OF THE STATIC AND QUASI-STATIC FIELDS AND APPLICATIONS TO THE SCATTERING FROM SMALL BODIES OF ARBITRARY SHAPE

0. Introduction

The calculation of static fields and some functionals of such fields, for example electrical capacitance or tensor polarizability, is of great interest in many applications. In particular, it is of basic interest for wave scattering by small bodies of arbitrary shape. Since the theory was initiated by Rayleigh [1] in 1871, very many papers have been published on this topic. Nevertheless the theory seemed incomplete in the following respect. Though wave scattering by a small body is a well understood process from the physical point of view there were no analytical formulas for the scattered field, scattering matrix, etc. In this chapter we obtain analytical formulas for the scattering matrix for the problems of scalar and vector wave scattering by a small body of arbitrary shape and by a system of such bodies. Analytical formulas for the calculation of the capacitance and polarizability of bodies of arbitrary shape with the needed accuracy are obtained. Two-sided variational estimates for the capacitance and polarizability are given.

The formulas mentioned above are of immediate use in applications. Some numerical examples are presented and some problems of interest in applications are solved. In Section 1, we present some new mathematical results, which form a basis for the rest, and some physical results. Proofs are given in other sections. The results are due to Ramm [18]-[47], [101].

1. Statement of the Problems and Main Results

1. Let A be a linear operator on a Hilbert space H with discrete spectrum. Let us assume for simplicity that A is a compact operator on H. We denote the characteristic values of A by λ_j: $\lambda_n A\Phi_n = \Phi_n$, $|\lambda_1| < |\lambda_2| \leq \ldots$, where λ_1 is semisimple. This means that the root subspace corresponding to λ_1 coincides with the eigenspace corresponding to λ_1 and is finite-dimensional. This case occurs in potential theory, including the applications of potential theory to elasticity and hydrodynamics (Parton-Perlin [1], Odquist [1]). We denote by $\{\phi_j\}$ an orthonormal basis of the eigenspace of the operator A corresponding to λ_1, $\lambda_1 A\phi_j = \phi_j$, $1 \leq j \leq m$, by $G = N(I - \overline{\lambda}_1 A^*)$, by $\{\psi_j\}$ an orthonormal basis of G, $\psi_j = \overline{\lambda}_1 A^* \psi_j$, $1 \leq j \leq m$, so $\dim N(I - \lambda_1 A) = \dim N(I - \overline{\lambda}_1 A^*)$, by $G^\perp = R(I - \lambda_1 A)$ the orthogonal complement to G in H, by P the orthogonal projection of H onto G. Let us consider the equation

$$g = \lambda_1 Ag + f, \quad f \in G^\perp. \tag{1.1}$$

The classical integral equations of potential theory are of the form (1.1). We introduce the operator

$$B_\gamma g = Ag + \gamma \sum_{j=1}^{m} (g,\psi_j)\psi_j \tag{1.2}$$

and the number

$$r_\gamma = \min(|\lambda_2|, |\lambda_1(1 + \gamma\lambda_1)^{-1}|), \tag{1.3}$$

where γ is an arbitrary number and $(\cdot,\cdot)$ is the inner product in H. Later, we choose γ. Let us consider the equation

$$g = \lambda_1 B_\gamma g + f. \tag{1.4}$$

It is clear that equations (1.1) and (1.4) are equivalent on the set $G^\perp$. This means that any solution $g \in G^\perp$ of equation (1.1) is a solution of equation (1.4) and any solution $g \in G^\perp$ to equation (1.4) is a solution to equation (1.1).

Theorem 1. *The operator B_γ has no characteristic values in the disk $|\lambda| < r_\gamma$. If $|1 + \gamma\lambda_1| < 1$, then the iterative process*

$$g_{n+1} = \lambda_1 B_\gamma g_n + F, \quad g_o = F; \quad F \equiv \lambda_1 Af - f \tag{1.5}$$

converges no more slowly than a geometric series with ratio q, $0 < q \leq |\lambda_1| r_\gamma^{-1}$, to an element $g = \Phi - f$, where $\Phi \in N(I - \lambda_1 A)$ and $P\Phi = Pf$. If $\dim G = 1$, $\phi \in N(I - \lambda_1 A)$, $\psi \in G$, and $||\psi|| = ||\phi|| = 1$, then $\Phi = \phi(f,\psi)/(\phi,\psi)$. The process (1.5) is stable in the following sense: the sequence h_n:

$$h_{n+1} = \lambda_1 B_\gamma h_n + F + \varepsilon_n, \quad h_o = F, \quad ||\varepsilon_n|| \leq \varepsilon \tag{1.6}$$

satisfies the estimate

$$\overline{\lim_{n\to\infty}} ||g - h_n|| = O(\varepsilon), \tag{1.7}$$

where

$$g = \lim_{n\to\infty} g_n.$$

Theorem 2. If dim $G = 1$ then the iterative process

$$f_{n+1} = \lambda_1 A f_n, \quad f_o = f \tag{1.8}$$

converges at the rate of a geometric series with ratio $q \leq |\lambda_1/\lambda_2|$ to the element $a\phi$, $a = (f,\psi)/(\phi,\psi)$.

Theorems 1 and 2 are applicable to the integral equations of potential theory. We note that the number γ in (1.2) should be chosen so that r_γ will be maximal, i.e., $r_\gamma = |\lambda_2|$ In order for this to be true, it is sufficient to take γ sufficiently close to $-\lambda_1^{-1}$, in particular one can take $\gamma = -\lambda_1^{-1}$.

Remark 1. The operator (1.2) and equation (1.4) have nothing to do with the operator and equation used in the so-called Schmidt Lemma (see Vainberg-Trenogin [1, p. 132]). Theorem 1 is an abstract analog to an iterative solution to the Robin problem in electrostatics.

2. Solving the third boundary-value problems for the Laplace operator by means of the single layer potential one comes to the equation

$$g + Tg = \lambda Ag \tag{1.9}$$

with $\lambda = -1$ for interior problems and $\lambda = 1$ for exterior problems and

$$Ag = \int_\Gamma \frac{\partial}{\partial N_s} \frac{1}{2\pi r_{st}} g(t)dt, \quad Tg = h \int_\Gamma \frac{g(t)dt}{2\pi r_{st}} \tag{1.10}$$

where Γ is a smooth closed surface, $h = \text{const}$, $r_{st} = |s - t|$ N_s is the unit outward pointing normal to Γ at the point s, and $\Gamma \subset \mathbb{R}^3$.

Theorem 3. *Assume that* $h = h_1 + ih_2$, $h_1 \geq 0$, $h_2 \leq 0$, $|h_1| + |h_2| > 0$. *Then*

(i) *all the eigenvalues of problem* (1.9) *satisfy the inequality* $|\lambda| > 1$;

(ii) *Equation* (1.9) *can be solved by the iterative process*

$$g_{n+1} + Tg_{n+1} = \lambda Ag_n + F, \quad g_o = \Phi, \tag{1.11}$$

where $\Phi \in L^2(\Gamma)$ *is arbitrary; method* (1.1) *converges no more slowly than a convergent geometric series;*

(iii) *If* $h > 0$ *all the eigenvalues of problem* (1.9) *are real.*

Remark 2. It can be shown that the iterative process

$$g_{n+1} + Tg_n = Ag_n + F, \quad g_o = \Phi \tag{1.12}$$

converges if $0 \leq h \leq \kappa$ and, in general, diverges if $h > \kappa$, where

$$\kappa \equiv \min_u \frac{\int_D |\nabla u|^2 dx}{\int_\Gamma |u|^2 dt} .$$

Remark 3. Let $g = \lim_{n\to\infty} g_n$, where the sequence g_n is defined by formula (1.11). Then the function

$$u(x) = \int_\Gamma \frac{g(t)dt}{4\pi r_{xt}} \tag{1.13}$$

is the solution to the problem

$$\Delta u = 0 \quad \text{in} \quad D, \quad \frac{\partial u}{\partial N_i} + hu\Big|_\Gamma = f, \tag{1.14}$$

for $\lambda = 1$, $F = -2f$. Here $\partial u/\partial N_e(\partial u/\partial N_i)$ is the limiting value of the normal derivative on Γ from the exterior (interior), $D \cup \Omega = \mathbb{R}^r$, $\Gamma = \partial D$, and D is interior domain.

Theorem 3 can be used in the Zaremba type problems. Let

$$\Delta u = 0 \quad \text{in} \quad D, \quad u\Big|_{\Gamma_1} = f_1, \quad \frac{\partial u}{\partial N_i}\Big|_{\Gamma_2} = f_2, \quad \Gamma_1 \cup \Gamma_2 = \Gamma, \tag{1.16}$$

where $\Gamma_1 \cap \Gamma_2 = \emptyset$, $\Gamma_1 \neq \emptyset$. Consider the problem

$$\Delta v = 0 \quad \text{in} \quad D, \quad \frac{\partial v}{\partial N_i} + h(s)v\Big|_{\Gamma} = F$$

$$F = \begin{cases} hf_1 & \text{on } \Gamma_1, \\ f_2 & \text{on } \Gamma_2, \end{cases} \qquad h(s) = \begin{cases} h & \text{on } \Gamma_1, \\ 0 & \text{on } \Gamma_2, \end{cases} \tag{1.17}$$

$$h = \text{const} > 0.$$

Theorem 4. The solution v_h to problem (1.17) tends in the norm of the space H_1 as $h \to +\infty$ to the solution $u(x)$ to problem (1.16), $\|u - v_h\|_{H_1} \leq ch^{-1}$, $c = \text{const}$. For any strictly inner subdomain $\tilde{D} \subset D$, the inequality $\|u - v_h\|_{W_2^2(\tilde{D})} \leq ch^{-1}$ holds.

3. In the following theorem we give necessary and sufficient conditions for the stationary Schwinger representation to be extremal. This is useful for obtaining various variational estimates.

Theorem 5. Let A be a symmetric linear operator on a Hilbert space. For the representation

$$(Af,f) = \max_{\phi \in D(A)} \frac{|(Af,\phi)|^2}{(A\phi,\phi)} \tag{1.18}$$

to hold it is necessary and sufficient that $A \geq 0$, i.e., $(A\phi,\phi) \geq 0$, for all $\phi \in D(A)$. By definition, we set $|(Af,\phi)|^2/(A\phi,\phi) = 0$ if $(A\phi,\phi) = 0$.

In applications this theorem is of use for the following reason. The value (Af,f) often has physical significance, e.g., electrical capacitance or polarizability. For a fixed denote by f the solution of equation $Af = g$. The value $X = (f,g)$ is called, in the terminology of physics, a linear functional of the field f. It is clear that $X = (Af,f)$. In the literature (Dolph [1], Dolph-Ritt [1]) the question of whether the Schwinger stationary representations for the scattering amplitude in quantum mechanics were extremal was discussed and a negative answer was obtained by means of relatively complicated arguments. In this theory $A = -\nabla^2 - k^2$ in $L^2(\mathbb{R}^3)$, and the scattering amplitude is some value of the form (Af,f). From Theorem 5 we obtain the negative answer immediately since $-\nabla^2 - k^2$, $k^2 > 0$, is not nonnegative.

Remark 4. For the representation

$$(Af,f) = \min_{\phi \in D(A)} \frac{|(Af,\phi)|^2}{(A\phi,\phi)} \tag{1.19}$$

to hold it is necessary and sufficient that $A \leq 0$.

4. Consider the equation

$$Kf = \int_D K(x,y)f(y)dy = g(x), \quad X \in D, \tag{1.20}$$

where $D \subset \mathbb{R}^r$ is a bounded domain, the operator $K: L^2(D) \to L^2(D)$ is compact, and $K(x,y) > 0$. Suppose there exists a function $h(x) > 0$ such that $Kh \leq c$ and $\int_D a(x)dx < \infty$, where $a(x) \equiv h(x)/((Kh)(x)) > 0$. Let $\phi = f(x)/a(x)$ and by $H_\pm$ denote the spaces $L^2(D;a^{\pm 1}(x))$ with the norms $\|g\|_\pm^2 = \int_D |g|^2 a^{\pm 1}(x)dx$. Let us rewrite equation (1.20) in the form

$$K_1\phi = g, \quad K_1\phi \equiv \int_D K(x,y)a(y)\phi(y)dy. \tag{1.21}$$

Let $Q = I - K_1$ and let $K_1\Phi_j = \lambda_j\Phi_j$, $\lambda_1 > |\lambda_2| \geq \dots$.

Here we use the theorem of Entsch which says that the maximal eigenvalue of an integral equation with positive kernel is positive and simple (Entsch [1]). An eigenvalue is called simple if the eigenspace is one-dimensional and the root subspace coincides with the eigenspace.

Theorem 6. *Assume the following statements hold:*

(i) $g \in H_+$ *and*

$$0 < C_1(\tilde{D}) \leq \int_{\tilde{D}} K(x,y)a(y)dy \leq C_2(\tilde{D}), \quad x \in D \tag{1.22}$$

for any measurable subset $\tilde{D}$, meas $\tilde{D} > 0$.

(ii) *Equation* (1.21) *has a solution in* H_+.

(iii) *The eigenfunctions* $\{\Phi_j\}$ *of the operator* K_1 *form a Riesz basis in* H_+ *and* $|\arg \lambda| \leq \pi/3$, $\lambda_j \neq 0$.

Then the iterative process

$$\phi_{n+1} = Q\phi_n + g, \quad \phi_o = g, \quad Q = I - K_1 \tag{1.23}$$

converges in H_+ *to a solution* $\phi(x)$ *of equation* (1.21), *and equation* (1.20) *has the solution* $f(x) = a(x)\phi(x)$ *in* H_-.

Remark 5. A sequence $\{\Phi_j\}$ forms a Riesz basis of the Hilbert space H if the system $\{\Phi_j\}$ is complete in H and for any numbers $c_1,\dots c_n$ the inequality

$$a \sum_{j=1}^{n} |c_j|^2 \leq |\sum_{j=1}^{n} c_j\Phi_j|^2 \leq b \sum_{j=1}^{n} |c_j|^2, \quad a > 0,$$

holds, where the coefficients a,b do not depend on n.

5. In this section we present some results of interest in the theory of static and quasi-static field. Consider a homogeneous body D with dielectric constant ε surrounded by a medium with dielectric constant ε_o. The body is in a

homogeneous initial electrostatic field $\mathscr{E} = -\nabla\phi$, $\phi = -(\mathscr{E},x)$, where x is the radius vector. The boundary Γ of the body D is assumed to be smooth, though it can actually be piecewise smooth without cusps. The problem is to find the field outside the body and the dipole moment of the body. The reader can find definitions of physical values in Landau-Lifschitz [1]. Denote by $u = \phi + v$ the scalar potential of the field. Then

$$\begin{aligned} &\Delta v = 0 \quad \text{in } \mathbb{R}^3, \quad v(\infty) = 0, \\ &\qquad \varepsilon \frac{\partial v}{\partial N_i} = \varepsilon_0 \frac{\partial v}{\partial N_e} + (\varepsilon_0 - \varepsilon)\frac{\partial \phi}{\partial N} \quad \text{on } \Gamma. \end{aligned} \tag{1.24}$$

We look for v of the form

$$v(x) = \int_\Gamma \frac{\sigma(t)\,dt}{4\pi\varepsilon_0 r_{xt}}, \quad r_{xt} \equiv |x - t|.$$

Substituting this formula for v into the boundary condition (1.24), we obtain the following integral equation for σ:

$$\sigma = -\gamma A\sigma - 2\gamma\varepsilon_0 \frac{\partial \phi}{\partial N}, \quad \gamma = \frac{\varepsilon - \varepsilon_0}{\varepsilon + \varepsilon_0}, \tag{1.25}$$

where $\varepsilon_0 > 0$,

$$A\sigma \equiv \int_\Gamma (2\pi)^{-1}\psi(t,s)\sigma(s)\,ds, \quad \psi(t,s) = \frac{\partial}{\partial N_t}\frac{1}{r_{st}}. \tag{1.26}$$

If $\varepsilon = +\infty$, $\gamma = 1$, the body D is an ideal conductor; if $\varepsilon = 0$, $\gamma = -1$, we can think of it as a superconductor in a magnetic field $\mathscr{H} = \mathscr{E}$. Equation (1.25) is to be solved with the following condition on σ:

$$\int_\Gamma \sigma(t)\,dt = 0 \tag{1.27}$$

which means the body is electrically neutral. Equation (1.25) has a unique solution satisfying condition (1.27). The right-hand side of equation (1.25) also satisfies (1.27). So we can apply Theorem 1, noting that now the set $G^{\perp}$ from Theorem 1 is described by equality (1.27). From Theorem 1 it follows that the solution of equation (1.25) including the case $\gamma = 1$, can be found by means of the iterative process

$$\begin{cases} \sigma_{n+1} = -\gamma A\sigma_n - 2\gamma\varepsilon_0 \frac{\partial\phi}{\partial N}, \quad \sigma_0 = \Phi \\ \sigma = \lim_{n\to\infty} \sigma_n, \end{cases} \tag{1.28}$$

where $\Phi \in L^2(\Gamma)$ is arbitrary if $\gamma < 1$, and Φ must satisfy condition (1.27) if $\gamma = 1$. We note that if $\varepsilon < \infty$ then $|\gamma| < 1$. In the disk $|\gamma| < 1$ there is no characteristic value of the operator A defined by formula (1.26). So for $|\gamma| < 1$, the iterative process (1.28) converges for any initial data. It is only the case $\gamma = 1$ that requires Theorem 1. It is well-known from potential theory (see, for example, Zabreiko et al. [1]) that $\lambda = -1$ modulo the minimal characteristic value of A. It is reasonable to take $\Phi = -2\gamma\varepsilon_0(\partial\phi/\partial N)$. Condition (1.27) is satisfied by such a choice. The dipole moment P is defined by

$$P = \int_\Gamma t\sigma(t)dt \tag{1.29}$$

and can be calculated as

$$P = \lim_{n\to\infty} \int_\Gamma t\sigma_n(t)dt,$$

where t is the radius vector of a point $t \in \Gamma$ and σ_n is defined by formula (1.28). Let us define the electrical polarizability tensor $\alpha_{ij}(\gamma)$ and magnetic polarizability tensor β_{ij} by the formulas

$$P_i = \alpha_{ij}(\gamma)\varepsilon_0 V \mathscr{E}_j; \tag{1.29'}$$

$$M_j = \alpha_{ij}(\tilde{\gamma})\mu_o V \mathscr{A}_j + \beta_{ij}\mu_o V \mathscr{A}_j. \tag{1.29''}$$

Here and below one must sum over the repeated indices, V is the volume of the body D, μ is the magnetic permeability of D, μ_o is the magnetic permeability of the medium, $\tilde{\gamma} \equiv (\mu-\mu_o)/(\mu+\mu_o)$, $\mathscr{A}$ is the homogeneous initial magnetic field, M and P are the magnetic and electric dipole moments acquired by the body in the field $\mathscr{A}$ and $\mathscr{E}$ respectively, and $\beta_{ij} = \alpha_{ij}(-1)$ is the magnetic polarizability tensor. Formulas (1.29'), (1.29") are valid also in the quasi-static theory, when the wavelength of the initial electromagnetic field is considerably greater than the characteristic dimension a of the body. If the depth δ of the skin layer is small, $\delta << a$, then both terms in formula (1.29'') are of the same order. If $\delta >> a$ the second term in the right-hand side of equality (1.29") is negligible in comparison to the first term if $\mu - \mu_o$ is not very small. Let us introduce the values

$$\alpha_{ij} = \alpha_{ij}(1), \tag{1.30}$$

$$\alpha_{ij}^{(n)}(\gamma) = \frac{2}{V}\sum_{m=0}^{n}\frac{(-1)^m}{(2\pi)^m}\frac{\gamma^{n+2}-\gamma^{m+1}}{\gamma-1}b_{ij}^{(m)}, \quad n > 0, \tag{1.31}$$

$$b_{ij}^{(0)} = V\delta_{ij}, \quad b_{ij}^{(1)} = \int_\Gamma\int_\Gamma \frac{N_i(t)N_j(s)}{r_{st}}\,dsdt,$$
$$\delta_{ij} = \begin{cases} 0, & i \neq j, \\ 1, & i = j. \end{cases} \tag{1.32}$$

$$b_{ij}^{(m)} = \int_\Gamma\int_\Gamma dsdt\, N_i(t)N_j(s)\underbrace{\int_\Gamma\cdots\int_\Gamma}_{m-1}\frac{1}{r_{st}}\psi(t_1,t)\psi(t_2,t_1)\cdots$$
$$\psi(t_{m-1},t_{m-2})dt_1dt_2\ldots dt_{m-1}, \quad \psi(t,s) = \frac{\partial}{\partial N_t}\frac{1}{r_{st}}. \tag{1.33}$$

In particular,

$$\alpha_{ij}^{(1)}(\gamma) = 2(\gamma + \gamma^2)\delta_{ij} - \frac{\gamma^2}{\pi V} b_{ij}^{(1)} \tag{1.34}$$

$$\alpha_{ij}^{(1)} = 4\delta_{ij} - \frac{1}{\pi V} b_{ij}^{(1)}, \quad \beta_{ij}^{(1)} = -\frac{1}{\pi V} b_{ij}^{(1)}. \tag{1.35}$$

Theorem 7. The following estimate is valid:

$$|\alpha_{ij}(\gamma) - \alpha_{ij}^{(n)}(\gamma)| \leq Aq^{n+1}, \quad 0 < q < 1, \ -1 \leq \gamma \leq 1, \tag{1.36}$$

where the constants A and q depend on the shape of the body and on γ.

Consider electromagnetic wave scattering by the body D. Let $a << \lambda$, $k = 2\pi/\lambda$, where λ is the wavelength of the initial field. From some model examples, such as scattering by a ball, it follows that the condition $a << \lambda$ is fulfilled if $a \leq \lambda/10$. If E is the field scattered by D then the scattering amplitude f_E is defined by the formula

$$E \sim \frac{\exp(ikr)}{r} f_E(k,n) \quad \text{as} \quad r \to \infty,$$

where r is the distance between D and the point x of observation and $n = x/r$. The scattering amplitude f_H for the magnetic field can be defined similarly. The following formulas are valid (Landau-Lifschitz [1]):

$$f_E = \frac{k^2}{4\pi\varepsilon_0}[n[P,n]] + \frac{k^2}{4\pi}\sqrt{\frac{\mu_0}{\varepsilon_0}}[M,n],$$
$$f_H = \sqrt{\frac{\varepsilon_0}{\mu_0}}[n,f_E]. \tag{1.37}$$

These formulas permit the calculation of the scattering amplitude for a small body of arbitrary shape if the dipole moments P and M induced by the initial fields are known. These moments can be calculated by formulas (1.29) and (1.29') in which one should replace $\mathscr{E}$, $\mathscr{H}$ by E, H, where E, H are the values of the initial electromagnetic field at the point where the body D is placed. The tensors $\alpha_{ij}(\gamma)$, β_{ij} in formulas (1.29') and (1.29") can be calculated with the prescribed accuracy by formulas (1.30) - (1.33), according to Theorem 7. These results allow one to write an explicit formula for the scattering matrix. Let the origin be inside D, the plane YOZ contain the orts s,n of the propagation directions for initial and scattered waves (this plane is called the plane of scattering), θ be the angle between the orts (the angle of scattering), $\cos\theta = (\nu,n)$, E_1 and E_2 be the projections of E on the axes OX and OY, f_1 and f_2 be the projections of the far distance scattered electrical field on the axes OX', OY'. The axis OZ' of the primed coordinate system lies in the plane of scattering and is directed along the ort n, and the axis OY' lies in the plane of scattering. The scattering matrix is defined by the formula

$$f_E = SE.$$

In Hulst [1] the following notations are used for the scattering matrix:

$$\begin{pmatrix} f_2 \\ f_1 \end{pmatrix} = \begin{pmatrix} S_2 & S_3 \\ S_4 & S_1 \end{pmatrix} \begin{pmatrix} E_2 \\ E_1 \end{pmatrix}.$$

The formula for the scattering matrix is

$$S = S(\theta) = \begin{pmatrix} S_2 & S_3 \\ S_4 & S_1 \end{pmatrix} \tag{1.38}$$

$$= \frac{k^2V}{4\pi} \begin{pmatrix} \mu_o\beta_{11}+\alpha_{22}\cos\theta-\alpha_{32}\sin\theta & \alpha_{21}\cos\theta-\alpha_{31}\sin\theta-\mu_o\beta_{12} \\ \alpha_{12}-\mu_o\beta_{21}\cos\theta+\mu_o\beta_{31}\sin\theta & \alpha_{11}+\mu_o\beta_{22}\cos\theta-\mu_o\beta_{32}\sin\theta \end{pmatrix}$$

Knowing the scattering matrix for a single small particle, one can calculate the refraction index tensor of a rarefied medium consisting of many such particles. If n_{ij} is the refraction index tensor, N is the number of particles in a unit volume then $n_{ij} = 1 + 2\pi Nk^{-2}S_{ij}(0)$. Knowing the tensor n_{ij} one can calculate the double refraction, absorption, rotation of the polarization plane, Stokes parameters, i.e., to answer the majority of the questions of practical importance for the theory of wave propagation in such a medium.

6. Let D be a conductor with the surface charge Q. The charge density on the boundary Γ satisfies the equation

$$\sigma = -A\sigma \tag{1.39}$$

and the condition

$$\int_\Gamma \sigma(t)dt = Q, \tag{1.40}$$

where the operator A is defined by formula (1.26). By Theorem 2, we conclude that the iterative process

$$\sigma_{n+1} = -A\sigma_n, \quad \sigma_0 = \frac{Q}{S}, \quad S = \text{meas}\,\Gamma, \quad \sigma = \lim_{n\to\infty} \sigma_n \tag{1.41}$$

converges no more slowly than a convergent geometric series. The process (1.41) is not stable in the sense of Theorem 1. A stable process for solving equation (1.39) is

$$\sigma_{n+1} = -A\sigma_n - \frac{1}{S}\int_\Gamma \sigma_n(t)dt + F, \quad \sigma_0 = F, \tag{1.42}$$

where

$$F = -\frac{1}{\sqrt{S}} - \frac{1}{2\pi\sqrt{S}}\int_\Gamma \psi(t,s)ds. \tag{1.43}$$

This can easily be obtained from Theorem 1. Let C be the capacitance of D. We introduce the values

$$C^{(n)} = 4\pi\varepsilon_0 S^2 \left\{ \frac{(-1)^n}{(2\pi)^n} \int_\Gamma \int_\Gamma \frac{dsdt}{r_{st}} \underbrace{\int_\Gamma \cdots \int_\Gamma}_{n} \psi(t,t_1) \cdot \psi(t_1,t_2) \cdots \psi(t_{n-1},t_n) dt_1 \cdots dt_n \right\}^{-1} \tag{1.44}$$

<u>Theorem 8</u>. <u>The estimate</u>

$$|C - C^{(n)}| \le A_1 q^{n+1}, \quad 0 < q < 1, \tag{1.45}$$

<u>holds, where</u> A_1 <u>and</u> q <u>are constants depending on the shape of</u> D.

For $n = 0$ we obtain

$$C^{(0)} = 4\pi\varepsilon_0 S^2 \left\{ \int_\Gamma \int_\Gamma \frac{dsdt}{r_{st}} \right\}^{-1}, \quad C^{(0)} \le C. \tag{1.46}$$

The approximation $C^{(0)} \approx C$ had been known in the literature for more than 65 years as Howe's empirical formula. No refinements of this empirical formula seem to be known. One can view formula (1.44) as such a refinement of Howe's empirical formula. The inequality in (1.46) can be obtained from the variational principle of type (1.18):

$$C = \max_\phi \left\{ \left(\int_\Gamma \phi dt \right)^2 / \int_\Gamma \int_\Gamma \phi(t)\phi(s)(4\pi\varepsilon_0 r_{st})^{-1} dsdt \right\}. \tag{1.47}$$

Indeed, if $\phi = 1$ is substituted into (1.47) then (1.46) is obtained. Formula (1.44) is of practical use because it makes it possible to work out a standard program for the calculation of the capacitance of a capacitor of arbitrary shape. Numerical examples show that accuracy 2-3% can be obtained by formula (1.44) for $n = 0,1$. The capacitances of the following conductors were calculated. First, a table of the capacitances of various parallelepipeds with arbitrary edges was calculated. Formula (1.44) for $n = 0$ gave an accuracy of ~ 3.5%. For the cube with unit edge we obtained $C^{(0)} = 4\pi\varepsilon_o \cdot 0.649$. In Pólya-Szegö [1] many papers dealing with the calculation of the cube's capacitance are mentioned, in which the best results obtained are $C \simeq 0.65 \cdot 4\pi\varepsilon_o$ and the two-sided estimate is $0.639 \leq C(4\pi\varepsilon_o)^{-1} \leq 0.667$. We can sharpen this lower estimate to $0.649 \leq C(4\pi\varepsilon_o)^{-1}$. The least accurate capacitance was obtained for the quadratic plate. The capacitance a of parallelepiped had not been previously tabulated in the literature. Second, a table of the capacitance of a circular cylinder was calculated. The table was obtained from formula (1.44) with $n = 0,1$, with the accuracy ~ 3%. We obtained a smooth transition to the asymptotic values of the capacitance for a long cylinder and for very short one (a disk).

7. Consider a flaky-homogeneous body D. Let the surface Γ_{j-1} surround the surface Γ_j, the dielectric constant of the layer between Γ_{j-1} and Γ_j be ε_j, $1 \leq j \leq p$, and $\mathscr{E} = -\nabla\phi$ be the initial field. We look for a potential of the form

$$u = \phi + v, \quad v = \sum_{j=1}^{p} \int_{\Gamma_j} \frac{\sigma_j(t)\,dt}{4\pi\varepsilon_o r_{xt}}. \tag{1.48}$$

Using the boundary conditions

$$\varepsilon_j \frac{\partial u}{\partial N_i} = \varepsilon_{j-1} \frac{\partial u}{\partial N_e} \quad \text{on} \quad \Gamma_j,$$

we obtain the system of integral equations

$$\sigma_j(t_j) = -\gamma_j \sum_{m \neq j, m=1}^{p} T_{jm}\sigma_m - \gamma_j A_j \sigma_j - 2\gamma_j \varepsilon_o \frac{\partial\phi}{\partial N_{t_j}}, \quad 1 \leq j \leq p, \tag{1.49}$$

where $\gamma_j = (\varepsilon_j - \varepsilon_{j-1})/(\varepsilon_j + \varepsilon_{j-1})$, the operator A_j is defined by formula (1.26) with Γ replaced by Γ_j, and

$$T_{jm}\sigma_m = \int_{\Gamma_m} \frac{\partial}{\partial N_{t_j}} \frac{1}{2\pi r_{t_j t}} \sigma_m(t)dt. \tag{1.50}$$

The system (1.49) can be written in the form

$$\sigma = -\mathscr{B}\sigma + f, \tag{1.51}$$

$$\sigma = (\sigma_1, \ldots, \sigma_p), \quad \int_{\Gamma_j} f_j(t)dt = 0, \quad 1 \leq j \leq p,$$

where the meanings of $\mathscr{B}$ and f are clear from (1.49).

Theorem 9. *The solution to system* (1.51) *does exist and is unique in* $L^2(\Gamma)$ *if* $\int_{\Gamma_j} \sigma_j dt = 0$, $1 \leq j \leq p$. *The solution for* $p = 2$ *can be calculated by means of the iterative process*

$$\sigma^{(n+1)} = -\mathscr{B}\sigma^{(n)} + f, \quad \sigma^{(o)} = f, \tag{1.52}$$

which converges no more slowly than some convergent geometric series.

Let us define the polarizability tensor A_{ij} of the flaky body by the formula $P_i = A_{ij}\varepsilon_o \mathscr{E}_j$. The multiplier V is omitted because for a flaky body the polarizability tensor depends not only on the shape of the body, but also on the constants ε_j and the shape of the surfaces Γ_j, so that from the physical point of view there is no reason to single out this multiplier. We define the dipole moment of the flaky body by the formula $P = \sum_{j=1}^{p} \int_{\Gamma_j} t\sigma_j dt$ and its approximate value by $P^{(n)} = \sum_{j=1}^{p} \int_{\Gamma_j} t\sigma_j^{(n)} dt$. The first order approximation is

$$A_{iq}^{(1)} = \sum_{j=1}^{p} \alpha_{iq}^{(1)}(\gamma_j)V_j + \sum_{j=1}^{p} \sum_{m\neq j, m=1}^{p} \alpha_{iq}^{(j,m)}, \tag{1.53}$$

where V_j is the volume of the body with boundary Γ_j, the tensor $\alpha_{iq}^{(1)}(\gamma_j)$ is defined by formula (1.34), and

$$\alpha_{iq}^{(j,m)} \equiv \begin{cases} -\dfrac{\gamma_j\gamma_m}{\pi} b_{iq}^{(j,m)}, & j > m, \\ -\dfrac{\gamma_j\gamma_m}{\pi} b_{iq}^{(j,m)} + 4\gamma_j\gamma_m V_m \delta_{iq}, & j < m, \end{cases} \tag{1.54}$$

where

$$b_{iq}^{(j,m)} = \int_{\Gamma_j}\int_{\Gamma_m} \frac{N_i(t_j)N_q(t_m)}{r_{t_j t_m}} dt_j dt_m. \tag{1.55}$$

8. Consider a system of p nonoverlapping homogeneous bodies with dielectric constants ε_j and boundaries Γ_j situated in an initial field $\mathscr{E} = -\nabla\phi$. Let us look for a potential in the form (1.48). For the functions σ_j in this representation of the potential we obtain the system of equations

$$\sigma_j = -\kappa_j \sum_{m\neq j, j=1}^{p} T_{jm}\sigma_m - \kappa_j A_j \sigma_j - 2\kappa_j \varepsilon_o \frac{\partial\phi}{\partial N_{t_j}}, \quad 1 \leq j \leq p, \tag{1.56}$$

where $\kappa_j = (\varepsilon_j - \varepsilon_o)/(\varepsilon_j + \varepsilon_o)$, and the operators T_{jm}, A_j are defined as in equation (1.44).

Theorem 10. The system (1.56) has a solution $\sigma \in L^2(\Gamma)$, $\sigma = (\sigma_1,\dots,\sigma_p)$, $\Gamma = \bigcup_{j=1}^{p} \Gamma_j$, satisfying the conditions $\int_{\Gamma_j} \sigma_j dt = 0$, $1 \leq j \leq p$. This solution is unique and can be calculated by an iterative process similar to (1.52). The process converges no more slowly than a convergent geometric series.

Let $P_i = B_{iq}\varepsilon_o \mathscr{E}_q$ denote the polarizability tensor of the system of bodies and let $P^{(n)} = \sum_{j=1}^{p} \int_{\Gamma_j} t\sigma_j^{(n)} dt$ denote the approximate value of P, where $\sigma_j^{(n)}$ is the n-th approximation for σ_j which is obtained by means of the iterative process for solving equation (1.56). The formula for the first order approximation $B_{iq}^{(1)}$ for the polarizability tensor can be obtained from formula (1.53) if one replaces $A_{iq}^{(1)}$ by $B_{iq}^{(1)}$, $\alpha_{iq}^{(1)}(\gamma_j)$ by $\alpha_{iq}^{(1)}(\kappa_j)$, $\alpha_{iq}^{(j,m)}$ by $\tilde{\alpha}_{iq}^{(j,m)}$, where $\tilde{\alpha}_{iq}^{(j,m)} = -(\kappa_j\kappa_m/\pi)b_{iq}^{(j,m)}$ and $b_{iq}^{(j,m)}$ are defined by formula (1.55).

9. Consider the basic problems of static field theory for bodies of zero volume, i.e., for metallic screens or thin ideal magnetic films. The surface F of the screen is not closed. Denote by $\mathscr{L}$ the edge of F, by $\mathscr{E} = -\nabla\phi$ the initial field. The charge density $\sigma(t)$ satisfies the equation

$$T\sigma = u_0 - \phi, \quad T\sigma \equiv \int_\Gamma \frac{\sigma(t)dt}{4\pi\varepsilon_0 r_{xt}} = u_0 - \phi, \tag{1.57}$$

and the condition

$$\int_F \sigma(t)dt = 0, \tag{1.58}$$

where $u_0 = \text{const}$ is the potential to which the screen was charged in the initial field. We must find σ and u_0 from (1.57) and (1.58) in the class of functions satisfying the edge condition

$$\sigma(t) \sim \frac{1}{\sqrt{\rho(t)}}, \quad \rho(t) \equiv \min_{s\in\mathscr{L}} |t-s|. \tag{1.59}$$

The problem of finding the free distribution of charge Q on the screen F can be reduced to solving the equation

$$T\omega = u, \quad u = \text{const}, \quad \int_F \omega dt = Q. \tag{1.60}$$

If $u = 1$ then $Q = C$, where C is the capacitance of the screen. It can be proved that problem (1.60) has a unique solution and this solution satisfies the edge condition (1.59). Define the dipole moment, polarizability tensor, and capacitance of the screen by the formulas

$$P = \int_F t\sigma(t)dt, \quad P_i = \tilde{\alpha}_{ij}\varepsilon_o \mathscr{E}_j, \quad C = \int_F \omega(t)dt, \tag{1.61}$$

where the potential u in (1.60) is assumed to be equal to 1. If F is a plane plate orthogonal to the axis X_3, then $\tilde{\alpha}_{i3} = \tilde{\alpha}_{3i} = 0$, $1 \le i \le 3$, so in this case the tensor $\tilde{\alpha}_{ij}$ is defined by four numbers $\tilde{\alpha}_{ij}$, $i \le i,j \le 2$ (actually by three numbers since $\tilde{\alpha}_{ij} = \tilde{\alpha}_{ji}$). Let us set

$$\begin{cases} h(t) = \dfrac{1}{\sqrt{\rho(t)}}, \quad a(t) = \dfrac{h(t)}{Th(t)}, \quad \psi(t) = \dfrac{\omega(t)}{a(t)}, \quad A_1\psi = T(a\psi), \\ u = 1. \end{cases} \tag{1.62}$$

Consider the iterative process of the type (1.23):

$$\psi_{n+1} = (I-A_1)\psi_n + 1, \quad \psi_o = 1. \tag{1.63}$$

By Theorem 6, the process converges in H_+ to a function ψ and $\omega = a\psi$ is a solution of the equation $T\omega = 1$ and $\int_F \omega dt = C$. In order to solve problem (1.57) - (1.58), we replace σ by a new unknown function ξ: $\xi(t) = -\sigma(t) + u_o\omega(t)$, where $\omega(t)$ is the function found above. It is easy to verify that $T\xi = \phi$, $\int_F \xi dt = Cu_o$. This equation for ξ is of the form (1.60) and can be solved by an iterative process as was shown above.

10. Here we formulate some variational principles of use in the calculation of capacitance and polarizability. We get two-sided estimates for these values. The principle (1.47) is convenient for obtaining estimates from below. To obtain estimates from above we use the well-known Dirichlet

principle (see Pólya-Szegö [1])

$$C = \min_{\Phi} \varepsilon_0 \int_\Omega |\nabla\Phi|^2 dx, \quad \Phi\Big|_\Omega = 1, \quad \Phi(\infty) = 0, \tag{1.64}$$

where, as usual, $\Omega = \mathbb{R}^3 \setminus D$. We note the estimate

$$C \leq \varepsilon_0 \int_0^\pi \int_0^{2\pi} \Big\{F^2(\theta,\phi) + |F'_\theta(\theta,\phi)|^2 + \frac{1}{\sin^2\theta} |F'_\phi(\theta,\phi)|^2\Big\} \frac{\sin\theta \, d\theta d\phi}{F(\theta,\phi)}, \tag{1.65}$$

where $r = F(\theta,\phi)$ is the equation of the surface Γ in a spherical coordinate system with origin inside D. In the following theorem, a variational principle for the polarizability tensor is given.

<u>Theorem 11</u>. <u>The formulas</u>

$$V\alpha_{jj} = \max_{\{\phi:\int_\Gamma \phi dt = 0\}} \frac{\left(\int_\Gamma t_j \phi(t) dt\right)^2}{\int_\Gamma \int_\Gamma \frac{\phi(t)\phi(s)}{4\pi\varepsilon_0 r_{st}} dsdt}, \tag{1.66}$$

<u>hold, where</u> V <u>is the volume of</u> D, $\phi \in L^2(D)$;

$$\begin{cases} V\alpha_{jj} = \min \int_\Omega |\nabla\phi_j|^2 dx, \quad \phi_j(\infty) = 0, \\ \phi_j\Big|_\Gamma = s_j + \text{const}, \end{cases} \tag{1.67}$$

<u>where</u> s_j <u>is the</u> j-<u>th coordinate of the point</u> $s \in \Gamma$, $\phi_j \in C^1(\Omega)$;

$$-V\beta_{jj} = V + \max \frac{\left(\int_\Gamma N_j(t) u(t) dt\right)^2}{\int_\Omega |\nabla u|^2 dx}, \quad u(\infty) = 0, \quad \Delta u = 0 \text{ in } \Omega \tag{1.68}$$

$$\begin{cases} -V\beta_{jj} = V + \min \int_\Omega |q_j(x)|^2 dx, \\ (q_j, N)\Big|_\Gamma = N_j(t), \quad \text{div } q_j(x) = 0 \text{ in } \Omega. \end{cases} \tag{1.69}$$

Knowing the diagonal elements of the polarizability tensor, one can estimate its nondiagonal elements since the tensor is diagonal in an appropriate coordinate system and by rotation the elements of the tensor are transformed according to the known transformation law. If we approach the limit as $V \to 0$ in formulas (1.66) - (1.69) and denote by $\tilde{\alpha}_{ij}$, $\tilde{\beta}_{ij}$ the limits of the left-hand sides of these formulas, then we obtain variational principles for the electrical polarizability tensor of thin screens and the magnetic polarizability tensor of thin magnetic films. We note in particular the case in which F is a plane plate. In this case only the elements $\tilde{\alpha}_{ij}$, $1 \le i,j \le 2$, and $\tilde{\beta} \equiv \tilde{\beta}_{33}$ are nonzero in the coordinate system with axis X_3 orthogonal to F.

11. Let F be an aperture in an infinite conducting plane, α_o be its coefficient of electrical polarizability, $\tilde{\beta}^o_{ij}$, $1 \le i,j \le 2$, be its tensor of magnetic polarizability, the third axis X_3 of the coordinate system be orthogonal to F, and e_1, e_2, e_3 be the orts of the coordinate axes. We assume that the electric field in the half-space $X_3 < 0$ is $\tilde{E}_o e_3$ and its asymptotics in the upper half-space $X_3 > 0$ is $\phi \sim (P,x)/4\pi\varepsilon_o|x|^3$, where $E = -\nabla\phi$ for $x_3 > 0$ and the vector P is called the electrical dipole moment of the aperture. The magnetic field in the lower half-space is $H_o = \tilde{H}_{o1}e_1 + \tilde{H}_{o2}e_2$, and its asymptotics in the upper half-space is $\psi \sim (M,x)/(4\pi\mu_o|x|^3)$, where $H = -\nabla\psi$ for $x_3 > 0$ and the vector M is the magnetic dipole moment of the aperture. The values α_o, β^o_{ij} are defined by the formula

$$P = \alpha_o \varepsilon_o \tilde{E}_o e_3, \quad M_i = \beta^o_{ij} M_o \tilde{H}_{oj}. \tag{1.70}$$

Theorem 12. The following formula holds:

$$\alpha_o = -\frac{1}{2}\tilde{\beta}, \quad \beta^o_{ij} = -\frac{1}{2}\tilde{\alpha}_{ij}. \tag{1.71}$$

Remark 7. This theorem gives a kind of duality principle in electrostatics.

12. Consider a conductor D placed in an inhomogeneous medium with dielectric tensor $\varepsilon_{ij}(x)$. The variational principle

$$C = \min_{u|_\Gamma=1, u(\infty)=0} \int_\Omega \varepsilon_{ij}(x)\frac{\partial u}{\partial x_i}\frac{\partial u}{\partial x_j}\,dx \tag{1.72}$$

allows one to estimate from above the capacitance in such a medium. To estimate the capacitance from below one can use the principle

$$C^{-1} = \min_{\substack{\operatorname{div}\varepsilon E=0 \\ \int_\Gamma(\varepsilon E,N)dt=1}} \int_\Omega \varepsilon_{ij}(x)E_i(x)E_j(x)\,dx. \tag{1.73}$$

We repeat that one must sum over repeated indices.

13. Formulas (1.28), (1.29'), (1.29"), and (1.30)-(1.35) allow us to solve the inverse problem of radiation theory. The problem consists of calculating the electromagnetic field at a point where a small probe is placed, from the field scattered by this probe and measured in the far distance zone.

14. Here we present some results concerning wave scattering by a system of small bodies of arbitrary shape. First we consider scalar wave scattering from a system of r small bodies with impedance boundaries. Let $u = u_o + v$, where u_o is the initial field, v is the scattered field, and

$$\begin{cases} (\Delta+k^2)v = 0 \quad \text{in} \quad \Omega, \quad \dfrac{\partial v}{\partial N} - hv\Big|_\Gamma = \left(-\dfrac{\partial u_o}{\partial N} + hu_o\right)\Big|_\Gamma, \\ |x|\left(\dfrac{\partial v}{\partial |x|} - ikv\right) \to 0 \quad \text{as} \quad |x| \to \infty. \end{cases} \tag{1.74}$$

Here $\Gamma = \bigcup_{j=1}^{r} \Gamma_j$, $D = \bigcup_{j=1}^{r} D_j$, $\Gamma_j = \partial D_j$, $\Omega = \mathbb{R}^3 \setminus D$, $D_j \cap D_m = \emptyset$ if $j \neq m$, N is the unit outward pointing normal to Γ, $h\Big|_{\Gamma_j} \equiv h_j = h_{1j} + ih_{2j}$, $h_{1j} \geq 0$, $h_{2j} \leq 0$, $|h_{1j}| + |h_{2j}| > 0$, and $k^2 > 0$. We look for v of the form

$$v = \sum_{j=1}^{r} \int_{\Gamma_j} \frac{\exp(ikr_{xt})}{4\pi r_{xt}} \sigma_j(t)\,dt. \tag{1.75}$$

The scattering amplitude can be found by the formula

$$f(n,k) = \sum_{j=1}^{r} \frac{1}{4\pi r_{xt}} \exp\{-ik(n,t_j)\}Q_j, \quad ka \ll 1, \tag{1.76}$$

where t_j is the radius vector of the j-th body, $a = \max_{1 \leq j \leq r} a_j$, a_j is the characteristic dimension of the j-th body, and Q_j is defined by the formula

$$Q_j = \int_{\Gamma_j} \exp\{-ik(n,t-t_j)\}\sigma_j(t)\,dt. \tag{1.77}$$

For $r = 1$, $ka \ll 1$, we obtain the approximation

$$f(n,k) = -\frac{hS}{4\pi(1+hSC^{-1})} u_{oo}, \tag{1.78}$$

for the scattering amplitude where C is the capacitance of the body, S is the surface area of the body, u_{oo} is the main term of the expansion in powers of ka of the initial field $u_{oo}(t_1,k)$ at the point t_1 where the body is, $u_o(t_1,k) = u_{oo} + O(ka)$. If $h = 0$ or is very small, one must take into account the terms up to the second order of smallness with respect to ka. If $h = 0$, $r = 1$, and $u_o = \exp\{ik(\nu,x)\}$, we obtain the formula

$$f(n,\nu,k) = \frac{ikV}{4\pi}\beta_{pq}n_p \frac{\partial u_o}{\partial x_q}\Big|_{x=0} + \frac{V}{8\pi}u_o\Big|_{x=0}, \tag{1.79}$$

for the scattering amplitude, where

$$\frac{\partial u_o}{\partial x_q} = ik\nu_q(1 + O(ka)), \tag{1.80}$$

V is the volume of the body, the origin of the coordinate system is inside the body, β_{pq} is the magnetic polarizability tensor defined in formula (1.29') and calculated from the formulas $\beta_{ij} = \alpha_{ij}(-1)$, (1.31), (1.36). Formula (1.79) is accurate up to the order of smallness $O((ka)^2)$. We note that the passage to the limit $h \to +\infty$ in formula (1.78) leads to the formula $f = (C/(4\pi))u_{oo}$ for the scattering amplitude for a small ideal conductor. This formula can be found in the literature (see Hönl et al. [1]). But formulas (1.78) and (1.79) are new.

Assume that $ka << 1$, $a << d$, where $d = \min_{i \neq j}|t_i - t_j|$. If $h_j \neq 0$ and is not too small, $1 \leq j \leq r$, then the scattering amplitude in the problem of scattering of the initial field u_o from the system of perfect conductors can be calculated with the accuracy $O(ka + (d/a))$ from the formula

$$f(n,k) = -\frac{1}{4\pi}\sum_{j=1}^{r}\exp\{-ik(n,t_j)\}h_jS_j(1+h_jS_jC_j)^{-1}u_{oj}, \tag{1.81}$$

where u_{oj} is the value of the initial field $u_o(x,k)$ at the point t_i where the i-th body is placed, and S_j is the area of Γ_j. If $h_j = 0$ we obtain the formula

$$f(n,k) = \frac{1}{4\pi}\sum_{j=1}^{r} ikV_j\beta_{pq}^{(j)}n_p \frac{\partial u_o(x,k)}{\partial x_q}\Big|_{x=t_j} + V_j\frac{\Delta u_o}{8\pi}\Big|_{x=t_j}\exp\{-ik(n,t_j)\}, \tag{1.82}$$

where V_j and $\beta_{pq}^{(j)}$ are the volume and the magnetic polarizability tensor of the j-th body. The theory developed can be applied to electromagnetic wave scattering.

15. Let us consider the scattering problem in the case where $r \sim 10^{23}$, so that we consider scattering in a medium formed by many small particles. If the particles are perfect conductors distributed in space with the density $f(x)$ and $d \gg a$, $d = \min_{i \neq j} |t_i - t_j|$, then the average field $\langle u(x,k) \rangle$ in the medium satisfies

$$\langle u(x,k)\rangle = u_0(x,k) - \int \frac{\exp(ikr_{xy})}{4\pi r_{xy}} f(y)C(y)\langle u(y,k)\rangle dy. \quad (1.83)$$

Here $\int = \int_{\mathbb{R}^3}$ and $C(x) = \int_0^\infty cg(x,c)dc$, where $g(x,c)$ is the probability density of the capacitance distribution of small bodies at the point x. From the point of view of acoustics, the Dirichlet boundary condition $u|_\Gamma = 0$ corresponds to soft bodies while the Neumann boundary condition $(\partial u/\partial N)|_\Gamma = 0$ corresponds to rigid bodies.

If the bodies are rigid then the equation for the average field is of the form

$$\langle u(x,k)\rangle = u_0(x,k) + \int \frac{\exp(ikr_{xy})}{4\pi r_{xy}} \left\{ B_{pq} \frac{\partial \langle u(y,k)\rangle}{\partial y_q} \frac{x_p - y_p}{r_{xy}} + \frac{1}{2} \Delta \langle u(y,k)\rangle b(y) \right\} dy, \quad (1.84)$$

where

$$B_{pq}(y) = ikV_{pq}N(y), \quad b(y) = VN(y), \quad (1.85)$$

V is the average volume of a single scatterer, β_{pq} is the magnetic polarizability tensor of the scatterer, $N(y)$ is the number of scatterers in a volume unit. If the impedance boundary condition is imposed on the boundaries of the scatterers, then for the average field equation (1.83) holds with $C(y)$ replaced by $q(y) = \langle hS/(1+hSC^{-1})\rangle$, where the averaging should be over a small volume around the point y. We note that equation (1.83) is equivalent to the Schrödinger equation

$$(-\Delta + f(y)C(y) - k^2)\langle u\rangle = 0, \tag{1.86}$$

where the field $\langle u\rangle - u_0$ satisfies the radiation condition, but equation (1.84) cannot be reduced to a differential equation. Equation (1.83) for the average field in a rarefied random medium often appeared in the literature (for example, in Morse-Feshbach [1]), but equation (1.84) is new. The same method can be applied to electromagnetic wave scattering in a rarefied medium and an integral equation for the Stokes vector can be deduced. Details will be presented in Section 6 below.

2. Investigation of a Class of Linear Operator Equations

Here we prove Theorems 1 through 6.

1. Proof of Theorem 1: Let $g = \lambda B_\gamma g$. If we take the inner product of this equality by ψ_j, we obtain $(g,\psi_j) = (\lambda/\lambda_1)(g,\psi_j) + \lambda\gamma(g,\psi_j)$, hence $(g,\psi_j)(1 - (\lambda/\lambda_1) - \lambda\gamma) = 0$. If $(g,\psi_j) \neq 0$ for some j, $1 \leq j \leq m$, then $\lambda = \lambda_1(1+\gamma\lambda_1)^{-1}$. If $(g,\psi_j) = 0$, $1 \leq j \leq m$, then $B_\gamma g = Ag$. Hence $\lambda \in \sigma(A)$. So there are no characteristic values of the operator B_γ in the disk $|\lambda| < r_\gamma$, where r_γ is defined by formula (1.3). Indeed, if $\lambda_1 \in \sigma(B_\gamma)$ then $(g,\psi_j) = 0$, $1 \leq j \leq m$, $g \neq 0$, so $\lambda_1 \in \sigma(A)$, i.e., $g = \lambda_1 Ag$, $g \in G^\perp$. Hence $g = (I - \lambda_1 A)h$, $(I - \lambda_1 A)^2 h = 0$. By assumption, λ_1 is semisimple, hence $(I - \lambda_1 A)h = 0$ and $g = 0$. This contradiction proves that $\lambda_1 \notin \sigma(B_\gamma)$. The spectrum of B_γ consists of $\sigma(A)\setminus\lambda_1$ and

possibly the number $\lambda_1(1 + \gamma\lambda_1)^{-1}$. We note that if $\Phi = \lambda B_\gamma \Phi$, $\lambda \neq \lambda_1$, then $\Phi \in G^\perp$. Hence $\Phi = \lambda A \Phi$, so the characteristic values of A other than λ_1 are characteristic values of B_γ. Let us prove that the process (1.5) converges. Let $|1 + \gamma\lambda_1| < 1$. Then $q = |\lambda_1| r_\gamma^{-1} < 1$, so there are no characteristic values of operator B_γ in the disk $|\lambda| < r$. Hence the process (1.5) converges no more slowly than a geometric series with ratio $|\lambda_1| r_\gamma^{-1}$. Taking into account that $F = \lambda_1 Af - f \in G^\perp$, so that $AF = B_\gamma F$ and setting $g = \sum_{j=0}^{\infty} \lambda_1^j B_\gamma^j F$, we have $Ag = B_\gamma g$, $g \in G^\perp$, $g = \lambda_1 B_\gamma g + F$. Hence $g + f = \lambda_1 A(g+f)$, so that $g + f = \Phi$, where $\Phi \in N(I - \lambda_1 A)$. Since $Pg = 0$ we see that $Pf = P\Phi$. If $\dim G = 1$ then $\dim N(I - \lambda_1 A) = 1$. Let $\phi \in N(I - \lambda_1 A)$, $\psi \in G$, $||\phi|| = ||\psi|| = 1$. Then $\Phi = c\phi$, where $c = \text{const}$. Taking the inner product of the equality $g + f = \Phi$ with ψ, we obtain $(f,\psi) = c(\phi,\psi)$ since $(g,\psi) = 0$. As λ_1 is semisimple, $(\phi,\psi) \neq 0$. So $c = (f,\psi)/(\phi,\psi)$. It remains to prove the stability of process (1.6), i.e., that the estimate (1.7) holds. We have

$$h_n = \sum_{j=0}^{n} (\lambda_1 B_\gamma)^j F + \sum_{j=0}^{n-1} (\lambda_j B_\gamma)^j \varepsilon_{n-1-j},$$

$$g = \sum_{j=0}^{\infty} (\lambda_1 B_\gamma)^j F, \quad ||\lambda_1 B_\gamma|| \leq q.$$

Hence

$$||g - h_n|| \leq \sum_{j=0}^{n-1} q^j \varepsilon + \sum_{j=n+1}^{\infty} q^j \, ||F|| \leq \frac{\varepsilon + ||F||\, q^{n+1}}{1 - q}$$

From here estimate (1.7) follows immediately. □

To prove Theorem 2 we need

Lemma 1. *Let $\mathscr{F}(\lambda)$ be a function with values in an operator Banach algebra which is analytic in the disk $|\lambda| < r$ and meromorphic in the disk $|\lambda| < r + \varepsilon$, $\varepsilon > 0$. Suppose that λ_1 is a simple pole of $\mathscr{F}(\lambda)$, $\operatorname{res}_{\lambda=\lambda_1} \mathscr{F}(\lambda) = C$, and $|\lambda_1| = r$. If there are no other poles in the disk $|\lambda| < r + \varepsilon$ and $\mathscr{F}(\lambda) = \sum_{j=0}^{\infty} a_n \lambda^n$ for $|\lambda| < r$, then $\lim_{n\to\infty} \lambda_1^{n+1} a_n = -C$.*

Proof of Lemma 1: The function $\mathscr{F}(\lambda) - \{C/(\lambda-\lambda_1)\}$ is analytic in the disk $|\lambda| < r + \varepsilon$. Hence $\mathscr{F}(\lambda) - \{C/(\lambda-\lambda_1)\} = \sum_{n=0}^{\infty} b_n \lambda^n$, $|\lambda| < r + \varepsilon$. If $|\lambda| < r$ we have the identity

$$\sum_{n=0}^{\infty} b_n \lambda^n = \sum_{n=0}^{\infty} (a_n + \frac{C}{\lambda_1^{n+1}}) \lambda^n,$$

which can be analytically continued into the disk $|\lambda| < r + \varepsilon$. So we conclude that $\lim_{n\to\infty} (\lambda_1^{n+1} a_n + C) = 0$. □

Proof of Theorem 2: Since the vector function $(I - \lambda A)^{-1} f = \sum_{j=0}^{\infty} \lambda^j A^j f$ is analytic in the disk $|\lambda| < |\lambda_1|$, has a simple pole at $\lambda = \lambda_1$, and has no other poles in the disk $|\lambda| < |\lambda_2|$, we use Lemma 1 and conclude that $\lim_{n\to\infty}(\lambda_1^{n+1} A^n f + C) = 0$ with rate of convergence $O(|\lambda_1/\lambda_2|)^n$. Since $f_n = \lambda_1^n A^n f$, we have $\lim_{n\to\infty} f_n = \eta$. Since $\eta = \lambda_1 A\eta$ and $\dim N(I - \lambda_1 A) = 1$, we have $\eta = a\phi$, $a = \text{const}$. We note that $(f_{n+1}, \psi) = \lambda_1 (Af_n, \psi) = (f_n, \overline{\lambda}_1 A^* \psi) = (f_n, \psi) = \ldots = (f, \psi)$, so that $a(\phi, \psi) = (f, \psi)$, $a = (f,\psi)/(\phi,\psi)$. □

Example 1. The Robin problem in electrostatics can be reduced to the equation

$$\phi = -A\phi, \quad \int_\Gamma \phi(s)ds \neq 0, \tag{2.1}$$

where the operator A is defined by formula (1.10). If Γ

is smooth (for example, of Lyapunov type) then the operator A is compact. It is known that $\lambda_1(A) = -1$, $|\lambda_2| > 1$, $\dim N(I+A) = 1$. Actually, it is known that all the $\lambda_j(A)$ are real but we need not use this fact. To solve equation (2.1) one can apply Theorems 1 and 2. In particular, one can find a solution of equation (2.1) by an iterative process. To make this process stable one should use process (1.6). We note that in our example, $\psi = S^{-1/2}$, $S = \text{meas}\,\Gamma$, so the iterative process is

$$\begin{cases} h_{n+1} = -A\phi_n - \frac{1}{S}\int_\Gamma h_n(t)dt + F + \varepsilon_n, \quad h_o = F, \\ F = -\psi - A\psi, \quad \psi = \frac{1}{\sqrt{S}}, \quad \|\varepsilon_n\|_{L^2(\Gamma)} < \varepsilon \end{cases} \tag{2.2}$$

and the sequence h_n satisfies the inequality

$$\overline{\lim_{n\to\infty}} \left\| \frac{\phi}{(\phi,\psi)} - \psi - h_n \right\| = O(\varepsilon).$$

We describe in detail the passage from the equation $\phi = -A\phi$ to the equation of the same type as (1.1). We look for ϕ of the form $\phi = \psi + h$, so that $h = -Ah - \psi - A\psi$. The right hand side of this equation $-(\psi + A\psi) \in G^\perp$ as in (1.1). Indeed, $\psi = -A^*\psi$, so $(\psi + A\psi, \psi) = 0$.

Proof of Theorem 3: The operator T in equation (1.9) is a positive definite operator in $L^2(\Gamma)$, while the operator A is compact if Γ is of Lyapunov type. We rewrite equation (1.9) in the form

$$(1 - \lambda)\frac{\partial u}{\partial N_i} + 2hu = (1 + \lambda)\frac{\partial u}{\partial N_e}, \quad u(x) = \int_\Gamma \frac{g(t)dt}{4\pi r_{xt}}. \tag{2.3}$$

Multiplying both sides of equation (2.3) by $\overline{u}$ we obtain the equality

$$\frac{1-\lambda}{1+\lambda} A + h \frac{B}{1+\lambda} = C, \tag{2.4}$$

where

$$A = \int_\Gamma \bar{u} \frac{\partial u}{\partial N_i} dt > 0, \quad B = 2 \int_\Gamma |u|^2 dt > 0,$$
$$C = \int_\Gamma \bar{u} \frac{\partial u}{\partial N_e} dt < 0. \tag{2.5}$$

Here the first and third inequalities can be easily obtained from Green's formula. If $A = 0$ or $C = 0$ then $u \equiv 0$. Let $\lambda = a + ib$. Taking real and imaginary part of equality (2.4) we obtain

$$\frac{(1 - a^2 - b^2)A + [h_1(1 + a) + h_2 b]B}{(1 + a)^2 + b^2} = C < 0, \tag{2.6}$$

$$\frac{-2bA + [h_2(1 + a) - h_1 b]B}{(1 + a)^2 + b^2} = 0. \tag{2.7}$$

Hence

$$(1 - |\lambda|^2)A + [h_1(1 + a) + h_2 b]B < 0, \tag{2.8}$$

$$h_2 = \frac{h_1 B + 2A}{1 + a} b. \tag{2.9}$$

As $h_2 \le 0$, $h_1 \ge 0$, we have $b \le 0$, $h_2 b \ge 0$. From here and (2.8) it follows that $|\lambda|^2 > 1$. Indeed, if $|\lambda|^2 < 1$ then $1 + a > 0$ and inequality (2.8) does not hold. If $|\lambda|^2 = 1$ then $h_1(1 + a) + h_2 b < 0$. This inequality does not hold if $h_2 < 0$. If $h_2 = 0$ then from (2.9) it follows that $b = 0$, so in this case the characteristic values of equation (1.9) are real, $b = 0$, $a^2 = 1$, $a = \pm 1$. Inequality (2.8) does not hold if $b = 0$, $|\lambda| = 1$. So we prove that $|\lambda_j| > 1$, and λ_j is real if h is real. It remains to prove point (ii) of Theorem 3. The equation $g + Tg = \lambda Ag + F$ is equivalent to the equation $g = \lambda Gg + QF$, where $Q = (I+T)^{-1}$, $G = QA$. There are no characteristic values of the operator G in the disk $|\lambda| \leq 1$. Hence the iterative process $g_{n+1} = \lambda G g_n + QF$, $g_o = \Phi$, converges in $L^2(\Gamma)$ no more slowly than a convergent geometric series where $\Phi \in L^2(\Gamma)$ is arbitrary. This process is equivalent to the process (1.11). □

<u>Proof of Remark 2</u>: Consider the equation $g = \mu(-T + A)g$, which can be written as a boundary condition for the potential $u(x)$:

$$(1 - \mu) \frac{\partial u}{\partial N_i} + 2\mu h u = (1 + \mu) \frac{\partial u}{\partial N_e} .$$

From here we obtain

$$(1 - \mu)A + \mu h B = (1 + \mu)C. \tag{2.10}$$

If $h \geq 0$ then arguments similar to those used in the proof of Theorem 3 show that μ is real and $\mu \in [0,1]$. If $-1 \leq \mu < 0$ then for h sufficiently large equality (2.10) does not lead to a contradiction. Let us find an interval such that if h is in the interval then $\mu \notin [-1,0)$. This will be the case if $(1 - \mu)A + \mu h B > 0$, or, equivalently, if

$$\frac{\int_D |\nabla u|^2 dx}{\int_\Gamma |u|^2 dt} > 2h|u|(1 + |\mu|)^{-1}.$$

Let

$$\kappa \equiv \min_{u \in W_2^1(D)} \frac{\int_D |\nabla u|^2 dx}{\int_\Gamma |u|^2 dt} .$$

Then the inequality

$$\kappa > \frac{2h|\mu|}{1+|\mu|} \tag{2.11}$$

is a sufficient condition for $\mu \notin [0,1)$. So if

$$0 \le h < \frac{1 + |\mu|}{2|\mu|} \kappa$$

then μ is not a characteristic value of the equation $g = \mu(-T + A)g$ and the iterative process $g_{n+1} = -Tg_n + Ag_n - 2f$, $g_o = \Phi$, converges at the rate of a convergent geometric series. Similarly, we can prove that for the equation $g = \mu(-T - A)g$ the condition analogous to (2.11) is

$$h < \frac{1+\mu}{2|\mu|} \kappa.$$

This inequality cannot be valid for all μ, $-1 \le \mu < 0$. Thus for the equation $g = (-T - A)g + 2F$, the iterative process $g_{n+1} = -Tg_n - Ag_n + 2f$, $g_o = 2f$, probably does not converge.

<u>Proof of Theorem 4</u>: We set $w_h = v_h - u$. Then

$$\Delta w_h = 0 \quad \text{in} \quad D, \qquad \left.\frac{\partial w_h}{\partial N}\right|_{\Gamma_2} = 0, \qquad \left.\frac{\partial w_h}{\partial N} + hw_h\right|_{\Gamma_1} = -\left.\frac{\partial u}{\partial N}\right|_{\Gamma_1}.$$

From here we obtain

$$\int_\Gamma w_h \frac{\partial w_h}{\partial N}\, dt + h \int_{\Gamma_1} |w_h|^2 dt = -\int_{\Gamma_1} w_h \frac{\partial u}{\partial N}\, dt.$$

According to Green's formula $\int_{\Gamma_1} w_h \frac{\partial w_h}{\partial N}\, dt = \int_D |\nabla w_h|^2 dx$. Hence

$$\int_D |\nabla w_h|^2 dx + h \int_{\Gamma_1} |w_h|^2 dt = -\int_{\Gamma_1} w_h \frac{\partial u}{\partial N}\, dt.$$

Therefore we obtain the estimates

$$h \|w_h\|_{L^2(\Gamma_1)} \le C, \quad C = \left\|\frac{\partial u}{\partial N}\right\|_{L^2(\Gamma_1)}.$$

From here it follows that

$$\|w_h\|_{L^2(\Gamma_1)} \le Ch^{-1}, \quad \int_D |\nabla w_h|^2 dx \le C^2 h^{-1}.$$

These estimates and the imbedding theorem imply the estimate

$$\|w_h\|_{L^2(D)} \le C_1\left(\|\nabla w_h\|_{L^2(D)} + \|w_h\|_{L^2(\Gamma_1)}\right),$$

where $C_1 = C_1(D,\Gamma_1)$. Only here do we use the fact that Γ_1 is nonempty. Hence

$$\|w_h\|_{W_2^1(D)} = O(h^{-1}) \quad \text{for} \quad h \to +\infty. \qquad \square$$

Proof of Theorem 5: If $A \ge 0$ then $|(Af,\phi)|^2 \le (A\phi,\phi)(Af,f)$ for all $f,\phi \in D(A)$. This is a variant of Cauchy's inequality. Hence $(Af,f) \ge |(Af,\phi)|^2(A\phi,\phi)^{-1}$ and equality holds for $\phi = \lambda f$. If $A \le 0$ then $-A \ge 0$, hence in this case the following representation holds:

$$(-Af,f) = \max\{|(-Af,\phi)|^2(A\phi,\phi)^{-1}\}.$$

This is equivalent to formula (1.19).

Suppose now that $(A\psi,\psi) < 0$, $(A\omega,\omega) > 0$, $\phi = \omega + \lambda\psi$, λ is a real number, and formula (1.18) is valid. Then

$$(Af,f) \ge \frac{|(Af,\omega)|^2 + 2\lambda \operatorname{Re}(Af,\omega)(\psi,Af) + \lambda^2|(Af,\psi)|^2}{(A\omega,\omega) + 2\lambda \operatorname{Re}(A\psi,\omega) + \lambda^2(A\psi,\psi)}$$

The denominator of this fraction has two real roots. Since the fraction is bounded from above its numerator has the same roots as the denominator. From here it follows that

$$\frac{|(Af,\omega)|^2}{|(Af,\psi)|^2} = \frac{(A\omega,\omega)}{(A\psi,\psi)}.$$

This is impossible because the right-hand side is negative while the left side is nonnegative. Hence $A \ge 0$. $\square$

Remark 1. Consider the stationary representation

$$(Af_i, f_j) = \text{st}\ \frac{(Af_i, \phi_j)(\phi_i, Af_j)}{(A\phi_i, \phi_j)} \tag{2.12}$$

where $A = A^*$, and st denotes stationary value. For $i = j$ the symbol st can be replaced by max if and only if $A \geq 0$. This fact is useful in applications.

<u>Proof of Theorem 6</u>: Under the assumptions of Theorem 6, we can use Entsch's theorem and conclude that $\lambda_1 > |\lambda_2| \geq \cdots$. Let $\phi(y)$ be a solution to equation (1.21), $\phi \in H_+$, $\delta_n \equiv \phi - \phi_n$. Then $\delta_n = Q^n \delta_o$. Let $\delta_o = \sum_{j=1}^{\infty} C_j \Phi_j$. This decomposition is possible in view of assumption (3) of the theorem. Then we have $\delta_n = \sum_{j=1}^{\infty} (1 - \lambda_j)^n C_j \Phi_j$. The operator K_1 has the eigenfunction $\Phi_1 = Ah \geq 0$ with eigenvalue $\lambda_1 = 1$. By Entsch's theorem $|\lambda_j(K_1)| < 1$. From here and from the assumption $|\arg \lambda_j| \leq \pi/3$ it follows that $|1 - \lambda_j| < 1$. Indeed, if $\lambda = r \exp(i\psi)$, $r < 1$, $|\psi| \leq \pi/3$, then $|1 - \lambda|^2 = 1 + r^2 - 2r \cos \psi \leq 1 + r^2 - r < 1$. Hence $|1 - \lambda_j|^n \to 0$ as $n \to \infty$. Therefore $||\delta_n||_+^2 \leq C \sum_{j=1}^{\infty} |1 - \lambda_j|^{2n} |C_j|^2 \to 0$ as $n \to \infty$. Thus the iterative process (1.23) converges in H_+. It is clear that $\lim_{n\to\infty} \phi_n = \phi$ is a solution to equation (1.21), $f = a(x)\phi(x)$ is a solution to equation (1.20), and $f \in H_-$.

3. Integral Equations of Static Field Theory for a Single Body and Their Applications. Explicit Formulas for the Scattering Matrix in the Problem of Wave Scattering of a Small Body of Arbitrary Shape

Here we prove the results presented in points 5-9 of Section 1.

Proof of Theorem 7: We set $\Phi = -2\gamma\varepsilon_o(\partial\phi/\partial N)$ in (1.28) and find

$$\sigma_n = \sum_{m=0}^{n} (-1)^m \gamma^m A^m (2\gamma(\mathscr{E},N))\varepsilon_o, \tag{3.1}$$

$$P_i^{(n)} = \int_\Gamma t_i \sigma_n(t)dt = \frac{2}{V}\sum_{m=0}^{n}(-1)^m\gamma^{m+1}\int_\Gamma t_j A^m(N_j)dtV\mathscr{E}_j\varepsilon_o, \tag{3.2}$$

where $P_i^{(n)}$ is the i-th coordinate of the n-th approximation to the dipole moment. We set $B = 2\pi A$ and find

$$\begin{cases} \alpha_{ij}^{(n)}(\gamma) = \dfrac{2}{V}\displaystyle\sum_{m=0}^{n}\frac{(-1)^m\gamma^{m+1}}{(2\pi)^m} J_{ij}^{(m)} \\ J_{ij}^{(m)} \equiv \displaystyle\int_\Gamma t_i B^m(N_j)dt. \end{cases} \tag{3.3}$$

Let us prove that

$$J_{ij}^{(m)} = b_{ij}^{(m)} - 2\pi J_{ij}^{(m-1)}.$$

We have

$$J_{ij}^{(0)} = \int_\Gamma t_i N_j(t)dt = \int_D \frac{\partial x_i}{\partial x_j}dx = \delta_{ij}V \equiv b_{ij}^{(0)}. \tag{3.4}$$

Furthermore, we obtain

$$\begin{aligned} J_{ij}^{(1)} &= \int_\Gamma s_i B(N_j)ds \\ &= \int_\Gamma dtN_j(t)\int_\Gamma s_i \frac{\partial}{\partial N_s}\frac{1}{r_{st}}ds = \int_\Gamma dtN_j(t)\Big(\int_\Gamma \frac{\partial s_i}{\partial N_s}\frac{ds}{r_{st}} - 2\pi \\ &= \int_\Gamma\int_\Gamma \frac{N_i(s)N_j(t)}{r_{st}}dsdt - 2\pi V\delta_{ij} \\ &= b_{ij}^{(1)} - 2\pi J_{ij}^{(0)}. \end{aligned} \tag{3.5}$$

In a similar manner, we obtain

$$J_{ij}^{(m)} = \int_\Gamma ds\, s_i B^m(N_j) = \int_\Gamma dt N_j(t) \int_\Gamma dt_1 \psi(t_1,t)\dots$$
$$\dots \int_\Gamma dt_{m-1}\psi(t_{m-1},t_{m-2})\left[\int_\Gamma \frac{N_i(s)ds}{r_{st}} - 2\pi(t_{m-1})_i\right]$$
$$= b_{ij}^{(m)} - 2\pi J_{ij}^{(m-1)}. \tag{3.6}$$

Using formula (3.6) we find

$$J_{ij}^{(m)} = \sum_{k=0}^{m} b_{ij}^{(k)}(2\pi)^{m-k}(-1)^{m-k}. \tag{3.7}$$

From (3.7) and (3.3) it follows that

$$\alpha_{ij}^{(n)}(\gamma) = \frac{2}{V}\sum_{m=0}^{n}\frac{(-1)^m\gamma^{m+1}}{(2\pi)^m}\sum_{k=0}^{n} b_{ij}^{(k)}(2\pi)^{m-k}(-1)^{m-k}$$
$$= \frac{2}{V}\sum_{k=0}^{n} b_{ij}^{(k)}\frac{(-1)^k}{(2\pi)^k}\frac{\gamma^{n+2}-\gamma^{k+1}}{\gamma - 1}. \tag{3.8}$$

The process (1.28) converges no more slowly than geometric series with ratio $0 < q < 1$, $q = q(\Gamma,\gamma)$. Hence the estimate (1.36) is valid. □

<u>Proof of Theorem 8</u>: Let Q be the surface charge on the conductor, $U^{(n)}$ be the average potential on its sufrace which is generated by the surface charge density $(Q/S)\omega_n(t)$, where $\omega_{n+1} = A\omega_n$, $\omega_o = 1$, and the operator A is defined by formula (1.26). By Theorem 2, the sequence ω_n converges to the free charge density on Γ with the surface charge S. If we set

$$C^{(n)} = \frac{Q}{U^{(n)}} = \frac{Q}{\frac{1}{S}\int_\Gamma ds \int_\Gamma dt \frac{Q\omega_n}{S}\frac{1}{4\pi\varepsilon_o r_{st}}}$$
$$= \frac{4\pi\varepsilon_o S^2}{\int_\Gamma\int_\Gamma \frac{\omega_n(t)dsdt}{r_{st}}} \tag{3.9}$$

and put $\omega_n = (-1)^n A^n 1$ in (3.9) we obtain formula (1.44) with the estimate (1.45) while the estimate (1.46) follows from (1.47). To end the proof we derive formula (1.47). Let the free surface charge distribution $\sigma(t)$ satisfy the equation

$$\int_\Gamma \frac{\sigma(t)dt}{4\pi\varepsilon_o r_{st}} = 1,$$

so that the potential on the surface is 1. Then $C = \int_\Gamma \sigma(t)dt$. Hence C is a linear functional of the type $(\sigma,1)$, where $(\cdot,\cdot)$ is the inner product in $L^2(\Gamma)$ (see Section 1, point 3). In this particular case formula (1.18) takes the form of (1.47). □

Remark 1. Formula (1.44) for $n = 0,1$, gives the capacitance of a conductor of arbitrary shape with an accuracy sufficient for practice. The following numerical results are of interest. Let C be the capacitance of a perfectly conducting cylinder of radius a, and length $2L$, $L_1 \equiv \frac{1}{2}C/L$, and $\ell \equiv L/a$. If $\ell \geq 10$, formula (1.44) for $n = 0$ gives C_1 with the accuracy of the asymptotic formula:

$$C_1 = 4\pi\varepsilon_o(\Omega^{-1} + 0.710\Omega^{-3}), \quad \Omega \equiv 2(\ell n\{4\ell\} - 1).$$

If $\ell \geq 5$, formula (1.44) for $n = 0$ gives C_1 which agrees with the numerical results of Wainstein [1] with the accuracy 1%, with 3% accuracy if $1 \leq \ell \leq 5$; if $0.1 \leq \ell \leq 1$, $n = 1$ formula (1.44) is accurate within 3% and for $\ell \leq 0.1$ it agrees with the asymptotic formula $C_1 = 4\varepsilon_o \ell^{-1}$ for C_1 as $\ell \to 0$. This formula can be obtained from the known formula $C = 8a\varepsilon_o$ for the capacitance of the disk with radius a, where $2LC_1 = 8a\varepsilon_o$. It can be shown (see Ramm [30]) that $C_{1hol}/C_{1sol} = 4.93/\ell n(16/\ell)$, where C_{1hol} is the capacitance per unit length of the hollow cylinder, i.e., the cylindrical tube, and C_{1sol} is the capacitance per unit length of the solid cylinder. A table of the capacitance of a parallelepiped with arbitrary edges calculated by formula (1.44) for $n = 0$ is given in Ramm [31].

We omit the proof of formulas in (1.37) because they are known (see Landau-Lifschitz [1, p. 72]). The proof in the form we need can be found in Appendix 3.

Proof of Formula (1.38) (Ramm [32]). All the elements of the scattering matrix are calculated by the same method. So we give in detail only the proof of the formula for S_2. We use the coordinate systems defined in Section 1, point 5. Let e_j, e_j' be the orts of the coordinate systems. Note that $(e_2',e_1) = 0$, $(e_2',e_2) = \cos\theta$, $(e_2',\nu) = -\sin\theta$. We have $f_2 = S_2E_2 + S_3E_1$. On the other hand,

$$f_2 = (f_E,e_2') = \frac{k^2}{4\pi\varepsilon_0}([n\ P,n)],e_2') + \frac{k^2}{4\pi}\sqrt{\frac{\mu_0}{\varepsilon_0}}([M,n],e_2'). \tag{3.10}$$

Furthermore, we have

$$([n[P,n]],e_2') = (P-n(P,n),e_2') = (P,e_2') = \varepsilon_0 V\alpha_{ij}E_j(e_i,e_2')$$

$$= \varepsilon_0 V\{(\alpha_{21}E_1 + \alpha_{22}E_2)\cos\theta - (\alpha_{31}E_1 + \alpha_{32}E_2)\sin\theta\}, \tag{3.11}$$

$$([M,n],e_2') = ([n,e_2'],M) = -(e_1,M) = -\mu_0 V(\beta H,e_1)$$

$$= -\mu_0 V(\beta_{11}H_1 + \beta_{12}H_2) = -\mu_0 V(-\beta_{11}E_2 + \beta_{12}E_1)\sqrt{\frac{\varepsilon_0}{\mu_0}}. \tag{3.12}$$

From formulas (3.10)-(3.12) we find S_2 as the coefficient of E_2:

$$S_2 = \frac{k^2 V}{4\pi}(\alpha_{22}\cos\theta - \alpha_{32}\sin\theta + \mu_o\beta_{11}).$$

In the calculation which led to formula (3.12) we used the equalities $H_1 = -\sqrt{\varepsilon_o/\mu_o}\, E_2$, $H_2 = \sqrt{\varepsilon_o/\mu_o}\, E_1$. □

Proof of Theorem 9: We consider the operator $\mathcal{B}$ in equation (1.51) to belong to the Hilbert space H of vector functions $\sigma = (\sigma_1,\ldots,\sigma_p)$ with the inner product $(\sigma,g) = \sum_{j=1}^{p} \int_{\Gamma_j} \sigma_j(t)\overline{g_j(t)}dt$. If Γ_j is smooth, $1 \le j \le p$, then $\mathcal{B}$ is compact in H. Let $\sigma = -\lambda\mathcal{B}\sigma$,

$$\mathcal{B} = \begin{pmatrix} \gamma_1 A_1 & \gamma_1 T_{12} & \cdots & \gamma_1 T_{1p} \\ \gamma_2 T_{21} & \gamma_2 A_2 & \cdots & \gamma_2 T_{2p} \\ \vdots & \vdots & & \vdots \\ \gamma_p T_{p1} & \gamma_p T_{p2} & \cdots & \gamma_p A_p \end{pmatrix}, \quad \gamma_j = \frac{\varepsilon_j - \varepsilon_{j-1}}{\varepsilon_j + \varepsilon_{j-1}}.$$

The function v defined in formula (1.48) satisfies the conditions

$$\begin{cases} \varepsilon_o \dfrac{\partial v}{\partial N_i}\Big|_{\Gamma_j} = \dfrac{A_j\sigma_j + \sigma_j}{2} + \dfrac{1}{2}\displaystyle\sum_{m\neq j, m=1}^{p} T_{jm}\sigma_m, \\ \varepsilon_o \dfrac{\partial v}{\partial N_e}\Big|_{\Gamma_j} = \dfrac{A_j\sigma_j - \sigma_j}{2} + \dfrac{1}{2}\displaystyle\sum_{m\neq j, m=1}^{p} T_{jm}\sigma_m. \end{cases} \tag{3.13}$$

We can rewrite the equality $\sigma = -\lambda\mathcal{B}\sigma$ in the form

$$\sigma_j = -\lambda\gamma_j(A_j\sigma_j + \sum_{m\neq j, j=1}^{p} T_{jm}\sigma_m). \tag{3.14}$$

From (3.13) and (3.14) we obtain

$$\varepsilon_o\left(\frac{\partial v}{\partial N_i} - \frac{\partial v}{\partial N_e}\right)\Big|_{\Gamma_j} = -\lambda\varepsilon_o\gamma_j\left(\frac{\partial v}{\partial N_i} + \frac{\partial v}{\partial N_e}\right)\Big|_{\Gamma_j}, \quad 1 \le j \le p. \tag{3.15}$$

Hence

$$(1+\lambda\gamma_j)\frac{\partial v}{\partial N_i}\bigg|_{\Gamma_j} = (1-\lambda\gamma_j)\frac{\partial v}{\partial N_e}\bigg|_{\Gamma_j}, \quad 1 \leq j \leq p. \tag{3.16}$$

Let D_o be the exterior domain with the boundary Γ_1, D_p be the interior domain with the boundary Γ_p, and D_j be the domain with the boundary $\Gamma_j \cup \Gamma_{j+1}$. Consider the identity

$$\begin{aligned} 0 = \sum_{j=0}^{p} a_j \int_{D_j} \bar{v}\, \Delta v dx &= -\sum_{j=0}^{p} a_j \int_{D} |\nabla v|^2 dx \\ &+ \sum_{j=1}^{p} \int_{\Gamma_j} \bar{v} \left(a_j \frac{\partial v}{\partial N_i} - a_{j-1} \frac{\partial v}{\partial N_e}\right) dt, \end{aligned} \tag{3.17}$$

where a_j are arbitrary constants. From (3.16) and (3.17) it follows that

$$\sum_{j=0}^{p} a_j \int_{D_j} |\nabla v|^2 dx = \sum_{j=1}^{p} \int_{\Gamma_j} \bar{v}\left(a_j - a_{j-1} \frac{1+\lambda\gamma_j}{1-\lambda\gamma_j}\right) \frac{\partial v}{\partial N_i}\, dt. \tag{3.18}$$

For $|\gamma_j| < 1$, $|\lambda| \leq 1$ we have $|\lambda\gamma_j| < 1$; for $\lambda = 1$, $|\gamma_j| < 1$ we have $(1+\lambda\gamma_j)/(1-\lambda\gamma_j) = \varepsilon_j/\varepsilon_{j-1}$. Let us set

$$a_o = \varepsilon_o, \quad a_j = a_{j-1} \frac{1+\lambda\gamma_j}{1-\lambda\gamma_j}, \quad 1 \leq j \leq p, \quad a_j = \varepsilon_j \quad \text{for} \quad \lambda = 1.$$

The identity (3.18) shows that $v = \text{const}$, and since $v(\infty) = 0$ we obtain $v \equiv 0$. Hence $\sigma = 0$. From here we conclude that equation (1.51) has a unique solution in H since $\mathscr{B}$ is compact. Consider now the case when $\lambda = 1$ and $\gamma_{j_o} = 1$ for some $1 \leq j_o \leq p$. In this case we assume that $p = j_o$, $\varepsilon_{jo} = \infty$, and $v|_{\Gamma_{jo}} = \text{const}$. Using an identity similar to (3.19) we prove that in this case the homogeneous equation (1.51) has only the trivial solution in H satisfying the conditions (*) $\int_{\Gamma_j} \sigma_j dt = 0$, $1 \leq j \leq p$, while the inhomogeneous equation (1.51) has a unique solution if the conditions (*) are valid for f, provided the solution belongs to H and satisfies the conditions (*). To prove the convergence of the process (1.52) it is sufficient to prove that the disk $|\lambda| \leq 1$ contains no characteristic values of the operator $\mathcal{B}$. First we assume that $|\gamma_j| < 1$ and set

$$a_o = \frac{1 - \lambda\gamma_1}{1 + \lambda\gamma_1}, \quad a_1 = 1, \quad a_2 = \frac{1 + \lambda\gamma_2}{1 - \lambda\gamma_2}, \quad p = 2.$$

Then the right-hand side of (3.18) is zero, Re $a_j > 0$, $0 \leq j \leq 2$. This can easily be verified. So $v = 0$, hence $\sigma = 0$. We proved the statement under the assumption that $|\gamma_j| < 1$. If $\gamma_2 = 1$, i.e., $\varepsilon_2 = \infty$, and $|\lambda| < 1$, $p = 2$, the proof is the same. Suppose that $\gamma_2 = 1$, $|\lambda| = 1$, $p = 2$. Then instead of the identity (3.18) we have

$$\sum_{j=0}^{1} a_j \int_{D_j} |\nabla v|^2 dx = \int_{\Gamma_1} \bar{v} \left(a_1 - a_o \frac{1 + \lambda\gamma_1}{1 - \lambda\gamma_1}\right) \frac{\partial v}{\partial N_i} dt - a_1 \int_{\Gamma_2} \bar{v} \frac{\partial v}{\partial N_e} dt, \tag{3.19}$$

and instead of the boundary condition (3.16) we have

$$(1 + \lambda) \left.\frac{\partial v}{\partial N_i}\right|_{\Gamma_2} = (1 - \lambda) \left.\frac{\partial v}{\partial N_e}\right|_{\Gamma_2}. \tag{3.20}$$

Let us prove that if $|\lambda| = 1$, $\lambda \neq 1$, then λ is not a characteristic value of $\mathcal{B}$. To this purpose, we set $a_o = (1-\lambda\gamma_1)/(1+\lambda\gamma_1)$, $a_1 = 1$, and note that

$$\int_{\Gamma_2} \bar{v} \frac{\partial v}{\partial N_e} dt = \frac{1 + \lambda}{1 - \lambda} \int_{\Gamma_2} \bar{v} \frac{\partial v}{\partial N_i} dt = \frac{1 + \lambda}{1 - \lambda} \int_{D_2} |\nabla v|^2 dx,$$

Re $(1+\lambda)/(1-\lambda) = 0$ if $|\lambda| = 1$, $\lambda \neq 1$. From here and (3.19) it follows that $v = \text{const}$ and hence $v \equiv 0$. If $\lambda = 1$, $\gamma_2 = 1$, $p = 2$, then from (3.20) we conclude that $\left.\frac{\partial v}{\partial N_i}\right|_{\Gamma_q} = 0$, $v|_{\Gamma_2} = \text{const}$, $v = \text{const}$ in D_2. In this case v is not necessarily zero, but is zero if $\int_{\Gamma_2} \frac{\partial v}{\partial N_e} dt = 0$. To prove this we use the identity (3.19) with $a_1 = 1$, $a_o = (1-\gamma_1)/(1+\gamma$

and take into account that $\int_{\Gamma_2} v\,(\partial v/\partial N_e)dt = 0$. Hence if $\varepsilon_2 = \infty$, the number $\lambda = -1$ can be a characteristic value of $\mathscr{B}$. Let us prove that $\lambda = -1$ can be only a semisimple characteristic value. Otherwise the equation $\psi = -\mathscr{B}\psi + \sigma$ is solvable, where $\sigma = -\mathscr{B}\sigma$, $\int_{\Gamma_2} \sigma_2 dt \neq 0$. But this leads to a contradiction. To show this we set $q_j = \int_{\Gamma_j} \psi_j dt$, $Q_j = \int_{\Gamma_j} \sigma_j dt$, $j = 1,2$, integrate the equations

$$\psi_1 = -\gamma_1 A_1 \psi_1 - \gamma_1 T_{12} \psi_2 + \sigma_1 \tag{3.21}$$

$$\psi_2 = -T_{12}\psi_1 - A_2\psi_2 + \sigma_2 \tag{3.22}$$

over Γ_1 and Γ_2, respectively, and take into account the equality

$$\int_\Gamma \frac{\partial}{\partial N_t} \frac{1}{2\pi r_{xt}}\, dt = \begin{cases} 0, & x \notin D, \\ -1, & x \in \Gamma, \\ -2, & x \in D. \end{cases} \tag{3.23}$$

As a result we obtain the system

$$\begin{cases} q_1 = \gamma_1 q_1 + 2\gamma_1 q_2 + Q_1, & (3.24) \\ q_2 = \qquad\quad q_2 \quad + Q_2. & (3.25) \end{cases}$$

Hence $Q_2 = 0$. Since $Q_2 = \int_{\Gamma_2} \sigma_2 dt \neq 0$ we have a contradiction. To end the proof we show that the subspace of functions satisfying the conditions

$$\int_{\Gamma_j} f_j\, dt = 0, \quad 1 \leq j \leq p \tag{3.26}$$

is invariant for the operator $\mathscr{B}$. To this end we note that

$$\int_{\Gamma_j} (\mathscr{B}f)_j dt = \gamma_j \int_{\Gamma_j} A_j f_j dt + \gamma_j \sum_{m \neq j} \int_{\Gamma_j} T_{jm} f_m dt = 0,$$

provided (3.26) holds. Here we use equality (3.23) and definitions (1.26) and (1.50). The restriction of the operator $\mathscr{B}$ to this invariant subspace has no characteristic value in the disk $|\lambda| \leq 1$. Thus the process (1.52) converges in H for any f satisfying conditions (3.26), and the limit $\sigma = \lim \sigma_n$ satisfies (3.26). □

Proof of Theorem 10: We write the system (1.56) in the form

$$\sigma = -\tilde{\mathscr{B}} + f, \tag{3.27}$$

where the matrix operator $\tilde{\mathscr{B}}$ is

$$\tilde{\mathscr{B}}_{jm} = \kappa_j \delta_{jm} A_j + \kappa_j (1-\delta_{jm}) T_{jm}, \quad \kappa_j = \frac{\varepsilon_j - \varepsilon_o}{\varepsilon_j + \varepsilon_o}. \tag{3.28}$$

Let us prove that the equation $\sigma = -\tilde{\mathscr{B}}\sigma$ has no nontrivial solutions if $|\lambda| \leq 1$, $|\kappa_j| < 1$. To this end, we write equation (3.27) in the form

$$(1 + \lambda\kappa_j) \left.\frac{\partial v}{\partial N_i}\right|_{\Gamma_j} = (1 - \lambda\kappa_j) \left.\frac{\partial v}{\partial N_e}\right|_{\Gamma_j}, \quad 1 \leq j \leq p, \tag{3.29}$$

and define the constants $\hat{\varepsilon}_j, \hat{\kappa}_j$ by the formulas

$$\begin{cases} \hat{\varepsilon}_j = \dfrac{\varepsilon_j + \varepsilon_o + \lambda(\varepsilon_j - \varepsilon_o)}{\varepsilon_j + \varepsilon_o - \lambda(\varepsilon_j - \varepsilon_o)} \varepsilon_o = \dfrac{1 + \lambda\kappa_j}{1 - \lambda\kappa_j} \varepsilon_o \\ \\ \hat{\kappa}_j = \dfrac{\hat{\varepsilon}_j - \varepsilon_o}{\hat{\varepsilon}_j + \varepsilon_o} = \lambda\kappa_j. \end{cases} \tag{3.30}$$

Then equalities (3.29) can be written as

$$\hat{\varepsilon}_j \left.\frac{\partial v}{\partial N_i}\right|_{\Gamma_j} = \varepsilon_o \left.\frac{\partial v}{\partial N_e}\right|_{\Gamma_j}, \quad 1 \leq j \leq p, \tag{3.31}$$

while σ satisfies the equation

$$\sigma = -\hat{\mathscr{B}}\sigma, \tag{3.32}$$

where the operator $\hat{\mathscr{B}}$ can be calculated from formula (3.28) in which κ_j should be replaced by $\hat{\kappa}_j$. We denote by

$D_o = \mathbb{R}^3 \setminus \bigcup_{j=1}^{p} D_j$ and consider the identity

$$0 = \sum_{j=0}^{p} \hat{\varepsilon}_j \int_{D_j} \bar{v}\Delta v dx = -\sum_{j=0}^{p} \hat{\varepsilon}_j \int_{D_j} |\nabla v|^2 dx$$
$$+ \sum_{j=1}^{p} \int_{\Gamma_j} \bar{v} \left(\hat{\varepsilon}_j \frac{\partial v}{\partial N_i} - \varepsilon_o \frac{\partial v}{\partial N_e}\right) dt. \tag{3.33}$$

Taking into account the conditions (3.31), from (3.33) we obtain

$$\sum_{j=0}^{p} \hat{\varepsilon}_j \int_{D_j} |\nabla v|^2 dx = 0. \tag{3.34}$$

If $|\kappa_j| < 1$, $|\lambda| \le 1$ it is easy to verify that $\mathrm{Re}\, \hat{\varepsilon}_j > 0$, $1 \le j \le p$. So from (3.34) and from the condition $v(\infty) = 0$ it follows that $v \equiv 0$, $\sigma = 0$. So we have proved that the operator $\tilde{\mathscr{B}}$ has no characteristic values in the disk $|\lambda| \le 1$, provided that $|\kappa_j| < 1$. Hence the iterative process

$$\sigma_{n+1} = -\tilde{\mathscr{B}}\sigma_n + f, \quad \sigma_o = \Phi \tag{3.35}$$

converges no more slowly than geometric series with the denominator $0 < q < 1$ for any $f, \Phi \in L^2(\Gamma)$. If $|\kappa_j| = 1$ for some j, $\lambda\kappa_j \ne 1$, then as above we can prove that the equation $\sigma = -\lambda\tilde{\mathscr{B}}\sigma$ has only trivial solution. If $\lambda\kappa_j = 1$, then it follows from (3.29) that $(\partial v/\partial N_i)|_{\Gamma_j} = 0$. For the sake of simplicity we assume that $\lambda\kappa_p = 1$. Then instead of the identity (3.33) we use the identity

$$\sum_{j=0}^{p-1} \hat{\varepsilon}_j \int_{D_j} |\nabla v|^2 dx - \varepsilon_o \int_{\Gamma_p} \bar{v} \frac{\partial v}{\partial N_e} ds = 0. \tag{3.36}$$

Since $\left.\frac{\partial v}{\partial N_i}\right|_{\Gamma_p} = 0$ it is clear that $v|_{\Gamma_p} = \text{const}$,

$\sigma_p = -\varepsilon_o \left.\frac{\partial v}{\partial N_e}\right|_{\Gamma_p}$. If $\int_{\Gamma_p} \sigma_p dt = 0$ we conclude from (3.36) that $v \equiv 0$ and $\sigma = 0$.

Remark 2. As in the proof of Theorem 9 it is possible to prove that $\lambda = -1$ can be a (necessarily semisimple) characteristic value of the operator $\tilde{\mathscr{B}}$.

Remark 3. If $\lambda\kappa_j = 1$ for $j = 1,2,\ldots k$, $\lambda\kappa_j \neq 1$ for $k + 1 \leq j \leq p$, then the equation $\sigma = -\lambda\tilde{\mathscr{B}}\sigma$ can have k linearly independent solutions.

Remark 4. The results of point 9 in Section 1 are immediate corollaries of Theorem 6. We note that the function $a(t)$ was determined in formula (1.62) in order that the edge condition (1.59) was satisfied. To verify that condition (iii) of Theorem 6 holds for the kernel of the operator A_1 defined by formula (1.62) we note that the kernel $a(y)/r_{xy}$, $a(y) > 0$, is symmetrizable, so its eigenfunctions form an orthonormal basis in H_+ and its eigenvalues are real.

4. Variational Principles for Calculation of the Electrical Capacitance and Polarizability Tensors for Bodies of Arbitrary Shape and Two-Sided Estimates of the Tensors

1. The proof of formula (1.47) was given near the end of the proof of Theorem 8. Formula (1.47) allows one to obtain estimates from below of the capacitance. An example of such an estimate is inequality (1.46). A proof of formula (1.64) can be found in Pólya-Szegö [1]. For convenience of the reader, we also give a proof. We start from the definition $Q = CV$, where Q is the total charge of the conductor D with boundary Γ, C is its capacitance, U is its potential, $\sigma = -\varepsilon_0(\partial u/\partial N_e)|_\Gamma$ is the surface charge density, and $u(x)$ is the potential out of the conductor. So

$$C = -\frac{\varepsilon_o}{U}\int_\Gamma \frac{\partial u}{\partial N_e}dt = -\frac{\varepsilon_o}{U^2}\int_\Gamma U\frac{\partial u}{\partial N_e}dt = \frac{\varepsilon_o}{U^2}\int_\Omega |\nabla u|^2 dx,$$

where $\Omega = \mathbb{R}^3 \setminus D$. If $U = 1$ then

$$C = \varepsilon_o \int_\Omega |\nabla u|^2 dx, \text{ where } \Delta u = 0 \text{ in } \Omega, \quad u|_\Gamma = 1, \quad u(\infty) = 0. \tag{4.1}$$

If $\Phi|_\Gamma = 1$, $\Phi(\infty) = 0$, $\Phi \in C^1(\Omega)$, then (*) $\int_\Omega |\nabla u|^2 dx \leq \int_\Omega |\nabla\Phi|^2 dx$. From here and (4.1) formula (1.64) follows. It remains to prove (*). If $\psi = \Phi - u$, then

$$\int_\Omega |\nabla\Phi|^2 dx = \int_\Omega |\nabla u|^2 dx + \int_\Omega |\nabla\psi|^2 dx + 2\int_\Omega \nabla\psi\nabla u dx. \tag{4.2}$$

Since $\Delta u = 0$ in Ω and $\psi|_\Gamma = 0$, we have

$$\int_\Omega \nabla\psi\nabla u dx = -\int_\Omega \psi\nabla u dx - \int_\Gamma \psi \frac{\partial u}{\partial N_e} dt = 0. \tag{4.3}$$

Hence (*) follows from (4.2). □

Setting $\Phi = F(\theta,\phi)/|x|$ in (1.64), where $r = F(\theta,\phi)$ is the equation of the surface Γ in spherical coordinates, we obtain (1.65).

Proof of Theorem 11: Let a conductor D be put in the homogeneous electrostatic field $E = \nabla x_j$. Then the surface charge density satisfies the equation

$$\int_\Gamma \frac{\sigma_j(t)dt}{4\pi\varepsilon_o r_{st}} = U_j + s_j \tag{4.4}$$

and the conditions

$$\int_\Gamma \sigma_j \, dt = 0, \quad U_j = \text{const}, \tag{4.5}$$

where s_j is the j-th coordinate of the radius vector of the point $s \in \Gamma$, U_j is the potential of the conductor, and the first condition (4.5) is the condition of electroneutrality.

We define the polarizability tensor by the equality

$$\varepsilon_0 V\alpha_{ij} = \int_\Gamma t_i\sigma_j(t)dt, \tag{4.6}$$

where V is the volume of the conductor. We obtain formula (1.66) from Theorem 5 and equation (4.4). Formula (1.67) holds, since the electrostatic energy of the real electrostatic field is minimal. The energy of the conductor D in the field $E = \nabla x_j$ is $\frac{1}{2}\varepsilon_0 V\alpha_{jj}$ (see Landau-Lifschitz [1, §2]) and the same energy is equal to $\frac{1}{2}\varepsilon_0\int_\Omega|\nabla\Phi_j|^2dx$, where Φ_j is the potential of the real field, so that $\Delta\Phi_j = 0$ in Ω, $\Phi_j(\infty) = 0$, $\Phi_j|_\Gamma = U_j + s_j$, $U_j = \text{const.}$ If $\phi \in C^1(\Omega)$, $\phi(\infty) = 0$, $\phi|_\Gamma = U_j + s_j$, then $\int_\Omega|\nabla\phi|^2dx \geq \int_\Omega|\nabla\phi_j|^2dx$. This can be proved as the similar inequality (*) was proved above. Hence formula (1.67) is valid. Formula (1.66) allows one to obtain estimates from below for the diagonal elements of the polarizability tensor, while formula (1.67) allows one to obtain estimates from above for these elements. By rotating the coordinate system, elements of the tensor are transformed according to the known transformation law, so we can obtain estimates of any element of the tensor from estimates of the diagonal elements. Let us prove formula (1.68). We have

$$\begin{aligned}\beta_{pj} &= \frac{1}{\mu_0 V}\int_\Gamma t_p\sigma_j(t)dt = \frac{1}{V}\int_\Gamma t_p\left(\frac{\partial\Phi_j}{\partial N_i} - \frac{\partial\Phi_j}{\partial N_e}\right)dt \\ &= \frac{1}{V}\left(\int_\Gamma \frac{\partial t_p}{\partial N}\Phi_j dt - \int_\Gamma t_p\frac{\partial t_j}{\partial N}dt\right) = \frac{1}{V}\int_\Gamma\frac{\partial\Phi_p}{\partial N_e}\Phi_j dt \\ &\quad - \delta_{pj} = -\frac{1}{V}\int_\Omega\nabla\Phi_p\nabla\Phi_j dx - \delta_{pj}.\end{aligned} \tag{4.7}$$

In particular,

$$-V\beta_{jj} - V = \int_\Omega|\nabla\Phi_j|^2dx, \tag{4.8}$$

$$\beta_{pj} = \frac{1}{V}\int_\Gamma \frac{\partial t_p}{\partial N}\Phi_j dt - \delta_{pj}. \tag{4.9}$$

Hence

$$\beta_{jj} < 0. \tag{4.10}$$

In formulas (4.7) - (4.9) the function Φ_j is the potential

$$\Phi_j(x) = \int_\Gamma \frac{\sigma_j(t)dt}{4\pi\mu_o r_{xt}}, \tag{4.11}$$

which is the solution to the problem

$$\Delta\Phi_j = 0 \quad \text{in} \quad \Omega, \quad -\frac{\partial\Phi_j}{\partial N_e} = -\frac{\partial t_j}{\partial N}\Big|_\Gamma = -N_j(t)\Big|_\Gamma, \quad \Phi_j(\infty) = 0, \tag{4.12}$$

while its density satisfies the equation

$$\sigma_j = A\sigma_j - 2\mu_o N_j(t), \quad A\sigma = \int_\Gamma \frac{\partial}{\partial N_s}\frac{1}{2\pi r_{st}}\sigma(t)dt. \tag{4.13}$$

The operator $-\partial/\partial N_e$ is positive definite on the set of functions which are boundary values on Γ of functions which are sufficiently smooth and harmonic in Ω and which vanish at infinity. Indeed,

$$-\int_\Gamma v\frac{\partial u}{\partial N_e}dt = \int_\Omega \nabla u\nabla v dx = -\int u\frac{\partial v}{\partial N_e}dt. \tag{4.14}$$

From (4.9), (4.14), (4.12), and Theorem 5, we obtain

$$-\beta_{jj} - 1 = \max_{\substack{\Delta u=0\\ u(\infty)=0}} \frac{1}{V}\frac{(\int_\Gamma N_j(t)u(t)dt)^2}{-\int_\Gamma u\frac{\partial u}{\partial N_e}dt}. \tag{4.15}$$

Applying Green's formula to the denominator of formula (4.15). we obtain formula (1.68). The maximum in this formula is attained by the solution to problem (4.12). The set of admissible functions in formula (1.68) can be extended to the set of functions $u(x) \in C^1(\Omega)$, $u(\infty) = 0$, $\int_\Omega|\nabla u|^2dx < \infty$. To prove this we note that

$$-V\beta_{jj} - V = \int_\Omega |\nabla\Phi_j|^2 dx \geq \frac{(\int_\Omega \nabla\Phi_j \nabla u dx)^2}{\int_\Omega |\nabla u|^2 dx} \tag{4.16}$$

and that inequality (4.16) holds for any $u(x)$ in the set defined. Let us prove formula (1.69). We have the identity

$$\int_\Omega |q_j - \nabla\Phi_j|^2 dx = \int_\Omega |q_j|^2 dx + V + V\beta_{jj}, \tag{4.17}$$

which holds for any $q_j(x) \in C^1(\Omega)$, such that

$$\operatorname{div} q_j = 0 \quad \text{in} \quad \Omega, \ (q_j, N)\Big|_\Gamma = N_j(t). \tag{4.18}$$

Equation (1.69) follows from (4.17). It remains to prove (4.17). We have

$$\int_\Omega |q_j - \nabla\Phi_j|^2 dx = \int_\Omega |q_j|^2 dx + \int_\Omega |\nabla\Phi_j|^2 dx - 2\int_\Omega q_j \nabla\Phi_j dx, \tag{4.19}$$

$$\begin{aligned}\int_\Omega q_j \nabla\Phi_j dx &= \int_\Omega \operatorname{div}(q_j\Phi_j) dx - \int_\Omega \Phi_j \operatorname{div} q_j dx = -\int_\Gamma (q_j, N)\Phi_j dt \\ &= -\int_\Gamma \Phi_j \frac{\partial\Phi_j}{\partial N_e} dt = \int_\Omega |\nabla\Phi_j|^2 dx.\end{aligned} \tag{4.20}$$

From (4.19), (4.20), and (4.8) we obtain (4.17). □

Remark 1. We can pass to the limit as $V \to 0$ in formulas (1.66) - (1.67). Denoting the limits in the left-hand side $\lim_{V\to 0} V\alpha_{jj}$ by $\tilde{\alpha}_{jj}$ we obtain from (1.66) and (1.67) variational principles for unclosed thin conducting screens. The admissible functions must satisfy the edge condition (1.59) in this case. It is not so easy to pass to the limit $V \to 0$ in formulas (1.68) and (1.69) because we cannot solve problem (4.12) for open surface by means of the potential (4.11). Indeed the normal derivative of (4.11) has a jump when crossing the open surface Γ, while the boundary condition in (4.12) shows that the normal derivative cannot have a jump. So for the open surface Γ we look for a solution of problem

(4.12) in the form

$$\psi_j = \int_\Gamma \eta_j(t) \frac{\partial}{\partial N_t} \frac{1}{4\pi\mu_o r_{xt}} dt. \qquad (4.21)$$

We note that $\partial\psi_j/\partial N$ has no jump when crossing Γ,

$$\psi_j \sim \frac{(M,x)}{4\pi\mu_o |x|^3} \quad \text{as} \quad |x| \to \infty, \quad M_j = \int_\Gamma \eta_j(t)N(t)dt \qquad (4.22)$$

where M is the magnetic dipole moment. In particular,

$$M_{jj} = \int_\Gamma \lambda_j(t)N_j(t)dt. \qquad (4.23)$$

By definition, $M_{jj} = \mu_o\tilde{\beta}_{jj}H_j = \mu_o\tilde{\beta}_{jj}$, as $H_j = 1$. Hence

$$\begin{aligned} \tilde{\beta}_{jj} &= \frac{1}{\mu_0}\int_\Gamma \eta_j(t)N_j(t)dt, \\ \tilde{\beta}_{pj} &= \tilde{\beta}_{jp} = \frac{1}{\mu_0}\int_\Gamma \eta_j(t)N_p(t)dt. \end{aligned} \qquad (4.24)$$

The operator $-\partial/\partial N_e$ is positive definite on the set of traces on Γ of functions ψ harmonic in Ω provided that the edge condition in the form

$$\lim_{\rho\to 0} \int_{S_\rho} \psi \frac{\partial\psi}{\partial N} dt = 0 \qquad (4.25)$$

is satisfied, where S_ρ is the surface of the torus generated by a circle with radius ρ whose center moves along the edge L of the screen while the plane of the circle is orthogonal to the line L. If condition (4.25) holds we can integrate over Γ as if it were closed and $\int_\Gamma = \int_{\Gamma_+} + \int_{\Gamma_-}$, where Γ_+ is the exterior side of the screen and Γ_- is the interior side. It makes no difference which side we call the exterior side. Now it is clear that as $V \to 0$ we obtain from (1.68)

$$-\beta_{jj} = \max\left\{\left(\int_\Gamma N_j(t)u(t)dt\right)^2 / \int_\Omega |\nabla u|^2 dx\right\}, \qquad (4.26)$$

where $u(\infty) = 0$, u satisfies condition (4.25), and Γ is the jump surface for $u(x)$. Such functions have a representation of the form (4.21). As $V \to 0$ formula (1.69) takes the form

$$-\tilde{\beta}_{jj} = \min \int_\Omega |q_j|^2 dx, \quad (q_j, N)\Big|_\Gamma = N_j, \quad \operatorname{div} q_j = 0 \text{ in } \Omega. \tag{4.27}$$

If Γ is a plane orthogonal to axis X_3 then it follows from formula (4.6) that $\tilde{\alpha}_{i3} = \tilde{\alpha}_{3i} = 0$, $1 \le i \le 3$, while it follows from formula (4.24) that only $\tilde{\beta}_{33} \equiv \tilde{\beta}$ is nonzero. We have from (4.26) and (4.27),

$$-\tilde{\beta} = \max\left\{\left(\int_\Gamma u dt\right)^2 \Big/ \int_\Omega |\nabla u|^2 dx\right\}, \tag{4.26'}$$

$$-\tilde{\beta} = \min \int_\Omega |q_j|^2 dx, \quad \operatorname{div} q_j = 0 \text{ in } \Omega, \quad q_3\Big|_\Gamma = 1. \tag{4.27'}$$

Admissible functions must satisfy the edge condition. From formula (4.15) we obtain as $V \to 0$,

$$-\tilde{\beta} = \max\left\{\left(\int_\Gamma \lambda(t) dt\right)^2 \Big/ \int_\Gamma\int_\Gamma \hat{\nabla}_t \eta(t) \hat{\nabla}_s \eta(s) \frac{ds dt}{4\pi r_{st}}\right\}, \tag{4.28}$$

where $\hat{\nabla} = e_1\partial_1 + e_2\partial_2$, $\partial_j = \partial/\partial x_j$, $\eta|_L = 0$. To prove (4.28) we note that $\int_\Gamma N_3(t) u(t) dt = \int_\Gamma \{u_+(t) - u_-(t)\} dt = \int_\Gamma \eta(t) dt$. Hence

$$\begin{aligned} -\tilde{\beta} &= \max\left\{\left(\int_\Gamma N_3 u dt\right)^2 \Big/ \int_\Gamma -\frac{\partial u}{\partial N_3} u dt\right\} \\ &= \max\left\{\left(\int_\Gamma \eta(t) dt\right)^2 \Big/ \int_\Gamma \frac{\partial^2}{\partial t_3^2} \int_\Gamma \frac{\eta(s) ds}{4\pi r_{st}} \eta(t) dt\right\} \\ &= \max\left\{\left(\int_\Gamma \eta dt\right)^2 \Big/ \int_\Gamma -\hat{\Delta}_t \int_\Gamma \frac{\eta ds}{4\pi r_{st}} \eta(t) dt\right\} \\ &= \max\left\{\left(\int_\Gamma \eta dt\right)^2 \Big/ \int_\Gamma\int_\Gamma \hat{\nabla}_t \eta(t) \hat{\nabla}_s \eta(s) \frac{ds dt}{4\pi r_{st}}\right\}. \quad \square \end{aligned}$$

Proof of Theorem 12: Let us formulate two principles

(A) Let there be an initial electrostatic field $\tilde{E}_o^{(2)} = E_o e_3$ in the half-space $X_3 < 0$ bounded by the conducting plane $X_3 = 0$. If we cut an aperture F in the plane $X_3 = 0$ then the field $E^{(2)}$ in the half-space $X_3 > 0$ can be calculated from the formula $E^{(2)} = H^{(1)} - H_o^{(1)}$, where $H^{(1)}$ is the magnetic field which is present when a magnetic plate F with $\mu = 0$ is placed in the initial field $H_o^{(1)} = -\frac{1}{2}\tilde{E}_o^{(2)} = -\frac{1}{2}E_o e_3$.

(B) Let there be a magnetostatic field $H_o^{(2)}$ parallel to the plane $X_3 = 0$ in the half-space $X_3 < 0$ bounded by the plane $X_3 = 0$ with $\mu = 0$. If we cut an aperture F in the plane then the field $H^{(2)}$ in the half-space $X_3 > 0$ can be calculated from the formula $H^{(2)} = -(E^{(1)} - E_o^{(1)})$, where $E^{(1)}$ is the electric field which is present when the metallic plate F is placed in the initial field $E_o^{(1)} = \frac{1}{2}H_o^{(2)}$.

Formula (1.71) follows immediately from these principles and from the definitions of α_o, $\tilde{\beta}$, β_{ij}^0, $\tilde{\alpha}_{ij}$. Both principles can be proved similarly. We give the proof of (A).

Let $S = \mathbb{R}^2 \setminus F$. We have $E^{(2)} = -\nabla u$, where

$$u = \begin{cases} \phi, & x_3 > 0, \\ -E_o x_3 + \phi, & x_3 < 0, \end{cases}$$

$\Delta\phi = 0$ outside S, $\phi|_S = 0$, $\phi(\infty) = 0$, and u, $\frac{\partial u}{\partial x_3}$ are continuous when crossing F, i.e., $(\partial\phi/\partial x_3)_+ = -E_o + (\partial\phi/\partial x_3)_-$. By symmetry we have $\phi(\hat{x},x_3) = \phi(\hat{x},-x_3)$, $\hat{x} = (x_1,x_2)$. Hence $(\partial\phi/\partial x_3)_- = -(\partial\phi/\partial x_3)_+$, $(\partial\phi/\partial x_3)_+ = -\frac{1}{2}E_0$. Here $(\partial\phi/\partial x_3)_\pm$ are the limiting values of $\partial\phi/\partial x_3$ on F for $x_3 \to \pm 0$. So $\Delta\phi = 0$ for $x_3 > 0$, $\phi|_S = 0$, $\phi(\infty) = 0$, $(\partial\phi/\partial x_3)_+ = -\frac{1}{2}E_o$.

$E^{(2)} = -\nabla\phi$ for $x_3 > 0$. The field $H^{(1)} - H_0^{(1)} = -\nabla\psi$ for $x_3 > 0$, where $\Delta\psi = 0$, $\psi(\infty) = 0$, and by symmetry $\psi(\hat{x},-x_3) = -\psi(\hat{x},x_3)$. The magnetostatic potential $v = \frac{1}{2} E_0 x_3 + \psi$ satisfies condition $(\partial v/\partial N)|_F = 0$, where N is the outward pointing normal to F. Hence $(\partial\psi/\partial x_3)_+ = -\frac{1}{2}E_0$. As ψ is odd in x_3, we conclude that $\psi|_{x_3=0} = 0$, $\psi|_S = 0$. Hence ϕ,ψ are the solutions of the same boundary-value problem in the half-space $x_3 > 0$. The solution of this problem is unique. Hence $\phi \equiv \psi$ for $x_3 > 0$. This means that $E^{(2)} = H^{(1)} - H_0^{(1)}$ for $x_3 > 0$. Principle (A) is proved. □

Example. For disk with radius a we have $\tilde{\beta} = -(8/3)a^3$, $\tilde{\alpha} = (16/3)a^3\delta_{ij}$, $1 \le i,j \le 2$, $\alpha_0 = (4/3)a^3$, $\beta^0_{ij} = -(8/3)a^3\delta_{ij}$, $1 \le i,j \le 2$, in SI units.

Proof of Formulas (1.72) and (1.73): We note that

$$\frac{1}{2}\int_\Omega \varepsilon_{ij}(x)\,\frac{\partial u}{\partial x_i}\,\frac{\partial u}{\partial x_j}\,dx$$

is the energy of the field with the potential $u(x)$, $u(\infty) = 0$, $u|_\Gamma = 1$. The minimum value of this energy is the energy of the real electrostatic field, i.e., $\frac{1}{2}\,Cu^2|_\Gamma = \frac{1}{2}C$. This gives formula (1.72). To prove formula (1.73), we note that if $\operatorname{div} \varepsilon E = 0$, $Q = \int_\Gamma(\varepsilon E,N)\,dt = 1$, the energy of the real electrostatic field is $\frac{1}{2}\,Q^2C^{-1} = \frac{1}{2}\,C^{-1}$. From here we obtain formula (1.73). □

5. Inverse Problem of Radiation Theory

For the sake of simplicity, we assume the probe material is such that its magnetic dipole radiation is negligible. So we have (see (1.37) and (1.29))

$$f_E = \frac{k^2}{4\pi\varepsilon_o}[n[P,n]], \quad P_i = \alpha_{ij}(\gamma)\varepsilon_o VE_j, \quad \gamma = \frac{\varepsilon'-\varepsilon_o}{\varepsilon'+\varepsilon_o}, \tag{5.1}$$

where ε' is the complex dielectric constant of the probe. According to Theorem 7, we can find the tensor $\alpha_{ij}(\gamma)$.

Set $f = f_E$, $b = k^2/4\pi\varepsilon_o$. The vector $f = bP - bn(P,n)$ is measured and vector E is to be calculated. Let n_1, n_2 be two noncollinear orts, $f_j = f(n_j)$, $j = 1,2$. Then

$$bP = f_1 + bn_1(P,n_1) = f_2 + bn_2(P,n_2). \tag{5.2}$$

Assume for simplicity that $(n_1,n_2) = 0$. Then from (5.2) we obtain

$$b(P,n_1) = (f_2,n_1). \tag{5.3}$$

Hence

$$P = \frac{f_1}{b} + \frac{n_1(f_2,n_1)}{b} = \frac{f_2}{b} + \frac{n_2(f_1,n_2)}{b} . \tag{5.4}$$

Knowing P, we can find E from the linear system

$$P_i = \alpha_{jj}\varepsilon_o VE_j, \quad 1 \le i \le 3. \tag{5.5}$$

The matrix α_{ij} is positive definite, so system (5.5) has a unique solution. So a practical method to solve the inverse problem can be described as: 1) measure f_1,f_2 in two orthogonal directions and find P from formula (5.4); 2) find E from system (5.5). □

6. Wave Scattering by a System of Small Bodies; Formulas for the Scattering Amplitude; and Determination of the Medium Properties from the Scattering Data

In this section we present a rigorous theory of scalar and vector wave scattering by a system of many small bodies of an arbitrary shape. We study the influence of boundary conditions on the scattering amplitude and give analytic formulas for the scattering amplitude. The equation for the average field in a medium consisting of many $(\sim 10^{23})$ small particles is obtained. The inverse problem of determining properties of the medium from the scattering data is discussed.

1. We consider scalar wave scattering by a body D with surface Γ. Let a be the characteristic dimension of the body, λ be the wavelength of the initial field, $k = 2\pi/\lambda$, and $ka << 1$ is the smallness condition for the body. Boundary conditions are important for the scattering process. For example, the scattering amplitude for a small acoustic soft body (the Dirichlet problem) is proportional to a, while for a rigid body (the Neumann condition) it is proportional to k^2a^3. The scattering is isotropic in the first case and has dipole character in the second. There are no formulas for the scattering amplitude, by a small body of an arbitrary shape for impedance or the Neumann boundary conditions.

Let $u_o(x,k)$ be the initial field, $u = u_o + v$ the total field, and v the scattered field. The scattering problem consists of solving the boundary-value problem

$$(\Delta + k^2)v = 0 \quad \text{in} \quad \Omega,$$

$$\frac{\partial v}{\partial N} - hv\Big|_\Gamma = \left(-\frac{\partial u_o}{\partial N} + hu_o\right)\Big|_\Gamma, \quad |x|\left(\frac{\partial v}{\partial |x|} - ikv\right) \to 0$$
$$\text{as} \quad |x| \to \infty, \quad (6.1)$$

where $\Gamma = \bigcup_{j=1}^{r} \Gamma_j$, $D = \bigcup_{j=1}^{r} D_j$, $\Gamma_j = \partial D_j$, $\Omega = \mathbb{R}^3 \setminus D$, $D_j \cap D_i = \emptyset$ for $i \neq j$, N is the unit outward pointing normal to Γ, $h|_{\Gamma_j} = h_j = h_{1j} + ih_{2j}$, $h_{2j} \leq 0$, $h_{1j} \geq 0$, $|h_{1j}| + |h_{2j}| > 0$, and $k^2 > 0$. We look for a solution of the form (1.75). It is not difficult to prove existence and uniqueness of the solution to the problem (6.1). The scattering amplitude which is defined by the formula

$$v \sim \frac{\exp(ik|x|)}{|x|} f(n,k), \quad n = \frac{x}{|x|}, \quad |x| \to \infty \qquad (6.2)$$

can be calculated from formula (1.76), where Q_j is defined by formula (1.77). Let a_j be the characteristic dimension of D_j, $a = \max_{1 \leq j \leq r} a_j$, d_{ji} be the distance between D_j and D_i, $d = \min_{i \neq j} d_{ij}$, and $\ell = \max_{i,j} d_{ij}$. Let us expand Q_j in powers of ka up to $(ka)^2$. First we expand $\sigma_j = \sigma_{jo} + ik\sigma_{j1} + \frac{1}{2}(ik)^2\sigma_{j2} + \ldots$ and substitute this expression for σ_j in (1.77). Then we expand the exponent in (1.77).

As a result we obtain

$$\begin{aligned} Q_j = &\int_{\Gamma_j} \sigma_{jo} dt + ik \int_{\Gamma_j} \sigma_{j1} dt - ik\left(n, \int_{\Gamma_j} \sigma_{jo}(t-t_j)dt\right) \\ &+ k^2\left(n, \int_{\Gamma_j} \sigma_{j1}(t-t_j)dt\right) - \frac{k^2}{2}\int_{\Gamma_j} \sigma_{j2} dt \\ &- \frac{k^2}{2}\int_{\Gamma_j} \sigma_{jo}(n, t-t_j)^2 dt + \ldots . \end{aligned} \qquad (6.3)$$

2. Here we consider the scattering amplitude when $r = 1$. For $\sigma_1 = \sigma$ we have the integral equation

$$\frac{A(k)\sigma - \sigma}{2} - \frac{h}{2} T(k)\sigma = \left(hu_o - \frac{\partial u_o}{\partial N}\right)\Big|_\Gamma, \tag{6.4}$$

where

$$\begin{aligned} A(k)\sigma &= \int_\Gamma \frac{\partial}{\partial N_s} \frac{\exp(ikr_{st})}{2\pi r_{st}} \sigma(t)dt, \\ T(k)\sigma &= \int_\Gamma \frac{\exp(ikr_{st})}{2\pi r_{st}} \sigma(t)dt. \end{aligned} \tag{6.5}$$

We set

$$u_0 = u_{00} + iku_{01} + \frac{(ik)^2}{2} u_{02} + \cdots \tag{6.6}$$

$$A(k) = A + ikA_1 + \frac{(ik)^2}{2} A_2 + \cdots \tag{6.7}$$

$$T(k) = T + ikT_1 + \frac{(ik)^2}{2} T_2 + \cdots . \tag{6.8}$$

If we substitute (6.6)-(6.8) into (6.4) and equate coefficients of like powers of ka, we obtain

$$\sigma_0 = A\sigma_0 - hT\sigma_0 = 2hu_{00}|_\Gamma + 2 \frac{\partial u_{00}}{\partial N}\Big|_\Gamma \tag{6.9}$$

$$\sigma_1 = A\sigma_1 + A_1\sigma_0 - hT\sigma_1 - hT_1\sigma_0 - 2hu_{01} + 2 \frac{\partial u_{01}}{\partial N}\Big|_\Gamma \tag{6.10}$$

$$\begin{aligned} \sigma_2 = A\sigma_2 - hT\sigma_2 + A_2\sigma_0 + 2A_1\sigma_1 - hT_2\sigma_0 - 2hT_1\sigma_1 \\ - 2hu_{02} + 2 \frac{\partial u_{02}}{\partial N}\Big|_\Gamma . \end{aligned} \tag{6.11}$$

We assume that the initial field is the plane wave $u_0 = \exp\{ik(\nu,x)\}$. Then

$$\begin{aligned} u_{00} = 1, \quad \frac{\partial u_{00}}{\partial N} = 0; \quad u_{01} = (\nu,t), \quad \frac{\partial u_{01}}{\partial N}\Big|_\Gamma = (\nu,N); \\ u_{02} = (\nu,t)^2, \quad \frac{\partial u_{02}}{\partial N}\Big|_\Gamma = 2(\nu,t)(\nu,N). \end{aligned} \tag{6.12}$$

We note that the following formulas hold:

$$A\sigma = \int_\Gamma \frac{\partial}{\partial N_s} \frac{1}{2\pi r_{st}} \sigma(t)dt, \quad A_1\sigma = 0,$$
$$A_2\sigma = \int_\Gamma \frac{\cos(r_{ts}, N_t)\sigma(t)dt}{2\pi}, \quad \int_\Gamma A\sigma dt = \int_\Gamma \sigma dt, \tag{6.13}$$
$$r_{ts} = s - t, \quad T\sigma = \int_\Gamma \frac{\sigma(t)}{2\pi r_{st}} dt$$
$$T_1\sigma = \int_\Gamma \frac{\sigma(t)dt}{2\pi}, \quad T_2\sigma = \int_\Gamma \frac{r_{st}\sigma(t)dt}{2\pi}.$$

In order to calculate the scattering amplitude we must find $Q_j = Q$ from formula (6.3), i.e., find the integrals in formula (6.3). To do so we integrate equation (6.9) over Γ and take into consideration (6.13). We obtain

$$\int_\Gamma \sigma_o dt = -hS - \frac{h}{4\pi} \int_\Gamma\int_\Gamma \frac{\sigma_o(t)dtds}{r_{st}}, \quad S = \text{meas } \Gamma. \tag{6.14}$$

The exact calculation of the integral in the right-hand side of (6.14) requires knowledge of the function $\sigma_o(t)$. This function can be obtained by an iterative process as in Theorem 3. But here we give an approximation using instead of the integral its average value

$$\int_\Gamma \frac{ds}{r_{st}} \approx \frac{1}{S} \int_\Gamma dt \int_\Gamma \frac{ds}{r_{st}} \equiv \frac{J}{S}, \quad J \equiv \int_\Gamma\int_\Gamma \frac{dsdt}{r_{st}}. \tag{6.15}$$

By formula (1.46) $C \approx C^{(0)} = 4\pi S^2/J$, $\varepsilon_o = 1$ in our case. So we obtain from (6.14) the following approximation

$$\int_\Gamma \sigma_o dt \approx -\frac{hS}{1 + \frac{hJ}{4\pi S}} \approx -\frac{hS}{1 + hSC^{-1}}, \tag{6.16}$$

where C is the capacitance of the capacitor with the same shape as D. Hence

$$f(n,k) \approx -\frac{hS}{4\pi(1+hSC^{-1})} u_{00} \tag{6.17}$$

is an approximate formula for the scattering amplitude of the initial field $u_o(x,k) = u_{00} + iku_{01} + \dots$ for the small body with the impedance boundary condition. This scattering is isotropic. If $h \to +\infty$ we obtain from (6.17) the known formula (see Hönl et al. [1])

$$f = -\frac{C}{4\pi} u_{00}. \tag{6.18}$$

Consider the case $h = 0$. Then equation (6.9) is homogeneous, $\sigma_o = 0$. Equation (6.10) takes the form

$$\sigma_1 = A\sigma_1 + 2(\nu,N) = A\sigma_1 + 2\,\frac{\partial u_{01}}{\partial N}. \tag{6.19}$$

Integrating equation (6.19) over Γ and using (6.13) we obtain $\int_\Gamma \sigma_1 dt = 0$. So only the terms of order k^2 remain in formula (6.3). Equation (6.11) can be written in our case as

$$\sigma_2 = A\sigma_2 + 2(\nu,t)(\nu,N) = A\sigma_2 + 2\,\frac{\partial u_{02}}{\partial N}. \tag{6.20}$$

Integrating over Γ we obtain $\int_\Gamma \sigma_2 dt = \int_\Gamma (\partial u_{02}/\partial N) dt = (V/V) \int_D \Delta u_{02} dx$,

$$\int_\Gamma \sigma_2 dt = \int_\Gamma (\nu,t)(\nu,N) dt = \int_D \{\operatorname{div}(\nu,x)\nu\} dx = |D| = V, \tag{6.21}$$

where $V = \text{meas } D$. For one body we can take $t_j = 0$ in (6.3), chosing the origin inside D.

Now

$$\left(n, \int_\Gamma t\sigma_1(t) dt\right) = -V\beta_{pq}\nu_q n_p, \tag{6.22}$$

where β_{pq} is the magnetic polarizability tensor, which is defined in Section 1, point 5 by the formula

$$\beta_{pq} = \frac{1}{V} \int_\Gamma t_p \sigma_q(t) dt, \tag{6.23}$$

where $\sigma_q = A\sigma_q - 2N_q$. Theorem 7 allows one to calculate tensor β_{pq} with the prescribed accuracy. So if $h = 0$, $r = 1$ we have the formula for the scattering amplitude

$$f(n,\nu,k) = -\frac{k^2V}{4\pi}\beta_{pq}\nu_q n_p - \frac{k^2V}{8\pi}, \quad f \sim k^2a^3. \tag{6.24}$$

Hence the scattering is anisotropic. Here we end the study of wave scattering by a single body.

3. Consider scattering by r bodies.

The integral equations for the problem are

$$\begin{aligned}\sigma_j &= A_j(k)\sigma_j - h_jT_j(k)\sigma_j + \sum{}'A_{jp}(k)\sigma_p - h_j\sum{}'T_{jp}(k)\sigma_p \\ &+ 2\frac{\partial u_o}{\partial N} - 2h_ju_0, \quad 1 \le j \le r, \quad \sum = \sum_{p=1}^{r}, \ \sum{}' = \sum_{p\neq j},\end{aligned} \tag{6.25}$$

the operators $A_j(k)$ and $T_j(k)$ are defined by formula (6.5) in which Γ should be replaced by Γ_j,

$$\begin{aligned}A_{jp}(k)\sigma_p &= \int_{\Gamma_p}\frac{\partial}{\partial N_{sj}}\frac{\exp(ikr_{s_jt_p})}{2\pi r_{s_jt_p}}\sigma_p(t_p)dt_p, \\ T_{jp}(k)\sigma_p &= \int_{\Gamma_p}\frac{\exp(ikr_{s_jt_p})}{2\pi r_{s_jt_p}}\sigma_p(t_p)dt_p.\end{aligned} \tag{6.26}$$

To simplify the calculations we assume that $d \gg a$ so that the terms A_{jp} and T_{jp} in (6.25) are negligible. Then for σ_{jo} we obtain an equation similar to (6.9). Hence for $h_j \neq 0$, $1 \le j \le r$, we obtain from (1.76) and (6.17) the formula

$$f(n,k) = -\frac{1}{4\pi}\sum_{j=1}^{r}\exp\{-ik(n,t_j)\}\frac{h_jS_j}{1+h_jS_jC_j^{-1}}u_{0j}, \tag{6.27}$$

where $u_{0j} = u_0(t_j,k)$.

If $h_j = 0$, $1 \le j \le r$, then from (1.76) and (6.24) we obtain

$$f(n,\nu,k) = -\frac{1}{4\pi}\sum_{j=1}^{r}\exp\{-ik(n,t_j)\}\left\{k^2V_j\beta_{pq}n_p\frac{\partial u_{01}(t_j,k)}{\partial x_q}\right.$$

$$\left.+ V_j\frac{1}{2}\frac{1}{V_j}\int_{D_j}\Delta u_{02}(x,k)dx\right\}. \tag{6.28}$$

Here we used the generalization of formula (6.24) for an arbitrary initial field $u_0(x,k)$. In this case the right-hand side of equation (6.19) is $2(\nabla u_{01},N)$, while the right-hand side of equation (6.20) is $2(\nabla u_{02},N)$. So we replace ν_q in (6.24) by $\partial u_{01}/\partial x_q$ and take into account that instead of (6.21) we have

$$\int_\Gamma \sigma_2(t)dt = \int_\Gamma (\nabla u_{02},N)dt = \int_D \Delta u_{02}dx. \tag{6.21'}$$

Hence instead of (6.24) we have

$$f(n,\nu,k) = -\frac{k^2V}{4\pi}\left\{\beta_{pq}n_p\frac{\partial u_{01}}{\partial x_q} + \frac{1}{2V}\int_D \Delta u_{02}dx\right\}. \tag{6.24'}$$

It is essential to calculate $\partial u_{01}/\partial x_q$ correctly in formulas (6.28) and (6.24'). To this end we note that the initial field $u_0(x,k)$ should be expanded in powers of ka, not k. For example, if $u_0 = \exp\{ik(\nu,x)\}$, then $u_0 = \exp\{ik(\nu,t_j)\}\{1+ik(\nu,x-t_j)+\cdots\}$, $k|x-t_j| \sim ka$ and

$$\frac{\partial u_{01}}{\partial x_q} = \left.\frac{\partial u_o(x,k)}{ik\partial x_q}\right|_{x=t_j}, \quad u_{02} = \left.\frac{\Delta u_{02}(x,k)}{(ik)^2}\right|_{x=t_j}. \tag{6.24''}$$

4. We can now obtain the integral equation for the average field in a medium consisting of many $(\sim 10^{23})$ small particles. Assume that $d \gg a$ and suppose that every particle is in a self-consistent field $u(x,k)$. If $\mathcal{F}_{jp}$ is an operator which allows us to calculate the field scattered by the p-th particle at the point t_j, then

$$u_j = u_{0j} + \sum{}' \frac{\exp(ikr_{jp})}{r_{jp}} \mathscr{F}_{jp} u_p. \tag{6.29}$$

Here u_j is the self-consistent field $\langle u(x_j)\rangle$ and u_{0j} is the initial field. If $((\partial u/\partial N) - hu)|_\Gamma = 0$ then from (6.29) and (6.17) we obtain

$$\langle u(x)\rangle = u_0(x) - \int \frac{\exp(ikr_{xy})}{4\pi r_{xy}} q(y) \langle u(y)\rangle dy, \tag{6.30}$$

$\int = \int_{\mathbb{R}^3}$, where $q(y)$ is the average value of $h_j S_j (1 + h_j S_j C_j^{-1})^{-1}$ over the volume dy in the neighborhood of the point y. If the particles are identical and $N(y)$ their distribution density then $q(y) = N(y)hS(1 + hSC^{-1})^{-1}$, where C is the capacitance of a particle and S is its surface area. If $h = 0$ then from (6.29), (6.24'), and (6.24") it follows that

$$\langle u(x,k)\rangle = u_0(x,k) + \frac{1}{4\pi} \int \frac{\exp(ikr_{xy})}{r_{xy}} \left\{ B_{pq}(y) \frac{\partial \langle u(y,k)\rangle}{\partial y_q} \frac{x_p - y_p}{r_{xy}} + \frac{1}{2} b(y) \Delta \langle u(y,k)\rangle \right\} dy. \tag{6.31}$$

where

$$b(y) = VN(y), \tag{6.32}$$

$$B_{pq}(y) = ikV\beta_{pq} N(y), \quad n_p = \frac{x_p - y_p}{r_{xy}}, \tag{6.33}$$

V is the volume of a particle, $N(y)$ is the number of particles in the unit of volume around point y, and β_{pq} is the magnetic polarizability tensor for a particle. We used the approximation

$$\frac{1}{V} \int_D \Delta u dy \approx \Delta u(y,k). \tag{6.34}$$

Equation (6.30) is equivalent to the Schrödinger equation

$$\Delta u + k^2 u - q(x)u = 0, \quad u = u_0 + v, \quad |x|\left(\frac{\partial v}{\partial |x|} - ikv\right) \to 0 \quad \text{as} \quad |x| \to \infty, \tag{6.35}$$

with

$$q(x) = N(x)\,\frac{hS}{1+hSC^{-1}} \,. \tag{6.36}$$

Equation (6.31) is nonlocal. It demonstrates once again the cardinal difference between scattering by rigid and nonrigid particles.

5. Here we discuss the inverse problem of determining properties of the medium from the scattering data. If the medium is rarefied enough, i.e., the inequality $d \gg a$ is so strong that $q(y)$ in equation (6.30) is small enough, we can solve equation (6.30) by an iterative process. It is sufficient to assume that

$$\max_x \int \frac{|q(y)|\,dy}{4\pi r_{xy}} < 1$$

in order that the iterative process converge in $C(\mathbb{R}^3)$. The scattering amplitude can be written as

$$f = -\frac{1}{4\pi}\int \exp\{-ik(n,y)\}q(y)u(y)\,dy. \tag{6.37}$$

Assume that $u_0(x,k) = \exp\{ik(\nu,x)\}$. Using the approximation $u \approx u_0$, which is justified for the rarefied medium, we obtain

$$f = f(n,\nu,k) = -\frac{1}{4\pi}\int \exp\{ik(\nu-n,y)\}q(y)\,dy. \tag{6.38}$$

From (6.38) we see that $q(y)$ is uniquely determined if $f(n,\nu,k)$ is known for $0 < k < \infty$ and $\ell = \nu - n \in S^2$, where S^2 is the unit sphere in $\mathbb{R}^3$. Usually in practice $q(y)$ is a compactly supported function. In this case f is an entire function of three complex arguments k_1, k_2, k_3, where the k_j

are the components of the vector $k\ell$, so that we can measure f only in some interval $0 < a \le k \le b$, $\ell \in S^2$ and then define $f(\ell,k)$ for all k by analytic continuation. This process is unstable. Actually, $q(y)$ is the inverse Fourier transform of $f(\ell,k)$. If $q(y)$ is a function depending only on a few parameters, it can be easily found from (6.38). For example, if $q(y) = \text{const}$ or $q(y) = c_1a_1(x) + c_2a_2(x)$, where the $a_j(x)$, $j = 1,2$, are some known functions, then the parameters c_1, c_2 can be easily calculated from (6.38) if we know $f(k,\ell_1)$, $f(k,\ell_2)$ for a fixed k and two directions ℓ_1, ℓ_2. Similar arguments are valid for equation (6.31). If we know $q(x)$ and properties of a single particle, i.e., h,S,C, we can find $N(x)$ from (6.36).

6. Consider electromagnetic wave scattering by a system of r small particles. Assume that the magnetic dipole scattering is negligible. This is so, for example, when light is scattered by small nonconducting particles. The scattering amplitude for a single particle can be found from the formula

$$f_E = \frac{k^2}{4\pi\varepsilon_0}[n,[P,n]], \quad P_j = \alpha_{jq}(\gamma)\varepsilon_0 VE_q, \quad \gamma = \frac{\varepsilon-\varepsilon_0}{\varepsilon+\varepsilon_0}, \tag{6.39}$$

(see formulas (1.37) and (1.29')). If $d \gg a$ we obtain the equation similar to (6.30) for average field $E(x,k)$:

$$E(x,k) = E_0(x,k) + k^2 \int \frac{\exp(ikr_{xy})}{4\pi\varepsilon_0 r_{xy}^3}[\mathbf{r}_{yx}[\alpha E(y,k),\mathbf{r}_{yx}]]dy \tag{6.40}$$

where $\mathbf{r}_{yx} = x-y$, $\alpha = \alpha(y) = N(y)\alpha_{jq}\varepsilon_0 V$, and $N(y)$ is the number of particles in the unit volume around y. Equation (6.40) can be used to solve the inverse problem discussed in the previous section. For example, if the particles are balls of radius a then $\alpha_{jq} = 3((\varepsilon-\varepsilon_o)/(\varepsilon+2\varepsilon_o))\delta_{jq}$, $V = (4/3)\pi a^3$. Suppose that $N(y) = N = \text{const.}$ Then the Born approximation for the scattering amplitude is

$$f(n,k) = k^2 N a^3 \frac{\varepsilon-\varepsilon_o}{\varepsilon+2\varepsilon_o} \int \exp\{-ik(n,y)\}[n[E_o(y,k),n]]dy \tag{6.41}$$

We can find Na^3 if we know $f(k,n)$ for some direction $n = n_1$. If N is known the value a can be calculated. Actually, one measures the absorption cross-section $d\sigma/d\Omega$. If $E_o(y,k) = E_o\exp\{ik(\nu,y)\}$, then from (6.41) we obtain, for $N = N(y)$,

$$\begin{aligned}\frac{d\sigma}{d\Omega} &= |f(n,k)|^2 \\ &= k^4 a^6 \left(\frac{\varepsilon-\varepsilon_o}{\varepsilon+2\varepsilon_o}\right)^2 |[n,[E_o,n]]|^2 \left|\int \{\exp ik(\nu-n,y)\}N(y)dy\right|^2\end{aligned} \tag{6.42}$$

If $n = \nu$, i.e., in the case of forward scattering, the integral in (6.42) is equal to N_Σ where N_Σ is the total number of particles in the scattering volume of the medium. If $n \neq \nu$ we average the result relative to random distribution of particles. Denote by line this averaging. Then

$$\begin{aligned}\overline{\frac{d\sigma}{d\Omega}} = k^4 a^6 \left|\frac{\varepsilon-\varepsilon_o}{\varepsilon+2\varepsilon_o}\right|^2 |[n,[E_o,n]]|^2 \iint \exp\{ik(n-\nu,y-y')\}\cdot \\ \cdot\, \overline{N(y)N(y')}dydy'.\end{aligned} \tag{6.43}$$

If the covariance $K(y,y') = \overline{N(y)N(y')}$ is known we can find $\overline{d\sigma/d\Omega}$. In many cases it is assumed that $K(y,y') = K^2(y)\delta(y-y$ (noncorrelated scatterers). If $K(y) = K = \text{const.}$ then the integral in (6.43) can be easily calculated. Let us take $K(y,y') = N^2\exp(-\rho^{-1}|y-y'|)$, where $\rho \ll 1$ is the correlation radius. Setting $y-y' = z$, $n-\nu = m$, $\theta = n\hat{\nu}$, $y+y' = z'$, we have

$$\frac{N^2}{8}\iint \exp\{ik(m,z)\}e^{-z/\rho}dzdz' = \frac{NN_\Sigma}{8}4\pi\int_0^\infty e^{-z/\rho}\,\frac{\sin(k|m|r)}{k|m|r}\,r^2dr$$

$$= \frac{\pi NN_\Sigma}{2k|m|}\int_0^\infty e^{-z/\rho}\sin(k|m|r)rdr = \frac{\pi NN_\Sigma\rho}{(1+4k^2\rho^2\sin^2(\theta/2))}. \quad (6.44)$$

Here we took into account that $|m|^2 = 4\sin^2(\theta/2)$, $\partial(y,y')/\partial(z,z') = 1/8$. Hence from (6.44) and (6.43) we obtain

$$\overline{\frac{d\sigma}{d\Omega}} = k^4a^6\left|\frac{\varepsilon-\varepsilon_0}{\varepsilon+2\varepsilon_0}\right|^2|[n,[E_0,n]]|^2\,\frac{\pi NN_\Sigma\rho}{1+4k^2\rho^2\sin^2(\theta/2)}\,. \quad (6.45)$$

7. Let us discuss the equation for the Stokes vector in a random medium consisting of small particles. The main assumption $d \gg a$ still holds. We use the notations adopted in Newton [1], which unfortunately differ from the notations in Hulst [1]. If the plane of scattering contains orts ν, n of the incident and scattered waves, $e_{||}$, $e_\perp$ are orts parallel and perpendicular to the plane of scattering, $e_{||}$ is the ort of axis OY in notations of Section 1, point 5, $E = E_{||}e_{||} + E_\perp e_\perp$, then the Stokes vector is the set of four numbers $\mathcal{J} = (J,Q,U,V)$, $J = |E_{||}|^2 + |E_\perp|^2$, $Q = |E_{||}|^2 - |E_\perp|^2$, $U = -2\,\mathrm{Re}(E_{||}E_\perp^*)$, $V = -2\,\mathrm{Im}(E_{||}E_\perp^*)$. Let the incident wave with Stokes vector $\mathcal{J}_0$ scattered by a particle have the Stokes vector $r^{-2}\mathcal{F}\mathcal{J}_0 = \mathcal{J}$ at distance r from the scattering particle, where $\mathcal{F}$ is some 4×4 matrix. Then for the Stokes vector in the medium we have the equation

$$\mathcal{J}(x) = \mathcal{J}_0(x) + \int \frac{1}{r_{xy}^2}\mathcal{J}(y)N(y)dy,$$

where $N(y)$ is the number of identical particles in a unit volume. In Newton [1] the following formulas for $\mathcal{F}$ are given:

$$2\mathcal{F}_{\substack{11\\21}} = |S_{11}|^2 \pm |S_{12}|^2 \pm |S_{21}|^2 + |S_{22}|^2,$$

$$2\mathcal{F}_{\substack{12\\22}} = |S_{11}|^2 \mp |S_{12}|^2 \pm |S_{21}|^2 - |S_{22}|^2,$$

$$\mathcal{F}_{\substack{13\\23}} = -\mathrm{Re}(S_{11}S_{12}^* \pm S_{22}S_{21}^*), \quad \mathcal{F}_{\substack{14\\24}} = -\mathrm{Im}(S_{11}S_{12}^* \mp S_{22}S_{21}^*)$$

$$\mathcal{F}_{\substack{31\\32}} = -\mathrm{Re}(S_{11}S_{22}^* \pm S_{22}S_{12}^*), \quad \mathcal{F}_{\substack{33\\44}} = \mathrm{Re}(S_{11}S_{22}^* \pm S_{12}S_{21}^*),$$

$$\mathcal{F}_{\substack{34\\43}} = \mathrm{Im}(S_{12}S_{21}^* \mp S_{11}S_{22}^*), \quad \mathcal{F}_{\substack{41\\42}} = \mathrm{Im}(S_{21}S_{11} \pm S_{22}S_{12}^*)$$

where $S_{11} = S_2$, $S_{22} = S_1$, $S_{12} = -S_3$, $S_{21} = S_4$, and S_j, $1 \le j \le 4$ are defined by formula (1.38).

Knowing the scattering matrix (1.38) for a single particle, one can find formulas for the refraction index tensor $n_{ij} = \delta_{ij} + 2\pi N k^{-2} S_{ij}(0)$, the coefficient of absorption $\chi = N\sigma = 4\pi N k^{-1} \mathrm{Im}\, S(0)$ for isotropic scattering by a particle, the cross-section $\sigma = 2\pi k^{-1}\, \mathrm{tr}\mathrm{Im}\, S(0)$ for anisotropic scattering by a single particle in the wave scattering, the rotation of the polarization plane, etc., so we can answer most of the questions which are important in practice.

7. Research Problems

1. To work out a standard program for calculation of the capacitance and the polarizability tensor for a body of arbitrary shape. According to formulas (1.44), (1.46), and (1.31)-(1.35) the program is to calculate integrals over Γ of functions with weak singularities. For example,

$$\int_\Gamma\int_\Gamma \frac{dsdt}{r_{st}}, \quad \int_\Gamma\int_\Gamma \frac{N_i(s)N_j(t)dsdt}{r_{st}},$$

$$\int_\Gamma\int_\Gamma \frac{dsdt}{r_{st}} \int_\Gamma \frac{\partial}{\partial N_t} \frac{1}{r_{t_1 t}} dt_1.$$

It is of considerable independent interest to find a good method to calculate multiple integrals of functions with weak singularities. The program must be tested on some special examples (e.g., cube, cylinder, ellipsoid, disk).

2. It is interesting to study how deviation from the spherical shape influences the scattering amplitude. For example, a falling drop of liquid first has a spherical shape at first, but its shape gradually tends to a hemisphere. There are photographs of the shapes in the geophysical literature. How does this drop scatter radio waves? To solve the problem one can use formulas (1.37), (1.29)-(1.29"), and (1.30)-(1.36). Actually if Problem 1 is solved there is no difficulties in solving Problem 2.
3. Solve Problem 1 for several bodies. Study the interaction of scattered waves.
4. Apply the iterative processes (1.6), (1.12), and (1.23) to some practical problems.
5. Prove Theorem 9 without the assumption $p = 2$ (see formula (3.19) et seq.).

8. Bibliographical Note

For many problems it is interesting to calculate static fields and linear functionals of such fields. In numerical analysis, variational methods and finite difference methods were the techniques primarily studied and applied to these problems. Many difficulties arise if the domain in which the field is to be determined is unbounded or does not have a smooth boundary, or the dimension of the space is large. Methods of integral equations can be helpful, giving iterative processes for calculating solutions and approximate analytical formulas for fields and linear functionals of static fields. It would be impossible even to mention all the books on the subject. Potential theory and its applications to mathematical physics are presented in Gunter [1], variational methods are described in Mihlin [1], [2], calculation of static fields is described in Buhgolz [1], a reference book on calculation of capacity is Jossel et al. [1]. Formulas for the polarizability tensor of an ellipsoid can be found in Landau-Lifschitz [1]. Wave scattering by small bodies was studied first by Rayleigh (1871), who continued this work until his death in 1919. Thomson [1] found that magnetic dipole scattering by small conducting particles is of the same order as electric dipole scattering. In many works scattering in dielectrics or in aqueous solutions was studied from the point of view of the theory of wave scattering by small bodies (Debye [1], Eskin [1], Brown [1]). Scattering from small apertures was studied by Bethe [1], Levine-Schwinger [1]. In Hulst [1] a review of the physics literature on light scattering by small particles is given. The problem of finding formulas for scattering amplitude for particles of an arbitrary shape was pointed out in many papers (see Hulst [1]). The results presented in Chapter 2 were obtained by the author Ramm ([18]-[45]). From the physical point of view, the main result of this chapter is formula (1.38) for the scattering matrix and Theorem 7 for the polarizability tensor. Now it is possible to calculate the field scattered by a small particle of arbitrary shape by means of explicit formulas. So work of Rayleigh is futhered here about one hundred years after it was initiated by Rayleigh [1]. We point out the book Marchenko-Hruslov [1], in which boundary-value problems in domains with granular boundary are under study. For selfadjoint integral equations of potential theory, the iterative process (1.23) was used in Tsyrlin [1]. In Stevenson [1], the solution of the electromagnetic scattering problem is represented as a power series in ka, where a is the characteristic dimension of the scatterer. A review paper on this subject is Kleinman [1]. The results of this chapter were summarized in Ramm [101]. Formulas (73), (75), (77), (78) from this paper (Section 2 in this Chapter) are improved: in these formulas in the cited paper an isotropic term similar to the term $\frac{1}{2} b(y)\Delta u(y,k)$ in formula (6.31) of this chapter was not taken into account.

CHAPTER III

INVESTIGATION OF A CLASS OF NONLINEAR INTEGRAL EQUATIONS AND APPLICATIONS TO NONLINEAR NETWORK THEORY

0. Introduction

Nonlinear oscillations in some networks can be described by the operator equation $Au + Fu = J$, where A is an unbounded linear operator on a Hilbert or Banach space, F is a nonlinear operator, and $B = A + F$ is monotone. This is true for the general passive one-loop network, consisting of an e.m.f. $E(t)$, and arbitrary linear passive stable one-port L, and a nonlinear one-port N with a monotone voltage-current characteristic $i = Fu$, where i is the current through N and u is the voltage on N. The theory presented below makes it possible to study nonlinear oscillations qualitatively, including questions of existence, uniqueness, stability in the large, convergence, and calculation of stationary regimes by means of an iterative process. Our assumptions concerning the network are more general than those usually adopted in the literature. No assumptions concerning the "smallness" of the nonlinearity or filter property of the linear one-port are made. Our results for nonlinear networks of the class defined above are final in the sense that if we omit the assumption concerning passivity of the network the results will not hold. Though the literature of the subject is vast, the theory and results presented here are new (see Ramm [46]-[53]).

1. Statement of the Problems and Main Results

1. Consider the equation

$$Bu = Au + Fu = J \tag{1.1}$$

under the following assumptions:

(1) The operator $F: H \to H$ on the Hilbert space H is bounded, defined everywhere on H, hemicontinuous, i.e., the function $(F(f + \lambda g), h)$ is continuous in $\lambda \in \mathbb{R}^1$ for all $f, g, h \in H$.

(2) The operator $A: H \to H$ is linear, closed, densely defined in H, $D(B) = D(A)$, and

$$\operatorname{Re}(Bu - Bv, u - v) = 0 \Rightarrow u = v, \quad u,v \in D(B) \tag{1.2}$$

There exists a sequence of bounded linear operators A_n such that

$$A_n u \to Au, \quad u \in D(A); \quad A_n^* u \to A^* u, \quad u \in D(A^*) \tag{1.3}$$

$$\operatorname{Re}(B_n u - B_n v, u - v) \geq 0, \quad u,v \in H, \quad B_n \equiv A_n + F \tag{1.4}$$

$$\operatorname{Re}(B_n u, u) \geq \gamma(\|u\|)\|u\|, \quad \gamma(t) \geq 0, \quad \gamma(t) \to +\infty \text{ as } t \to +\infty, \tag{1.5}$$

where $\|u\|$ is the norm in H, $\to$ denotes convergence in H, and $\rightharpoonup$ denotes weak convergence in H.

(3) $$\operatorname{Re}(Bu - Bv, u - v) \geq \nu_R(\|u-v\|)\|u-v\| \tag{1.6}$$

for $\|u\| \leq R$, $\|v\| \leq R$, where $\nu_R(t)$ is continuous in t, $\nu(0) = 0$, and $\nu_R(t) > 0$ for $t > 0$.

We note that (1.6) implies (1.2), but it is convenient to have the above assumptions because we shall sometimes use (1) and (2) without (3).

Instead of assumptions (1)-(3) we shall also use the following assumptions:

$$\text{(4)} \quad \mathrm{Re}(Au,u) \geq \delta\|u\|^2, \quad \delta > 0, \quad u \in D(A) \text{ and } R(A) = H. \tag{1.7}$$

$$\text{(5)} \quad \|Fu\| \leq \varepsilon\|u\| + c(\varepsilon), \quad \varepsilon > 0, \quad u \in H, \quad c(\varepsilon) = \text{const}, \tag{1.8}$$

$$\mathrm{Re}(Fu - Fv, u - v) \geq 0, \quad u,v \in H, \tag{1.9}$$

$$\text{(6)} \quad \|Fu - Fv\| \leq C(\rho)\|u-v\|, \quad \|u\| \leq \rho, \quad \|v\| \leq \rho, \quad 0 < \rho < \infty, \quad C(\rho) = \text{const}. \tag{1.10}$$

If A is the generator of a strongly continuous semigroup in H then a sequence A_n with property (1.3) exists.

If (1.7) holds we denote by H_A the Hilbert space which is the completion of $D(A)$ with respect to the metric generated by the form $\mathrm{Re}(Au,u)$. The sesquilinear form $[u,v] = \frac{1}{2}\{(Au,v) + (u,Av)\}$ is the inner product in H_A, $[u,u]$ $\mathrm{Re}(Au,u)$. By R_λ we denote the operator $(A+\lambda I)^{-1}$, $\lambda > 0$, where I is the identity on H.

Let us explain now what kinds of problem in nonlinear network theory can be reduced to equation (1.1). Let S be a nonlinear system consisting of a linear one-port L, a nonlinear one-port N, and an e.m.f. $E(t)$. Let the voltage-current characteristic of N be $i = Fu$, where i is the current through N, u is the voltage on N. The linear one-port is described by the equality $i = Au_1$, where u_1 is the voltage on L and A is the admittance operator. The operator $Z = A^{-1}$ is the impedance operator of the one-port, $u_1 = Zi$. We assume that

$$Zi = \int_{-\infty}^{t} g(t,\tau)i(\tau)d\tau. \tag{1.11}$$

The function $g(t,\tau)$ is called the weight function or the impulse response of the two-port. Kirchhoff's equation for the circuit is

$$u + ZFu = E, \tag{1.12}$$

or

$$Bu \equiv Au + Fu = J, \quad J = AE. \tag{1.13}$$

In network theory the passage from (1.12) to (1.13) is called the replacement of the voltage generator $E(t)$ by the equivalent current generator J. Passivity of the network from the mathematical point of view means that

$$\mathrm{Re}(Bu - Bv, u - v) > 0, \quad u,v \in D(A) = D(B). \tag{1.14}$$

If $J(t) = J(t+T)$ then we take as H the space $L^2[0,T]$ of T-periodic functions with inner product $(u,v) = T^{-1} \int_0^T uv^* dt$. If $J(t)$ is an almost periodic function we take as H the Besicovich space B_2 which is the completion of the set of trigonometric polynomials in the metric generated by the inner product $(u,v) = \lim_{T\to\infty} (2T)^{-1} \int_{-T}^{T} uv^* dt$. Inequality (1.14) means that power consumed by the one-port L-N is positive. So the network S is passive if (1.14) holds. If F is monotone then N is passive. The greater δ is in condition (1.7), the less assumptions we need concerning F. For example, if F is not monotone but $F + aI$, $a > 0$, is monotone, then for

$\delta > a$ the one-port L-N can be transformed into an equivalent one-port L_a-N_a where N_a is described by the monotone operator $F + aI$ and L_a is described by the operator $A - aI$ satisfying inequality (1.7) with $\delta_a = \delta - a > 0$. In order to study the stationary regime of the network S under the condition that $E(t)$ is periodic, we use the equation

$$u + QFu = E \tag{1.15}$$

or

$$A_p u + Fu = J, \quad J = A_p E, \quad A_p = Q^{-1}, \tag{1.16}$$

where

$$\begin{cases} Qi = \int_0^T \phi(t,\tau) i(\tau) dt, & (1.17) \\ \phi(t,\tau) = T^{-1} \sum_{n=-\infty}^{\infty} \exp\{in\,\omega(t-\tau)\} Z(in\,\omega, t), & (1.17') \end{cases}$$

and

$$\begin{gathered} Z(i\lambda,t) = \int_0^{\infty} g(t,t-s)\exp(-i\lambda s)ds, \quad \omega = \frac{2\pi}{T}, \\ Z(i\lambda,t+T) = Z(i\lambda,T). \end{gathered} \tag{1.18}$$

If the one-port L is time invariant, i.e., $g(t,\tau) = g(t-\tau)$ then $Z(i\lambda,t) = Z(i\lambda)$, where

$$Z(p) = \int_0^{\infty} \exp(-pt) g(t) dt, \quad p = \sigma + i\lambda, \tag{1.19}$$

where $Z(p)$ is the usual operator impedance of the one-port. It is usually assumed in the literature that a linear one-port is time invariant. $Y(p) = Z^{-1}(p)$ is called the operator admittance. If $E(t)$ is almost periodic then the stationary regime satisfies equations (1.12) and (1.13), and A, F are operators on B_2. Transient regimes in the network can be studied with the help of the equation

$$u(t) = E(t) + m(t) - \int_0^t g(t,\tau)f(\tau,u(\tau))d\tau, \quad t \geq 0, \tag{1.20}$$

where $m(t)$ is a reaction on initial condition and $i = f(t,u(t))$ is the nonlinear characteristic of N. The following questions are to be studied:

(1) Does equation (1.16) have a unique periodic solution if J is periodic? Is this solution stable in the large, i.e., under arbitrary perturbations of the initial conditions? Is it stable under small periodic perturbations of $E(t)$ (so-called stability under permanently acting perturbations)? Is the network convergent? (A network is called convergent if there exists only one stationary regime in the network and for any initial data the transient regime converges to the stationary regime.) From the mathematical point of view this means that equation (1.12) has a unique solution in the class of functions uniformly bounded on $I = (-\infty,\infty)$ and every solution of equation (1.20) tends as $t \to +\infty$ to the uniformly bounded solution of equation (1.12) uniformly or in some other sense.

(2) The same question for almost periodic $J(t)$ or $E(t)$.

(3) Let $B(I)$ be the space of measurable functions bounded on I with the norm $|u| = \sup_{t\in I}|u(t)|$. $E \in B(I)$ can for example be a sequence of random impulses. Does a solution of equation (1.12) exist in $B(I)$? Is it unique?

<u>Remark 1</u>. In applications to network theory condition (1.8) is not restrictive because in practice the nonlinear characteristic is considered to be bounded at infinity so condition (1.8) is satisfied even for $\varepsilon = 0$.

Remark 2. We consider equation (1.1) in Hilbert space but our results, assumptions (1)-(6), and proofs are valid for operators $B: X \to X^*$ acting from a Banach space X into its conjugate X^*, provided that X is reflexive, i.e., $X^{**} = X$.

Theorem 1. Let conditions (1) and (2) be satisfied. Then equation (1.1) has a unique solution in H. If, in addition, condition (3) holds then the map B^{-1} is continuous on H.

Theorem 2. Let conditions (4), (5), and (6) be satisfied. Then for any $u_o \in H$ and sufficiently large $\lambda > 0$ (see formula (2.4) below) the sequence

$$u_{n+1} = \lambda R_\lambda u_n - R_\lambda F u_n + R_\lambda J, \quad u_o \in H \tag{1.21}$$

converges in H_A to a solution of equation (1.1) no more slowly than a convergent geometric series. The solution of equation (1.1) is unique in H and the map $B^{-1}: H \to H_A$ is continuous.

Remark 3. Surjectivity in Theorem 2 is known from monotonicity theory and uniqueness is obvious. Conditions (5) and (6) are used for constructing the solution. The map T^{-1} from H to $D(A)$ equipped with the norm $(\|u\|^2 + \|Au\|^2)^{\frac{1}{2}}$ is continuous.

2. In the following theorem about equation (1.12) we assume all the functions are real-valued, $Fu = f(t,u(t))$, Z is defined by formula (1.11), and $\dot{E} = dE/dt$.

Theorem 3. Let $|E| < \infty$, $|\dot{E}| < \infty$, $f(t,u)$ be measurable in t, uniformly bounded if $|u| \leq R$, for all $R > 0$, and uniformly continuous in u for $t \in I$. If

$$\sup_{t\in I} \int_I |g(t,\tau)|dt \equiv G < \infty, \tag{1.22}$$

$$|f(t,u)| \leq \varepsilon|u| + c(\varepsilon), \quad \varepsilon > 0, \quad c(\varepsilon) = \text{const}, \tag{1.23}$$

then all uniformly bounded solutions of equation (1.12) are a priori bounded on I. If, moreover, $0 \leq \Delta f(t,u)/\Delta u \leq \mu$ and for sufficiently large numbers ℓ the inequality holds

$$\varepsilon \int_{-\ell}^{\ell} u^2 dt \leq \mu^{-1} \int_{-\ell}^{\ell} u^2 dt + \int_{-\ell}^{\ell} dt u(t) \int_{-\infty}^{t} g(t,\tau)u(\tau)d\tau, \quad \varepsilon > 0, \tag{1.24}$$

where $u(t)$ is an arbitrary bounded measurable function, then equation (1.12) has no more than one solution $U(t)$ uniformly bounded on I. If, moreover, the function $g(t,\tau)$ is continuously differentiable in t for $t \geq \tau$, $\tau \in I$,

$$\sup_{t\in I}\left(|g(t,t)| + \int_{-\infty}^{t} |\dot{g}(t,\tau)|d\tau\right) < \infty, \tag{1.25}$$

then there exists a unique solution $U(t)$ of equation (1.12) which is uniformly bounded on I. If, moreover, the following inequalities hold:

$$\int_0^\infty dt\left(\int_{-\infty}^0 |g(t,\tau)|d\tau\right)^2 < \infty, \quad m(t) \in L^2(0,\infty)), \quad m(t) \to 0 \quad \text{as} \quad t \to +\infty, \quad \sup_{t\geq 0}|m(t)| < \infty, \tag{1.26}$$

$$\sup_{s\geq 0} \int_0^\infty\int_0^\infty |g(t,\tau)g(t,s)|d\tau dt < \infty, \tag{1.27}$$

$$\sup_{t\geq 0} \int_0^\infty |g(t,\tau)|^2 d\tau < \infty, \quad |g(t,\tau)| \underset{|t-\tau|\to\infty}{\to} 0, \quad \int_{-\infty}^0 |g(t,\tau)|d\tau \underset{t\to+\infty}{\to} 0, \tag{1.28}$$

then every solution of equation (1.20) satisfies

$$\lim_{t\to+\infty} |U(t) - u(t)| = 0, \tag{1.29}$$

i.e., the network S is convergent.

Remark 4. Conditions (1.22) and (1.24)-(1.28) hold if, for example, $g(t,\tau) = g(t-\tau)$, $|\dot{g}(t)| + |g(t)| \leq c \exp(-\alpha t)$, $\alpha > 0$, $\mu^{-1} + \text{Re } Z(i\lambda) \geq \varepsilon > 0$ for $\lambda \in I$, where $Z(p)$ is defined by formula (1.19). These assumptions about linear one-ports are usually adopted in the literature.

Remark 5. Theorem 3 remains valid if $g(t,\tau) = R\delta(t-\tau) + g_1(t,\tau)$, where $R = \text{const} > 0$, $\delta(t)$ is the delta function, and $g_1(t,\tau)$ satisfies the assumptions of Theorem 3.

Remark 6. The solution of equation (1.12) is unique in $B(I)$ if

$$0 < \mu^{-1} \int_{-\ell}^{\ell} u^2 dt + \int_{-\ell}^{\ell} dt u(t) \int_{-\infty}^{t} g(t,\tau)u(\tau)d\tau, \quad \ell > T_0, \tag{1.30}$$

where T_0 is a sufficiently large number and $u(t)$ is an arbitrary bounded measurable function.

Remark 7. The monotonicity of the operator F is not necessary for equation (1.15) to have a solution. If the operator QF is compact in H and Q is bounded in H,

$$(Qu,u) \geq 0, \quad (Fu,u) \geq \gamma\|Fu\| - C(\gamma), \quad \gamma > 0,\ C(\gamma) \geq 0, \tag{1.31}$$

then equation (1.15) has a solution in H.

Condition (1.31) allows arbitrary growth of the nonlinearity at the infinity. If QF is compact in H, Q is a linear bounded operator in H,

$$\|Fu\| \leq A\|u\|^{\alpha} + B, \quad A > 0, \quad B = \text{const}, \quad 0 \leq \alpha < 1 \tag{1.32}$$

then equation (1.15) has a solution in H. It is essential that in Theorems 1 and 2 there is no assumption concerning the compactness of ZF, because this operator is not compact in the problem of finding an almost periodic stationary regime.

3. Consider the equation

$$u(t) = h(t) - \int_0^t g(t-\tau)f(u(t))dt, \quad t \geq 0, \tag{1.33}$$

where $f(u)$ is a piecewise continuous function,

$$uf(u) \geq 0, \quad u \in I, \tag{1.34}$$

and $g(t)$ is the impulse response of a linear passive stable two-port so that the following conditions hold (see Kontorovich [1, p. 219]): $Z(p)$ is analytic for $\sigma \geq 0$, and

$$\sup_{\sigma \geq 0} \int_{\sigma - i\infty}^{\sigma + i\infty} |Z(\sigma + i\lambda)|^2 d\lambda < \infty,$$
$$\operatorname{Re} Z(\sigma + i\lambda) \geq \frac{\sigma}{\sqrt{\sigma^2 + \lambda^2}} |Z(\sigma + i\lambda)|, \quad \sigma \geq 0. \tag{1.35}$$

Very often the one-port is assumed to be exponentially stable, so that conditions (1.35) hold for $\sigma > -\gamma$, $\gamma > 0$. In particular,

$$\operatorname{Re} Z(\sigma + i\lambda) \geq \frac{\sigma + \gamma}{\sqrt{(\sigma + \gamma)^2 + \lambda^2}} |Z(\sigma + i\lambda)|, \quad \sigma \geq -\gamma.$$

Hence

$$\operatorname{Re} Z(i\lambda) \geq \frac{\gamma}{\sqrt{\gamma^2 + \lambda^2}} |Z(i\lambda)|. \tag{1.36}$$

The impedance $Z(i\lambda)$ satisfies (Kontorovich [1], p. 251)

$$-\lim_{\lambda \to +\infty} \lambda \operatorname{Im} Z(i\lambda) = \frac{2}{\pi} \int_0^\infty \operatorname{Re} Z(i\lambda) d\lambda. \tag{1.37}$$

Let

$$\tilde{u}(\lambda) = \int_0^\infty \exp(-i\lambda t)u(t)dt, \quad u(t) = 0 \quad \text{for} \quad t < 0, \tag{1.38}$$

$$N_1(u) \equiv \left\{\gamma \int_{-\infty}^{\infty} \frac{|\tilde{u}(\lambda)|^2}{|Z(i\lambda)|\sqrt{\lambda^2+\gamma^2}} d\lambda\right\}^{1/2}, \tag{1.39}$$

$$N_2(h) = \left\{\gamma^{-1} \int_{-\infty}^{\infty} \frac{|\tilde{h}|^2 \sqrt{\gamma^2+\lambda^2}}{|Z(i\lambda)|} d\lambda\right\}^{1/2} + N_1(h). \tag{1.40}$$

Theorem 4. Let conditions (1.34) and (1.36) hold and assume $N_2(h) < \infty$. Then the solution of equation (1.33) satisfies the inequality

$$N_1(u) \leq N_2(h). \tag{1.41}$$

Remark 8. If $0 < C_1 \leq \sqrt{\lambda^2 + \gamma^2}|Z(i\lambda)| \leq C_2 < \infty$ then the norm N_1 defined by (1.39) is equivalent to the norm of $L^2([0,\infty))$, while the norm N_2 defined by (1.40) is equivalent to the norm of $W_2^1([0,\infty))$.

Remark 9 Denote $\exp(at)h(t)$ by $h_2(t)$. If $a < \gamma$ and $N_2(h_a) < \infty$ then inequality (1.41) is valid provided that the substitutions $u \to u_a$, $h \to h_a$, $\gamma \to \gamma-a$, $i\lambda \to i\lambda-a$ have been done. Inequality (1.41) means that solutions of equation (1.33) are stable under small (in the norm N_2) perturbations of the function $h(t)$.

Remark 10. In the literature on stability in the large and absolute stability (see Lefschetz [1]), the stability of the system

$$\dot{x} = Px + q\phi(\sigma) + f(t), \quad \sigma = r\cdot x,$$

is studied, where $x \in \mathbb{R}^n$, P is a Hurwitz matrix, and $\phi(\sigma)$ is a nonlinear function. This problem can be easily reduced to the study of solutions of equation (1.33). But even if it is possible (for example, if $Z(p)$ is rational), reduction of integral equation (1.33) to the system of differential equations is rarely advisable.

4. The following theorem is useful for the study of stability under continuously acting perturbations.

Theorem 5. *Let* A *be a continuous compact map of a Banach space* X *into itself*, $T = I + A$, *where* I *is the identity operator in* X. *If* T *is injective and* T^{-1} *is bounded then* T *is a homeomorphism of* X *onto* X.

5. The following theorems are useful in numerical analysis.

Theorem 6. *Let* F *be a Fréchet differentiable operator on a Hilbert space* H, *and assume* B *is linear injective operator*, $D(B) \supset R(F'(u))$, $D(B) \supset R(F(u))$, $\mathrm{Re}BF'(u) \geq a > 0$, $\|BF'(u)\| \leq b$ *for all* $u \in H$, $b > a$, $\gamma = ab^{-2}$, *and* $q = (1 - a^2b^{-2})^{1/2}$. *Then the iterative process* $u_{n+1} = u_n - \gamma BFu_n$, *with* $u_o \in H$ *arbitrary*, *converges to the unique solution of the equation* $Fu = 0$ *at the rate* $\|u-u_n\| = O(q^n)$.

Theorem 7. *Let* $Au + Fu = 0$, $A \geq d > 0$, $0 \leq F'(u) \leq M$ *for all* $u \in H$, *and* $B = A^{-1/2}$. *Then the iterative process* $v_{n+1} = v_n - b^{-2}BF_1(v_n)$, $v_0 \in H$, *converges to the unique solution of the equation* $F_1(v) = 0$, $F_1(v) \equiv A^{1/2}v + F(Bv)$, $b = 1 + Md^{-1}$, *and* $u = Bv$ *is the unique solution of equation* $Au + Fu = 0$. *If* $u_n = Bv_n$ *then* $\|u - u_n\| = O(b^{-2n})$.

2. Existence, Uniqueness and Stability of Solutions of Some Nonlinear Operator Equations and an Iterative Process to Solve the Equations

Here we prove Theorems 1, 2, and 5 - 7.

Proof of Theorem 1: Consider the sequence of equations

$$B_n u_n = A_n u_n + Fu_n = J, \quad n = 1,2,\dots \,. \tag{2.1}$$

From the main theorem of monotone operator theory (see, for example, Lions [1], p. 182) and conditions (1.4) and (1.5) it follows that equation (2.1) has a solution for any $n = 1,2,\dots$. From (2.1) and (1.5) we get

$$\begin{aligned} \gamma(\|u_n\|)\|u_n\| &\le \mathrm{Re}[(A_n u_n, u_n) + (Fu_n, u_n)] \\ &= \mathrm{Re}(J, u_n) \le \|J\|\,\|u_n\| \,. \end{aligned} \tag{2.2}$$

Since $\gamma(t) \to +\infty$ as $t \to \infty$ we conclude that $\|u_n\| \le C = C(\|J\|)$. We denote by C various constant which do not depend on n. The operator F is bounded, hence $\|Fu_n\| \le C$. From here and (2.1) we obtain $\|A_n u_n\| \le C$. Since H is weakly compact (Kato [1]), there exist weakly convergent subsequences of $\{u_n\}$, $\{Fu_n\}$, $\{A_n u_n\}$. We also denote the subsequence by u_n. Then $u_n \rightharpoonup u$, $Fu_n \rightharpoonup v$, $A_n u_n \rightharpoonup w$. Let us prove that $u \in D(A)$, $Au + Fu = J$. We have

$$(u,A^*y) = (u-u_n,A^*y) + (u_n,A^*y-A_n^*y) + (A_n u_n,y), \quad \text{for all } y \in D(A^*) \tag{2.3}$$

From here, (1.3), and taking into account that $u_n \rightharpoonup u$ we find that $(u,A^*y) = (w,y)$, for all $y \in D(A^*)$. Hence $u \in D(A)$, $Au = w$. From (1.4) it follows that $0 \le \mathrm{Re}(x-u_n, B_n x-B_n u_n)$, for all $x \in D(A)$. From here, (2.1), and (1.3) we obtain as $n \to \infty$ the inequality $0 \leq \text{Re}(x-u, Bx-J)$, for all $x \in D(A)$. As F is hemicontinuous and A is linear, $B = A+F$ is hemicontinuous. We set $x - u = \lambda y \in D(A)$, where λ is a number. Then $0 \leq \text{Re}(y, B(u + \lambda y)-J)$, for all $y \in D(A)$. Since B is hemicontinuous we let $\lambda \to 0$ and obtain $0 \leq \text{Re}(y, Bu - J)$, for all $y \in D(A)$. Hence $Bu = J$. Condition (1.2) guarantees the uniqueness of the solution to equation (1.1). It remains to prove that B^{-1} is continuous provided (1.6) holds. Let $Bu_n = J_n$, $Bu = J$, and $J_n \to J$. From (1.6) we obtain

$$\nu_R(\|u_n - u_m\|)\,\|u_n - u_m\| \leq \text{Re}(Bu_n - Bu_m, u_n - u_m)$$

$$= \text{Re}(J_n - J_m, u_n - u_m) \leq \|J_n - J_m\|\,\|u_n - u_m\| .$$

Hence $\|u_n - u_m\| \to 0$ as $n,m \to \infty$. We set $u_o = \lim u_n$. We prove that B is closed below. Hence $u_o \in D(B)$, $Bu_o = J$. Since the solution of equation (1.1) is unique we conclude that $u_o = u$. Hence $J_n \to J$ implies $u_n \to u$, i.e., B^{-1} is continuous. Now we prove that B is closed. Suppose $Bu_n = Au_n + Fu_n \to J$, $u_n \to u$. Then $\|Fu_n\| \leq C$. Passing to a subsequence we can assume that $Fu_n \rightharpoonup v$, $Au_n \rightharpoonup w$. We have $(Au_n,y) = (u_n,A^*y)$, for all $y \in D(A^*)$. If $n \to \infty$ we get $(w,y) = (u,A^*y)$ for all $y \in D(A) = D(B)$. Hence $u \in D(B)$, $Au = w$. As above, we prove that $Bu = J$. □

<u>Proof of Theorem 2</u>: Let $\|u_n\| = a_n$, $u_n - u_{n-1} = \eta_n$, $\|\eta_n\| = b_n$, $Fu_n - Fu_{n-1} = \psi_n$, $(A+\lambda I)\eta_n = y_n$, $\|y_n\| = h_n$. From equation (1.21) we obtain

$$a_{n+1} \leq \lambda\|R_\lambda\|\, a_n + \|R_\lambda\|(\varepsilon a_n + c(\varepsilon)) + \|R_\lambda\|\,\|J\| .$$

From (1.7) it follows that $\|R_\lambda\| \leq (\lambda+\delta)^{-1}$. Hence $a_{n+1} \leq \gamma a_n + C$, where $0 < \gamma = (\lambda+\varepsilon)/(\lambda+\delta) < 1$ if $0 < \varepsilon < \delta$. Iterating the inequality $a_{n+1} \leq \gamma a_n + C$, we conclude that $\sup a_n \leq R$, where $R = \text{const}$. Let us prove that the sequence u_n converges. From (1.21) we obtain

$$\eta_{n+1} = \lambda R_\lambda \eta_n - R_\lambda \eta_n, \quad y_{n+1} = \lambda\eta_n - \psi_n.$$

Hence

$$\|y_{n+1}\|^2 = \lambda^2\|\eta_n\|^2 + \|\psi_n\|^2 - 2\lambda \text{Re}(\eta_n,\psi_n) \leq \lambda^2\|\eta_n\|^2 + \|\psi_n\|^2.$$

Here we took into account that F is monotone and thus $\text{Re}(\eta_n,\psi_n) = \text{Re}(u_n - u_{n-1}, Fu_n - Fu_{n-1}) \geq 0$. From (1.10) it follows that $\|\psi_n\| \leq C(R)b_n$. So $h_{n+1}^2 \leq b_n^2(\lambda^2 + C^2(R))$. But we have $\eta_n = R_\lambda y_n$, $\|\eta_n\| \leq (\lambda+\delta)^{-1}\|y_n\|$. Hence $h_{n+1}^2 \leq q^2 h_n^2$, $q = \{\lambda^2 + C^2(R)\}^{1/2}/(\lambda+\delta)$, $q = q_{min}$ if $\lambda = C^2(R)\delta^{-1}$, and

$$q_{min} = C(R)(C^2(R) + \delta^2)^{-1/2} < 1. \tag{2.4}$$

Iterating the inequality $h_{n+1} \leq qh_n$ we see that $h_n = O(q^n)$, $\|(A+\lambda)\eta_n\| = O(q^n)$, $\|\eta_n\| \leq (\lambda+\delta)^{-1}O(q^n)$. Since $\eta_n = u_n - u_{n-1}$ we conclude that the sequence u_n converges in H no more slowly than a convergent geometric series. If $u = \lim u_n$ then passing to the limit in (1.21), we obtain $u = \lambda R_\lambda u - R_\lambda Fu + R_\lambda J$. Hence u is a solution of equation (1.1). The uniqueness of this solution follows from (1.7) and (1.9). Indeed, if $Bu = J$ and $Bv = J$ then $0 = \text{Re}(Bu - Bv, u - v) \geq \delta\|u-v\|^2$. Hence $u = v$. It remains to prove that $B^{-1}: H \to H_A$ is continuous. Let $Bu_n = J_n$. We have

$$\delta\|u_n-u_m\|^2 \le \mathrm{Re}(Au_n-Au_m,\ u_n-u_m) \le \mathrm{Re}(Bu_n-Bu_m,\ u_n-u_m)$$

$$= \mathrm{Re}(J_n-J_m,\ u_n-u_m) \le \|J_n-J_m\|\,\|u_n-u_m\| .$$

Thus $\|u_n-u_m\| \le \delta^{-1}\|Bu_n-Bu_m\|$, $\|u_n-u_m\|_{H_A} \le \delta^{-1/2}\|Bu_n-Bu_m\|$, where $\|u\|_A^2 = \mathrm{Re}(Au,u)$. □

Proof of Theorem 5: First we prove that T^{-1} is continuous on $R(T)$. Let $Tu_n = f_n$, $f_n \to f \in R(T)$. Since T^{-1} is bounded we have $\|u_n\| \le C$. Since A is compact, $Au_{n_k} \to v$ for some subsequence $\{u_{n_k}\}$ of $\{u_n\}$. Hence $u_{n_k} = f_{n_k} - Au_{n_k} \to u$. Passing to the limit in the equality $u_{n_k} + Au_{n_k} = f_{n_k}$ we obtain $u + Au = f$. Since T is injective, $u = T^{-1}f$. Thus any subsequence converges to the same limit $T^{-1}f$. This means that $u_n \to T^{-1}f$. Hence $R(T)$ is closed in X and T^{-1} is continuous on $R(T)$. If A is compact, $T = I+A$, and T^{-1} is continuous on $R(T)$, then T is an open map, i.e., T maps open sets into open sets (see Krasnoselskij-Zabreiko [1, p. 161]). Hence $R(T)$ is open, since it is an image of the open set X under the open mapping T. Therefore $R(T)$ is closed and open in X, implying $R(T) = X$. So T is a homeomorphism of X onto X. □

Proof of Remark 7: First we prove the second part of this remark.

If condition (1.32) holds, $R > 0$ is sufficiently large, and $\|u\| = R$, then $(u,u) \ge (E-QFu,u)$ for $\|u\| = R$. From here it follows that equation (1.15) has a solution. Here we used the following well known theorem (Krasnoselskij-Zabreiko [1], p. 339): If T is compact in H, $(Tu,u) < \|u\|^2$ for $\|u\| = R$ then equation $u = Tu$ has a solution in the ball $||u|| \leq R$. In our case $T = E\text{-}QF$.

To prove the first proposition of Remark 6 we consider the equation $u + \lambda QFu = E$, $0 \leq \lambda \leq 1$. From here we obtain $(u,Fu) + \lambda(QFu,Fu) = (E,Fu)$. Making use of (1.31) we have $\gamma||Fu|| \leq C(\gamma) + ||E|| \, ||Fu||$. Taking $\gamma > ||E||$, we obtain $||Fu|| \leq C$. Therefore $||u|| \leq ||E|| + ||QFu|| \leq C$, since Q is bounded. By the Leray-Schauder principle (Krasnoselskij-Zabreiko [1, p. 298]) we conclude that the equation $u + QFu = E$ has a solution. □

Proof of Theorem 6: Let $Q = BF'(V)$, $\mathrm{Re}\, Q \geq a$, $||Q|| \leq b$, $v \in H$ be arbitrary. We note that $||I-\alpha Q||^2 = ||I-2\alpha \mathrm{Re}\, Q + \alpha^2 QQ^*|| \leq 1 - 2\alpha a + b^2\alpha^2 = k^2(\alpha)$, $k^2_{min} = k^2(\gamma) \equiv q^2 = 1-a^2b^{-2}$, $\gamma = ab^{-2}$. The equation $F(u) = 0$ is equivalent to $u = u - \gamma BF(u) \equiv Tu$, and $||Tu - Tv|| \leq \sup_w ||I-\gamma BF'(w)|| \, ||u-v|| \leq q||u-v||$. Hence from the contraction mapping principle we obtain the statement of Theorem 6.

Proof of Theorem 7: The equation $Au + Fu = 0$ is equivalent to $F_1(v) = 0$, and $u = Bv$. Since $BF_1'(w)h = h+BF'(Bw)Bh$, for all $w \in H$, we obtain $BF_1'(w) \geq 1$, $||BF_1'(w)|| \leq 1 + Md^{-1} \equiv b$. Hence from Theorem 6, it follows that the iterative process $v_{n+1} = v_n - b^{-2}BF_1(v_n)$, with arbitrary $v_o \in H$, converges to the unique solution of the equation $F_1(v) = 0$ at the rate $||v - v_n|| = O(b^{-2n})$. Therefore $u = Bv$ is the solution of the equation $Au + Fu = 0$, the sequence $u_n = Bv_n$ converges to this solution u, $u = Bv$, at the rate $||u - u_n|| \leq d^{-1/2}||v_n - v|| \leq d^{-1/2}O(b^{-2n})$.

Remark 1. The iterative process given in Theorem 7 can be used for the calculation of stationary regimes in nonlinear networks (see Section 3).

Remark 2. In Section 1 it is stated that if A is a generator of a strongly continuous semigroup then a sequence A_n exists with property (1.3). One example of such a sequence is known as the Yosida approximation: $A_n = -\lambda_n I - \lambda_n^2(A-\lambda_n I)^{-1}$, where the λ_n are real and $\lambda_n \to +\infty$ (see S. G. Krein [1], p. 66). These operators A_n commute.

Remark 3. The results of Theorems 1 and 2 are close to final from the point of view of network theory. Indeed, consider a consecutively connected e.m.f. $e(t) = e(t+T)$, a nonlinearity with a voltage-current characteristic $f(u) > 0$, $f(u) > 0$, $|f(u)| < c_1$, for all $u \in R^1$ and a capacitor with capacitance C. If u is the voltage on the capacitor then $f(e - u) = Cdu/dt$. This equation has no T-periodic solutions since $C \int_0^T (du/dt)dt = 0$ if $u(0) = u(T)$, while $\int_0^T f(e - u)dt > 0$ because $f(u) > 0$, for all u. In this example all the conditions of Theorem 2 hold except $\delta > 0$ in formula (1.7) and all the conditions of Theorem 1 hold except coercivity: in formula (1.5), $\gamma(t)$ does not go to infinity as $t \to +\infty$. From the point of view of network theory, a capacitor is a conservative but not a passive one-port.

3. Existence, Uniqueness, and Stability of the Stationary Regimes in Some Nonlinear Networks. Stability in the Large and Convergence in the Nonlinear Networks

Here we prove Theorems 3 and 4.

1. Proof of Theorem 3: From conditions (1.22), (1.23), and equation (1.12) we obtain $|u| < \varepsilon G|u| + Gc(\varepsilon) + |E|$. Taking $\varepsilon G = \frac{1}{2}$ we have

$$|u| \leq 2(|E| + cG). \tag{3.1}$$

Hence all uniformly bounded solutions of equation (1.12) are a priori bounded on $I = (-\infty,\infty)$. Let us show that condition (1.24) implies the uniqueness of the uniformly bounded solution of equation (1.12). Suppose u, v are such solutions, and set $w = u - v$ and $\Phi = Fu - Fv$. Then

$$w + \int_{-\infty}^{t} g(t,\tau)\Phi(\tau)dt = 0.$$

Multiplying this equation by $\Phi(t)$ and integrating over $(-\ell,\ell)$, making use of inequality (1.24) and the inequality $\mu^{-1}\Phi^2 \leq \Phi w$, we obtain

$$0 \geq \mu^{-1}\int_{-\ell}^{\ell}\Phi^2 dt + \int_{-\ell}^{\ell} dt\Phi(t)\int_{-\infty}^{t} g(t,\tau)\Phi(\tau)d\tau \geq \varepsilon\int_{-\ell}^{\ell}\Phi^2 dt. \tag{3.2}$$

Hence $\Phi \equiv 0$, $u \equiv v$. We have proved the uniqueness of the uniformly bounded solution to equation (1.12). Actually Remark 5 has been also proved. Now we prove the existence of such a solution. Let

$$u_n(t) + \int_{-n}^{t} g(t,\tau)f(\tau,u_n(\tau))dt = E(t), \quad t \geq -n. \tag{3.3}$$

For any $n = 1,2,\ldots$ this equation has a unique solution on $(-n,\infty)$. This is so because (3.3) is a Volterra equation with sublinear nonlinearity: $|f(t,u)| \leq A + B|u|$, A, B = const, and (1.22) holds. For any n, an inequality similar to (3.1) holds, so the set $\{u_n(t)\}$ is bounded uniformly in n. Using (1.25) and differentiating (3.3), we obtain

$$\dot{u}_n + g(t,t)f(t,u_n(t)) + \int_{-n}^{t} \dot{g}(t,\tau)f(\tau,u_n(\tau))d\tau = \dot{E}. \tag{3.4}$$

Hence $|\dot{u}_n| \leq C$. So we proved that

$$|u_n| + |\dot{u}_n| \leq C,$$

where C does not depend on n. Therefore we can choose a subsequence (denoted again by u_n), which converges uniformly on any finite interval to a uniformly bounded function $U(t)$. We pass to the limit as $n \to \infty$ in equation (3.3). Using Lebesgue's theorem we obtain

$$U(t) + \int_{-\infty}^{t} g(t,\tau)f(\tau,U(t))dt = E(t), \quad t \in I. \tag{3.5}$$

We have already proved that equation (3.5) has only one uniformly bounded solution. Hence any subsequence of the sequence u_n tends to the same limit $U(t)$. Therefore $u_n(t) \to U(t)$ as $n \to \infty$, $t \in I$. It remains to prove (1.29). We have

$$\begin{cases} U(t) = E(t) + n(t) - \int_0^t g(t,\tau)f(\tau,U(\tau))d\tau, \\ n(t) \equiv -\int_{-\infty}^{0} g(t,\tau)f(\tau,U(\tau))d\tau. \end{cases} \tag{3.6}$$

Let us denote

$$v = U-u, \quad q = n(t) - m(t),$$
$$\psi(t) = \begin{cases} f(t,U(t)) - f(t,u(t)), & t \geq 0 \\ 0, & t < 0, \end{cases} \tag{3.7}$$

where $u(t)$ is the solution of equation (1.20) and $m(t)$

is the function appearing in that equation. It is clear that

$$v(t) = q(t) - \int_0^t g(t,\tau)\psi(\tau)d\tau. \tag{3.8}$$

Multiplying (3.8) by ψ, integrating over $(0,\infty)$, and using (1.24), we obtain

$$\begin{aligned} \varepsilon\int_0^\infty \psi^2 dt &\le \mu^{-1} \int_0^\infty \psi^2 dt + \int_0^\infty dt\psi(t) \int_0^t g(t,\tau)\psi(\tau)d\tau \\ &\le \left(\int_0^\infty q^2 dt\right)^{1/2}\left(\int_0^\infty \psi^2 dt\right)^{1/2}. \end{aligned} \tag{3.9}$$

Hence

$$\int_0^\infty \psi^2 dt \le C. \tag{3.10}$$

From (3.8), (1.27), (1.26), and (3.10) it follows that

$$\begin{aligned} \int_0^\infty v^2 dt &\le 2\left\{\int_0^\infty q^2 dt + \int_0^\infty\left(\int_0^t g(t,\tau)\psi(\tau)d\tau\right)^2 dt\right\} \\ &\le 4\int_0^\infty m^2 dt + 4\int_0^\infty n^2 dt + C \int_0^\infty \psi^2 dt \le C. \end{aligned} \tag{3.11}$$

Here we took into account that condition (1.27) implies the boundedness in $L^2(I_+)$, $I_+ = [0,\infty)$, of the integral operator $\int_0^t g(t,s)\psi(s)ds$. Indeed,

$$\begin{aligned} &\int_0^\infty dt \int_0^t\int_0^t g(t,s)g(t,\tau)\psi(s)\psi(t)dsd\tau \\ &\le \int_0^\infty dt \int_0^t\int_0^t |g(t,s)g(t,\tau)|\frac{\psi^2(s)+\psi^2(\tau)}{2} dsdt \\ &\le \int_0^\infty ds\psi^2(s) \int_0^\infty\int_0^\infty |g(t,s)g(t,\tau)|dtd\tau \\ &\le C \int_0^\infty ds\psi^2(s). \end{aligned}$$

We denote by C various constants. From (3.8) and (1.28) it follows that

$$v(t) \to 0 \quad \text{as} \quad t \to +\infty. \tag{3.12}$$

Indeed both terms in the right-hand side of (3.8) tend to zero as $t \to +\infty$. Let us prove this. We have

$$\begin{aligned}|q(t)| &\leq \int_{-\infty}^{0} |g(t,\tau)||f(\tau,U(\tau))|d\tau + |m(t)| \\ &\leq C\int_{-\infty}^{0} |g(t,\tau)|dt + |m(t)| \to 0 \quad \text{as} \quad t \to +\infty.\end{aligned} \tag{3.13}$$

Here we used conditions (1.26) and (1.28) and took into account that $|f(t,U(t))| \leq C$ since $|U(t)| \leq C$. Furthermore, we have

$$\begin{aligned}\left|\int_0^t g(t,\tau)\psi dt\right| &\leq \int_0^N |g(t,\tau)||\psi|dt \\ &+ \left(\int_N^t |g(t,\tau)|^2 dt\right)^{1/2}\left(\int_N^t \psi^2 dt\right)^{1/2}.\end{aligned} \tag{3.14}$$

We note that all the solutions of equation (1.20) are uniformly bounded on I_+ provided that $|E| + |m| < C$ and conditions (1.22) and (1.23) hold. This can be proved as inequality (3.1) was proved. Hence $|\psi(t)| \leq C$. Let us take N so large that the second term in the right-hand side of formula (3.14) will be less than $\alpha > 0$, where α is an arbitrary small given number. This is possible because of (3.10) and (1.22). Then we fix N and take t so large that the first term in the right-hand side of formula (3.14) will be less than α. This is possible because of condition (1.28) and the inequality $|\psi| \leq C$. Since $\alpha > 0$ is arbitrary small, we prove that

$$\left|\int_0^t g(t,\tau)\psi dt\right| \to 0 \quad \text{as} \quad t \to +\infty. \tag{3.15}$$

Formula (3.12) follows from (3.13) and (3.15). □

<u>Proof of Remark 3</u>: If $g(t,\tau) = g(t-\tau)$, $|g(t)| + |\dot{g}(t)| \leq c\exp(-\alpha t)$, and $\alpha > 0$, then conditions (1.25)-(1.28) are

certainly satisfied. Let $\tilde{u}(\lambda)$ be defined by formula (1.38),

$$u_\ell(t) \equiv \begin{cases} 0, & |t| > \ell, \\ u(t), & |t| \leq \ell. \end{cases}$$

Using Parseval's equality the right-hand side of inequality (1.24) can be written as

$$J_1 + J_2 \equiv \frac{1}{2\pi}\int_{-\infty}^{\infty} [\mathrm{Re}\ Z(i\lambda) + \mu^{-1}]|\tilde{u}_\ell|^2 d\lambda + \int_{-\ell}^{\ell} dt u(t) \int_{-\infty}^{-\ell} g(t,\tau)u(\tau)d\tau. \tag{3.16}$$

If $\mathrm{Re}\ Z(i\lambda) + \mu^{-1} \geq \varepsilon > 0$ then $J_1 \geq \varepsilon\int_{-\ell}^{\ell}|u|^2dt$. Let us prove that

$$|J_2| \leq \delta \int_{-\ell}^{\ell}|u|^2 dt, \tag{3.17}$$

where $\delta > 0$ is arbitrary small if ℓ is large enough. This completes the proof of Remark 3. If $u \in L^2(I)$ then $J_2 \to 0$ as $\ell \to \infty$. Indeed,

$$\begin{aligned} |J_2| &\leq \int_{-\ell}^{\ell} dt|u(t)| \int_{-\infty}^{-\ell} \exp\{-\alpha(t-\tau)\}|u|d\tau \\ &\leq \left(\int_{-\ell}^{\ell} dt|u(t)|\exp(-\alpha t)\right)\left(\int_{-\infty}^{-\ell} u^2 dt\right)^{1/2} \frac{\exp(-\alpha\ell)}{\sqrt{2\alpha}} \\ &\leq \left(\int_{-\ell}^{\ell} u^2 dt\right)^{1/2} \frac{\exp(\alpha\ell)}{\sqrt{2\alpha}} \left(\int_{-\infty}^{-\ell} u^2 dt\right)^{1/2} \frac{\exp(-\alpha\ell)}{\sqrt{2\alpha}} \to 0 \\ &\qquad \text{as } \ell \to \infty. \end{aligned}$$

If $u \notin L^2(I)$ then $\int_{-\ell}^{\ell} u^2 dt \to \infty$ as $\ell \to \infty$. Next we prove (3.17) using inequality $|u| \leq C$. We have

$$\begin{aligned} |J_2| &\leq C\int_{-\ell}^{\ell} dt|u(t)|\exp(-\alpha t)\alpha^{-1} \exp(-\alpha\ell) \\ &\leq C\alpha^{-1}\exp(-\alpha\ell)\left(\int_{-\ell}^{\ell} u^2 dt\right)^{1/2} \frac{\exp(\alpha\ell)}{\sqrt{2\alpha}} \\ &\leq C_1\left(\int_{-\ell}^{\ell} u^2 dt\right)^{1/2} = o\left(\int_{-\ell}^{\ell} u^2 dt\right) \text{ as } \ell \to \infty. \end{aligned} \tag{3.18}$$

Inequality (3.18) is more than we need. □

Proof of Remark 4: Let us write equations (1.12) and (1.20) in the form

$$u + Rf(t,u) + K_1Fu = E(t) \tag{3.19}$$

$$u + Rf(t,u) = E(t) + m_1(t) - \int_0^t g_1(t,\tau)f(\tau,u(\tau))d\tau. \tag{3.20}$$

We set $v = u + Rf(t,u)$ and note that the function $u = u(t,v)$ inverse to $v = v(t,u)$ increases monotonically.

We rewrite equation (3.19) as

$$v + K_1F_1v = E(t), \tag{3.21}$$

where

$$F_1v = f(t,u(t,v)) \equiv f_1(t,v), \tag{3.22}$$

and equation (3.20) as

$$v = E(t) + m_1(t) - \int_0^t g_1(t,\tau)f_1(\tau,v(\tau))d\tau. \tag{3.23}$$

Since

$$0 \leq \frac{\Delta f}{\Delta u} \leq \mu, \quad 1 \leq \frac{\Delta v}{\Delta u} \leq 1 + R\mu$$

we have

$$1 \geq \frac{\Delta u}{\Delta v} \geq (1 + R\mu)^{-1}, \quad 0 \leq \frac{\Delta f_1}{\Delta v} = \frac{\Delta f_1}{\Delta u}\frac{\Delta u}{\Delta v} \leq \mu.$$

Now we prove an inequality similar to (1.23) for the function $f_1(t,v)$. We have

$$|f_1(t,v)| \leq \varepsilon|u(t,v)| + c(\varepsilon);$$

$$|u| - R\varepsilon|u| - RC(\varepsilon) \leq |v| \leq |u| + R\varepsilon|u| + Rc(\varepsilon).$$

Hence

$$|u(t,v)| \leq (1 - R\varepsilon)^{-1}|v| + C_1(\varepsilon).$$

Therefore

$$|f_1(t,v)| \leq \varepsilon|v| + C(\varepsilon) \tag{3.24}$$

if $\varepsilon > 0$ is small enough. Equations (3.21) and (3.23) are similar to (1.12) and (1.20), respectively, so we can apply Theorem 3 to these equations. □

Proof of Theorem 4: Denote $\tilde{\psi} = f(u(t))$. From (1.31) we obtain

$$\tilde{u} = \tilde{h} - Z(i\lambda)\tilde{\psi}. \tag{3.25}$$

Multiplying (3.25) by $\tilde{\psi}^*$ and integrating over I, we obtain

$$\int_I \tilde{u}\tilde{\psi}^* d\lambda = \int_I \tilde{h}\tilde{\psi}^* d\lambda - \int_{-\infty}^{\infty} Z(i\lambda)|\tilde{\psi}|^2 d\lambda. \tag{3.26}$$

Using Parseval's equality and condition (1.34), we obtain

$$\int_I u\tilde{\psi}^* d\lambda = 2\pi \int_I uf(u)dt \geq 0. \tag{3.27}$$

From (1.36) and (3.26) it follows that

$$\gamma \int_I \frac{|Z(i\lambda)|}{\sqrt{\gamma^2 + \lambda^2}} |\tilde{\psi}|^2 d\lambda < \left\{ \int_I \frac{|\tilde{h}|^2\sqrt{\gamma^2 + \lambda^2}}{\gamma|Z(i\lambda)|} d\lambda \right\}^{1/2} \left\{ \int_I |\tilde{\psi}|^2 \frac{|Z(i\lambda)|\gamma}{\sqrt{\gamma^2 + \lambda^2}} d\lambda \right\}^{1/2} \tag{3.28}$$

From here we obtain

$$\gamma \int_I \frac{|Z(i\lambda)||\tilde{\psi}|^2 d\lambda}{\sqrt{\gamma^2 + \lambda^2}} \leq \int_I \frac{\sqrt{\gamma^2 + \lambda^2}|\tilde{h}|^2}{\gamma|Z(i\lambda)|} d\lambda. \tag{3.29}$$

Let us multiply the identity $|\tilde{u}|^2 = \tilde{h}\tilde{u}^* - Z(i\lambda)\tilde{\psi}\tilde{u}^*$ by the function $A(\lambda) = \gamma(\gamma^2 + \lambda^2)^{-1/2}|Z^{-1}(i\lambda)|$. We obtain

$$\int_I A(\lambda)|\tilde{u}|^2 d\lambda \le \int_I A(\lambda)|\tilde{h}\tilde{u}|d\lambda + \int_I A(\lambda)|Z(i\lambda)||\tilde{\psi}\tilde{u}|d\lambda$$
$$\le \left(\int_I A(\lambda)|\tilde{u}|^2 d\lambda\right)^{1/2}\left(\int_I A(\lambda)|\tilde{h}|^2 d\lambda\right)^{1/2}$$
$$+ \left(\int_I A(\lambda)|\tilde{u}|^2 d\lambda\right)^{1/2}\left(\int_I A(\lambda)|Z(i\lambda)|^2|\tilde{\psi}|^2 d\lambda\right)^{1/2}.$$

Inequality (1.41) follows from here immediately. □

2. Application to nonlinear networks. Equation (1.16) for a periodic stationary regime is an equation of type (1.1) in the Hilbert space $H = L^2[0,T]$ of periodic functions. By periodic, we mean T-periodic. If the nonlinear characteristic $f(u)$ is monotone, $f(u)u^{-1} \to 0$ as $|u| \to \infty$, $0 \le \Delta f(u)/\Delta u \le \mu$, and if the linear one-port is stable and passive, then conditions (1.7)-(1.10) are valid, with A replaced by A_p from equation (1.16). If the linear one-port is time invariant then the conditions

$$\operatorname{Re} Z^{-1}(i\lambda) \ge \delta > 0, \quad 0 \le \Delta f(u)/\Delta u \le \mu,$$
$$u^{-1}f(u) \to 0 \quad \text{as} \quad |u| \to \infty.$$

are sufficient for conditions (1.7)-(1.10) to be valid. It is easy to explain condition (1.7) from the physical point of view. The linear one-port consumes the amount of power

$$(A_p u,u) = \sum_{n=-\infty}^{\infty} \operatorname{Re} Z^{-1}(inw)|u_n|^2 \ge \delta(u,u),$$

where

$$u_n = T^{-1}\int_0^T \exp(-in\omega)u(t)dt.$$

We note that if $|\lambda Z(i\lambda)| < C$ then, using (1.36), one obtains

$$\operatorname{Re} Z^{-1}(i\lambda) = \operatorname{Re}\frac{Z(i\lambda)}{|Z(i\lambda)|^2} \ge \frac{\gamma}{\sqrt{\lambda^2 + \gamma^2}|Z(i\lambda)|} \ge \delta_1 > 0.$$

Under these assumptions we can apply Theorem 2 and conclude that there exists a unique periodic regime in the network. The regime is stable towards small periodic perturbations of $E(t)$.

Consider the case when $E(t) \in B_2$ is almost periodic and assume that operator F, $Fu = f(t,u(t))$, acts in B_2. For example, this is so if $Fu = f(u)$ and f is uniformly continuous on the set of values of the function $u(t)$. The integral equation for the statonary regime can be written in the form (1.12) or (1.13). It is essential that operator Z in equation (1.12) is not compact in B_2, so that the Leray-Schauder method cannot be applied. Nevertheless Theorems 1 and 2 are applicable. The iterative process (1.21) allows one to calculate the stationary regime. The conditions $0 \leq \Delta f(u)/\Delta u \leq \mu$, $u^{-1}f(u) \to 0$ as $|u| \to \infty$, $f(u)$ is uniformly continuous on I, $\operatorname{Re} Z^{-1}(i\lambda) \geq \delta > 0$, $J \in B_2$ are sufficient for Theorem 2 to be applicable.

Theorem 1 is also applicable under these conditions. Operators A_n from (1.3) can be chosen as

$$A_n u = \sum_{m=-n}^{n} \exp(im\omega t) Z^{-1}(im\omega) u_m$$

for periodic problem, and

$$A_n u = \sum_{m=1}^{n} \exp(i\lambda_m t) Z^{-1}(i\lambda_m) a_m,$$

where

$$a_m = \lim_{\ell \to \infty} (2\ell)^{-1} \int_{-\ell}^{\ell} u(t) \exp(-i\lambda_m t)dt,$$

for almost periodic problem, where $\{\lambda_m\}$ is the spectrum of almost periodic function $u(t)$. If A is normal the sequence

A_n from formula (1.3) does exist. For example if the spectral representation of the normal operator A is $A = \int_{-\infty}^{\infty}\int (x+iy)E(dxdy)$, where $E(\Delta)$ is the spectral measure, then $A_n = \iint_{x^2+y^2\leq n^2} (x+iy)E(dxdy)$ satisfies condition (1.3). If $E(t)$ is uniformly bounded but not periodic or almost periodic, for example if $E(t)$ is a sequence of random impulses, then Theorem 3 is applicable. If the nonlinear system S satisfies the hypotheses of Theorem 3, then there exists a unique forced regime in S which is uniquely determined by $E(t)$ and is stable under small (in the norm of the space $B(I)$) perturbations of $E(t)$. Unfortunately no reflexive Banach space containing all functions uniformly bounded on I is known so we cannot apply Theorem 2 in this case.

3. Discontinuous nonlinearities. If the nonlinear characteristic does not satisfy condition (1.10) and is discontinuous, then the iterative process (1.21) might be divergent. We describe here another iterative process for solving equation (1.1) with a discontinuous nonlinear operator F. Let us assume that $(Au,u) \geq d(u,u)$, $d > 0$, F is bounded in H, and condition (1.9) holds.

We rewrite equation (1.1) in the form

$$Tu \equiv u + A^{-1}Fu - E(t) = 0. \tag{3.30}$$

where the operator T is defined everywhere on H_A and $[Tu - Tv, u - v] = (ATu - ATv, u - v) = [u - v, u - v] + (Fu - Fv, u - v) \geq [u - v, u - v]$. Now we use the following proposition (see Perov-Jurgelas [1]).

Proposition. Suppose that $T: H \to H$ *satisfies the inequality* $(Tu - Tv, u - v) \geq \gamma \|u - v\|^2$. *Then the iterative process*

$$u_{n+1} = u_n - \beta_n h_n, \quad h_n = \frac{Tu_n}{\|Tu_n\|}, \quad \beta_n = \min(\rho_n, \frac{d_n}{2\rho_n}),$$

$$d_{n+1} = (1 - q_n)d_n, \quad q_n = (1 - \beta_n \rho_n d_n^{-1})\beta_n \rho_n^{-1} \tag{3.31}$$

$$d_1 = \frac{\|Tu_1\|}{\gamma}, \quad \rho_n = \frac{\|Tu_n\|}{2\gamma}, \quad n > 1$$

converges in H *to an element* u *at the rate* $\|u - u_n\| \leq d_n^{1/2} = O(n^{-1/2})$, *and* u *is the unique generalized solution of the equation* $Tu = 0$. *An element* u *is called a generalized solution of the equation* $Tu = 0$ *if* $(Tv, v - u) \geq \gamma \|v - u\|^2$, *for all* $v \in H$.

We omit the somewhat technical proof of the proposition and explain how to calculate with formula (3.30). We take an arbitrary $u_1 \in H$, find ρ_1, d_1 then β_1, q_1, d_2, h_1, and u_2. This is the first step of the calculations. Then we find ρ_2, β_2, q_2, h_2, d_3, u_3 and so on. Equation (3.30) can be solved under the assumptions made by means of process (3.31).

4. *Remark 1*. If condition (1.36) holds then $\operatorname{Re}\{(1 + i\lambda q)Z(i\lambda)\} > 0$ for some $q > 0$. Indeed for $|\lambda| < R$, R a fixed arbitrarily large number, the inequality holds if q is small enough. For $|\lambda| > R$, the inequality holds since $\operatorname{Re} Z(i\lambda) > 0$ (see (1.36)) and $\operatorname{Re} i\lambda q Z(i\lambda) = -q\lambda \operatorname{Im} Z(i\lambda) > 0$ (see (1.37)).

4. Stationary Regime in a Nonlinear Feedback Amplifier

Here we consider another problem of practical interest, the network on Figure 1 representing a feedback amplifier. The problem is to find u on the load Z_ℓ, where B is a linear feedback circuit, K is a linear quadripole in the amplifier. Using the theorem about equivalent generator we pass to the equivalent circuit given in Figure 2, where e_e is the equivalent e.m.f., i.e., the voltage on the open terminals a and b, and Z_e is the equivalent impedance. For simplicity we assume that the input impedance of K is infinite, as is often the case in practice. Then $e = ke$, where k is the amplifier coefficient of K by open terminals ab. The impedance Z_e is the impedance of the one-port Z_{ab} with $e = 0$ and without the diode ab. It can be shown that

$$Z_e = Z_\ell'(1 + kk_1) \tag{4.1}$$

where k_1 is the transmission coefficient of quadripole B, k is the coefficient of amplification of K for open terminals ab, $Z_\ell' = Z_\ell Z_B/(Z_\ell + Z_B)$, and Z_B is the impedance of B between its input terminals when its output terminals are open.

To the circuit on Figure 3 we can apply theory developed in Sections 1-3. We assume that the voltage-current characteristic $i = f(u)$ of the diode is monotone, $|f(u) - t(v)| \leq c|u - v|$. The conditions of Theorems 1 and 2 are satisfied if

$$\operatorname{Re} Z_e > 0. \tag{4.2}$$

Hence we have the theorem.

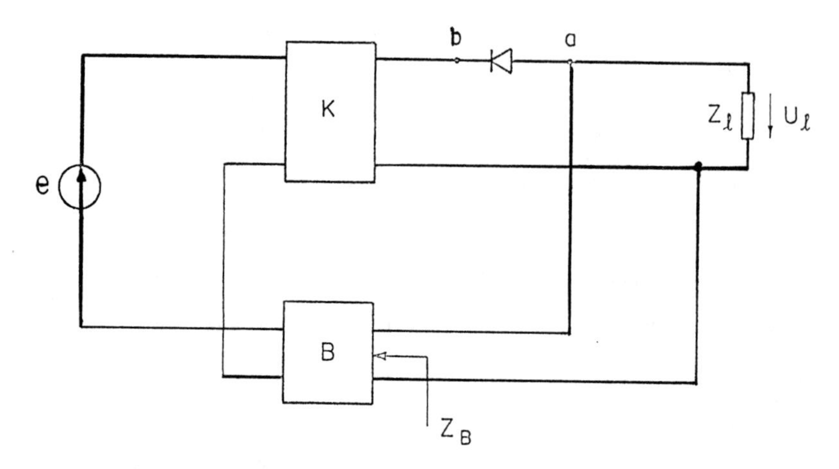

Figure 1

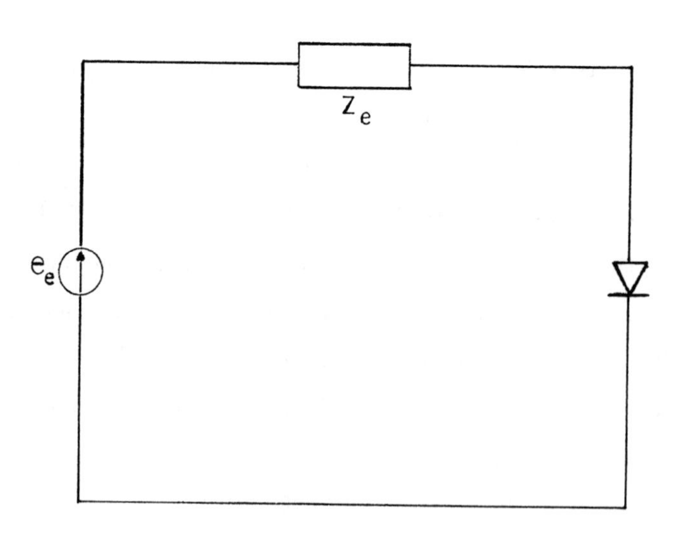

Figure 2

Theorem 8. _Condition_ (4.2) _is sufficient for existence, uniqueness, and stability in_ $H = L^2[0,T]$ _under small perturbations in_ H _of_ e(t) _of the periodic regime in the network of Figure_ 1. _This regime can be calculated by means of an iterative process of type_ (1.21).

Remark 1. Condition (4.2) is easy to verify in practice.

5. Research Problems

1. It would be interesting to generalize Theorem 2 by replacing equation (1.7) by the inequality $(Au,u) > 0$ for $\|u\| > 0$.
2. It would be interesting to find a practically important problem in which the nonlinearity is not small and the linear one-port does not satisfy the filter property, and to solve the problem by means of the process (1.21) and the iterative process given in Theorem 7.
3. In Ramm [95] it is proved that by perturbation by damping imaginary part of modulo maximal eigenfrequencies of a linear system without losses decreases. What can be said about other eigenfrequencies?
4. In Ramm [75] a criterion is given for all solutions of a system of nonlinear differential equations to exist on $(0,\infty)$ in terms of spectral properties of a certain linear operator. It would be interesting to study spectral properties of a differential operator using known results concerning global existence of solutions of simultaneous nonlinear differential equations.

5. Using an idea given in Ramm [75], is it possible to study infinite dimensional problems of type: $\dot{\phi} = F(\phi)$, $\phi(0) = \psi$, where F is a nonlinear operator in X? When is $\phi(t)$ defined on $I_+ = (0,\infty)$? The solution of this problem may be of interest in statistical mechanics, or the dynamics of infinitely many particles.

6. Bibliographical Note

The theory of nonlinear operator equations was intensely studied in the literature (Browder [1], Lions [1], Gajewski et al. [1], Brezis [1], Vainberg M. [1], Ortega-Rheinboldt [1] Petryshyn [1], Krasnoselskij-Zabreiko [1], Krasnoselskij [1], Krasnoselskij et al. [1]). There are still more publications on nonlinear oscillation theory in the literature (Demidovich [1], Aizerman-Gantmaher [1], Cesari [1], Rosenvasser [1], Hsu-Mayer [1], Malkin [1,2], Lefschetz [1], Bogolubov-Mitropolsky [1], Hale [1], Krasnoselskij [1], Krasnoselskij-Burd-Kolesov [1], and many others). In the mechanical and engineering literature nonlinear oscillations were studied mostly either by variants of the small parameter method, averaging method, or under filter hypotheses about the linear part of the network. The specialized literature deals with the absolute stability of automatic and control systems (Lefschetz [1], Hsu-Mayer [1]), for which the Popov frequency method (Popov V. [1]) and some results due to Kalman and Yakubovich are of use (Lefschetz [1]). The results of this chapter are due to the author Ramm [46-53]. From the mathematical point of view, assumption (1.3) is convenient and easy to verify. The proofs of Theorems 1 and 2 are very simple.

The statements of these theorems are convenient in applications as was shown in Sections 2-5. The method and the results of the study of stationary regimes in nonlinear networks given in Sections 2 and 3 are new and in a certain sense the results are final. Specifically, if the network of Figure 2 in Section 4 is not passive there can be more than one periodic regime, etc. Many of the previously published results (Hsu-Mayer [1]) are immediate corollaries or particular cases of the results of Sections 1-3. In Section 4 a problem of importance for practice is studied. The result of point 1, Section 4 (formula (4.1)) is due to G. S. Ramm. All the nonlinear oscillation problems are treated in this chapter in the large, without any assumptions concerning the filter property of the linear part or the smallness of nonlinearity. Equations of type (1.20) were studied under various assumptions by many authors. Londen [1] and Kiffe [1] are recent papers on this subject. The essential part of the proof of Theorem 5 is the theorem about open maps in a Banach space (Krasnoselskij-Zabreiko [1]). In the proof of Theorem 6 we use a variant of a well-known lemma (see Gajewski et al. [1, p. 104]).

For the problems studied in this chapter, the Newton-Kantorovich method (Kantorovich-Akilov [1]) is not convenient because our problems are nonlocal and it is not clear a priori how to choose initial approximation close enough to the desired solution so that Newton-Kantorovich method converges. The first statement of Remark 6 in Section 1 can be found in Brezis-Browder [1] and Ramm [49]. Theorems 3, 4 are proved in Ramm [50]. A summary of the results presented in Sections 1-3 is given in Ramm [101]. A recent monograph Dolezal [1] deals with the mathematics of network theory.

CHAPTER IV

INTEGRAL EQUATIONS ARISING IN THE OPEN SYSTEM THEORY

1. Calculation of the Complex Poles of Green's Function in Scattering and Diffraction Problems

In quantum mechanics, in potential scattering theory, and in diffraction theory it is important to know the complex poles of Green's functions. These poles determine energy losses in open systems, and are called resonances in quantum mechanics (see Baz et al. [1], Lifschitz [1]). Here we give a general method for numerical calculation of these complex poles. The method is described for quantum mechanics scattering problems and for diffraction problems.

1. Consider the Schrödinger equation

$$(-\Delta + V(x) - k^2)\psi = 0, \quad x \in \mathbb{R}^3, \quad V(x) = 0 \quad \text{for} \quad |x| > R_0. \tag{1.1}$$

For simplicity, we assume that $V(x) \in C^1(\mathbb{R}^3)$.

It can be proved (see, for example, Ramm [78, 87, 90]) that the Green's function for the operator in (1.1), i.e., the resolvent kernel of the Schrödinger operator with potential $V(x)$, can be analytically continued to the whole complex plane as a meromorphic function with argument k. If

$V(x)$ is real-valued then the complex poles of the Green's function lie in the lower half-plane $\operatorname{Im} k < 0$. If $V(x)$ is complex-valued, the spectral properties of the Schrödinger operator in $\mathbb{R}^3$ and in the exterior domain were studied in Ramm [82, 88, 94, 101]. We present here a general method for calculating the complex poles of Green's function. This problem is of great interest in physics because the knowledge of the location of the complex poles allows one to calculate the law of decay of the system described by equation (1.1). The reader can consult Newton [1] concerning this point.

We pass from equation (1.1) to the integral equation

$$\psi = T(k)\psi, \quad T(k)\psi = \int g(x,y,k)V(y)\psi(y,k)dy, \tag{1.2}$$

$$g(x,y,k) \equiv \frac{\exp(ik|x-y|)}{4\pi|x-y|}, \quad \int = \int_{\mathbb{R}^3}. \tag{1.3}$$

Every nontrivial solution of equation (1.1) generates a nontrivial solution of equation (1.2) and vice versa. Equation (1.2) we consider in the space $H = L^2_{R_0}$ of functions with the inner product $(u,v) = \int_{|x|\le R_0} uv^* dx$, $\|u\| = (u,u)^{1/2}$. Let $\{\Phi_j\}$ be an orthonormal basis in H,

$$\psi_N = \sum_{j=1}^{N} C_j \Phi_j, \tag{1.4}$$

where the C_j are constants and N is some integer. It is clear that k is a pole of the resolvent kernel of the Schrödinger operator if and only if equation (1.2) has a nontrivial solution in H. Substituting (1.4) into (1.2) and taking the inner product in H with Φ_ℓ, we obtain

$$\sum_{j=1}^{N} a_{\ell j}(k) C_j = 0, \tag{1.5}$$

where

$$a_{\ell j}(k) = \delta_{\ell j} - (T\Phi_j, \Phi_\ell). \tag{1.6}$$

The system (1.5) has a nontrivial solution if and only if

$$\det a_{\ell j}(k) = 0. \tag{1.7}$$

Let $k_m^{(n)}$, $1 \leq m < \infty$, denote the roots of equation (1.7), while k_m, $1 \leq m < \infty$, denote the complex poles of Green's function for equation (1.1).

Theorem 1. If $V(x) \in C^1(\mathbb{R}^3)$ and $V(x)$ is bounded, then $k_m^{(N)} \to k_m$ as $N \to \infty$ for all $1 \leq m < \infty$ uniformly on any compact subset of the complex k-plane. Moreover, all k_m can be obtained as limits of $k_m^{(N)}$.

Proof of Theorem 1: First we prove that $k_m^{(N)} \to k_m$ as $N \to \infty$. We fix a disk $K_R = \{k: |k| \leq R\}$, where $R > 0$ is an arbitrary large fixed number. Let $|k_m| < R$ for $1 \leq m \leq n$ and $|k_m| > R$ for $m > n$, $D_{\varepsilon,R} = \{k: |k-k_m| \geq \varepsilon, |k| \leq R\}$, $\varepsilon > 0$ is small enough, $m \leq n$. We have

$$\|(I - T(k))^{-1}\| \leq M, \quad k \in D_{\varepsilon,R}, \quad M = M_{\varepsilon,R}. \tag{1.8}$$

Let L_N be the linear span of $\Phi_1, \ldots, \Phi_N$, and P_N be the orthogonal projection of H onto L_N. Then equation (1.5) can be written as

$$\psi_N = P_N T(k) \psi_N, \quad \psi_N \in L_N. \tag{1.9}$$

Under our assumptions, $T(k)$ is compact in H and $P_N \to I$, where $\to$ denotes strong convergence of operators in H. Hence $\|(I - P_N)T(k)\| \to 0$ as $N \to \infty$. Therefore $\|I - T(k) - (I - P_N T(k))\| \to 0$ as $N \to \infty$. Since $I - T(k)$ is invertible for $k \in D_{\varepsilon,R}$ we conclude that $I - P_N T(k)$ is also invertible for $k \in D_\varepsilon$ if N is large enough. But for $k = k_m^{(N)}$, $m \leq n$, the operator $I - P_N T(k)$ has no inverse. Hence if N is large enough all $k_m^{(N)} \notin D_{\varepsilon,R}$, i.e., $k_m^{(N)}$ lies in the ε-neighborhood of k_m, $1 \leq m \leq n$. This means that $k_m^{(N)} \to k_m$ as $N \to \infty$, $1 \leq m \leq n$. It remains to prove that for any fixed k_m there exists a sequence $k_m^{(N)}$ of roots of equation (1.7) such that $k_m^{(N)} \to k_m$ as $N \to \infty$. Let $|k - k_m| = \varepsilon$ and $\varepsilon > 0$ be so small that there are no other $k_\ell, \ell \neq m$, inside the disk $|k - k_m| \leq \varepsilon$. Then $\|(I - T(k))^{-1}\| \leq M$ for $|k - k_m| = \varepsilon$. Hence for N large enough, $\|(I - P_N T(k))^{-1}\| \leq M_1$ for $|k - k_m| = \varepsilon$. Suppose that there are no $k_\ell^{(N)}$ in the disk $|k_m - k| \leq \varepsilon$. Then $\|(I - P_N T(k))^{-1}\| \leq M_1$ for $|k_m - k| \leq \varepsilon$, according to the maximum modulus principle for analytic operator-valued functions, M_1 does not depend on N, provided N is large enough. Hence $\|(I - T(k))^{-1}\|$ is also bounded for $|k - k_m| \leq \varepsilon$. This contradiction proves that for $N > N_\varepsilon$ there is a $k_\ell^{(N)}$ such that

$$|k_m - k_\ell^{(N)}| < \varepsilon. \qquad \square$$

Remark 1. We do not want to discuss practically important but much more traditional questions as to how to calculate the roots of equation (1.7) numerically (see Ortega-Rheinboldt [1]).

2. Now we consider the diffraction problem

$$(\Delta+k^2)\psi = 0 \quad \text{in} \quad \Omega \tag{1.10}$$

$$\frac{\partial\psi}{\partial N} - h\psi\Big|_\Gamma = 0 \tag{1.11}$$

$$\psi \sim \frac{\exp(ik|x|)}{|x|} f(n,k) \quad \text{as} \quad |x| \to \infty, \quad n = \frac{x}{|x|}, \tag{1.12}$$

where Γ is the smooth closed boundary of a bounded domain $D \subset \mathbb{R}^3$, $\Omega = \mathbb{R}^3 \setminus D$, $h = \text{const}$, and $f(n,k)$ is called the scattering amplitude. If we look for a solution of the form

$$\psi = \int_\Gamma g(x,y,k)\sigma(y)dy \tag{1.13}$$

and substitute this into (1.11) we obtain

$$\sigma = Q\sigma, \tag{1.14}$$

where

$$\begin{aligned} Q\sigma \equiv & \int_\Gamma \frac{\partial}{\partial N_t} \frac{\exp(ikr_{st})}{2\pi r_{st}} \sigma(s)ds \\ & - h \int_\Gamma \frac{\exp(ikr_{st})}{2\pi r_{st}} \sigma(s)ds, \quad r_{st} = |s-t|. \end{aligned} \tag{1.15}$$

Let $\{\phi_j\}$ be an orthonormal basis in $L^2(\Gamma)$, and

$$\sigma_N = \sum_{j=1}^N C_j \phi_j. \tag{1.16}$$

Substituting (1.16) into (1.14) and taking the inner product with ϕ_i in $L^2(\Gamma)$, we obtain

$$\sum_{j=1}^N b_{ij}(k)C_j = 0, \quad b_{ij}(k) = \delta_{ij} - (Q\phi_j, \phi_i), \tag{1.17}$$

where $(\phi,\psi) = (\phi,\psi)_{L^2(\Gamma)}$.

The system (1.17) has a nontrivial solution if and only if

$$\det b_{ij}(k) = 0, \quad 1 \le i,j \le N. \tag{1.18}$$

As in the proof of Theorem 1, it can be shown that the roots $k_j^{(N)}$ of equation (1.18) tend to complex poles of Green's function for problem (1.10)-(1.12) as $N \to \infty$ and all the complex poles can be obtained in this way. If boundary condition (1.11) takes the form $u|_\Gamma = 0$, then equation (1.14) takes the form

$$B\sigma \equiv \int_\Gamma \frac{\exp(ikr_{st})}{4\pi r_{st}} \sigma(s)ds = 0, \quad t \in \Gamma, \tag{1.19}$$

while equation (1.18) takes the form

$$\det \beta_{ij}(k) = 0, \quad 1 \le i,j \le N, \tag{1.20}$$

where

$$\beta_{ij}(k) \equiv (B\phi_j, \phi_i). \tag{1.21}$$

Remark 2. In the literature there has been no general approach to the calculation of the complex poles of Green's functions for scattering or diffraction problems. For some systems with special geometrical properties the complex poles were calculated (Wainstein [1]). For example, if Γ is a sphere and the ϕ_j are the spherical functions, then equation (1.21) allows one to calculate the precise values of the complex poles of Green's function for the exterior Dirichlet problem for the ball with boundary Γ.

We can approach the latter problem, i.e., the problem of finding the complex poles of Green's function of equation (1.10) with the boundary condition

$$\psi|_\Gamma = 0 \tag{1.22}$$

and the radiation condition (1.12) at infinity in a different way.

Consider the problem

$$B(k)\phi_n = \rho_n(k)\phi_n, \quad n = 1,2,\ldots, \tag{1.23}$$

where the $\rho_n(k)$ are the eigenvalues of the operator $B(k)$. From the arguments given above it follows that the roots k_{nj}

of the functions $\rho_n(k)$ are poles of Green's function. Indeed, if $\rho_n(k_{nj}) = 0$ then $B\phi_{nj}(t) = 0$. Hence $u(x) \equiv \int_\Gamma g(x,t,k_{nj})\phi_{nj}(t)dt$ is a solution to the problem (1.10), (1.22), and (1.12). If $u \not\equiv 0$ in Ω then k_{nj} is a pole of Green's function. If $u \equiv 0$ in Ω then $u \not\equiv 0$ in D otherwise $\phi_{nj} \equiv 0$. Hence if $u \equiv 0$ in Ω then k_{nj}^2 belongs to the spectrum of the inner Dirichlet problem for domain D, so that $k_{nj}^2 > 0$. It can be shown that poles of Green's function are the roots of the equations

$$\lambda_n(k) = -1, \quad n = 1,2,\ldots, \tag{1.24}$$

where the $\lambda_n(k)$ are the eigenvalues of the following operator:

$$\int_\Gamma \frac{\partial}{\partial N_s} \frac{\exp(ikr_{st})}{2\pi r_{st}} \psi_n(t)dt = \lambda_n(k)\psi_n(s). \tag{1.25}$$

2. Calculation of Diffraction Losses in Some Open Resonators

1. In the physics literature the following integral equation for the current on mirrors of a confocal resonator was obtained (see, for example, Wainstein [1]):

$$Af = \lambda f, \quad Af \equiv \frac{b}{2\pi} \int_S \exp\{-ib(x,u)\}f(u)du, \tag{2.1}$$

where S is the domain on the plane $\mathbb{R}^2$ in which a mirror is projected, $u,x \in \mathbb{R}^2$, $(x,u) = x_1u_1 + x_2u_2$, S is a centrally symmetric domain, and $b > 0$ is certain scalar parameter depending on the distance between the mirrors, wave number, and dimension of the mirror. The diffraction losses for the n-th mode are defined by the formula

$$\alpha_n = 1 - |\lambda_n|^2, \quad n = 0,1,2,\ldots . \tag{2.2}$$

So the problem consists of estimating the eigenvalues of the nonselfadjoint operator A in formula (2.1). In Popov, M [1] the estimate

$$\alpha_{0e} \leq \alpha_0 \leq \alpha_{0i}, \tag{2.3}$$

is proved, where α_{0i}, α_{0e} are the losses on the zero mode for mirrors S_i and S_e, respectively, $S_i \subset S \subset S_e$, the origin is a symmetry center for domains S_i, S, S_e, and inclusion is set-theoretic. It follows from (2.3) that we can estimate α_0 from above and below if we take as S_i a circle or rectangle inscribed in S and a circumscribed circle or rectangle as S_e. For a circle or a rectangle of large dimension, asymptotic formulas for α_0 are known (Wainstein [1]). Here we present a new method of studying of α_n and give a numerical variational approach for calculating of α_n.

2. In general our method is based on the observation that the operator A in (2.1) is normal in $H = L^2(S)$. This can be verified by direct calculation. The kernel of operator $AA^* - A^*A$ is equal to

$$\left(\frac{b}{2\pi}\right)^2 2i \int_S \sin\{(z-u,x)\}dx = 0$$

since S is centrally symmetric. For the compact normal operator A,

$$|\lambda_n| = s_n, \tag{2.4}$$

where the λ_n are eigenvalues of A and the s_n are its singular values, i.e., $s_n(A) = \lambda_n\{(A^*A)^{1/2}\}$. The operator A^*A is nonnegative in H. According to the proof of the first statement of Theorem 6 in Chapter 1 we have

$$s_n(S_1) \leq s_n(S_2) \quad \text{if} \quad S_1 \subset S_2, \tag{2.5}$$

where $s_n(S) = \lambda_n\{(A^*A)^{1/2}\}$. Hence we obtain

<u>Theorem 2</u>. <u>The following inequalities hold</u>

$$\alpha_{ne} \leq \alpha_n \leq \alpha_{ni}, \quad n = 0,1,2,\ldots . \tag{2.6}$$

<u>Remark 3</u>. We can use the well-known minimax definition of $|\lambda_n|^2$ in order to calculate α_n by the variational method

$$|\lambda_n|^2 = \min_{L_n} \max_{\substack{f \in L_n^\perp \\ \|f\|=1}} \|Af\|^2, \quad n = 0,1,2,\ldots, \tag{2.7}$$

where L_n is an n-dimensional subspace in H and $L_n^\perp$ is its orthogonal complement. For $n = 0$ formula (2.7) takes the form

$$|\lambda_0|^2 = \max_{\|f\|=1} \|Af\|^2. \tag{2.8}$$

<u>Remark 4</u>. The kernel of the operator $A^*A \geq 0$ is

$$\left(\frac{b}{2\pi}\right)^2 \int_S \exp\{ib(z-u,x)\}dx. \tag{2.9}$$

According to the second statement of Theorem 6 from Chapter 1 we conclude that

$$\lambda_0(S) \to 1 \quad \text{as} \quad S \to \mathbb{R}^2. \tag{2.10}$$

<u>Remark 5</u>. Since S is centrally symmetric it is easy to verify that the subspaces H_+ and H_- of symmetric and antisymmetric functions are invariant subspaces for A, $H_+ = \{f \in H: f(x) = f(-x)\}$, $H_- = \{f \in H: f(x) = -f(x)\}$, and $H = H_+ \oplus H_-$. Hence all the eigenfunctions of equation (2.1) can be chosen to be either symmetric or antisymmetric.

Remark 6. Let the area of S be fixed. For what shape of a centrally symmetric domain S with the fixed area will diffraction losses be minimal? Our conjecture is that the losses will be minimal for the disk.

We give some arguments to make the conjecture plausible. If S is a rectangle with smaller side $2a$ then the asymptotic formula for α_0 for large mirrors shows that the diffraction losses are defined asymptotically by the minimum distance from the origin to the boundary of the domain S. Hence these losses are minimal for the domain for which the aforementioned distance is maximal, i.e., for the disk of given area.

3. Some Spectral Properties of Nonselfadjoint Integral Operators of Diffraction Theory

1. In diffraction theory it is of interest to know spectral properties of some nonselfadjoint operators. The reader can learn why it is of interest from the book by Voitovich-Kacenelenbaum-Sivov [1]. Consider, for example, the operator

$$B\phi = \int_\Gamma \frac{\exp(ikr_{st})}{4\pi r_{st}} \phi(t)dt, \quad r_{st} = |s-t|, \quad k > 0, \tag{3.1}$$

and let ϕ_n be its eigenfunctions

$$B\phi_n = \lambda_n\phi_n, \quad |\lambda_1| \geq |\lambda_2| \geq \cdots . \tag{3.2}$$

We assume that $k > 0$, Γ is a closed smooth surface in $\mathbb{R}^3$, and $H = L^2(\Gamma)$. The first question of interest in diffraction theory is whether the set of eigenfunctions and associated functions of the operator B is complete in H. The second question is when are there no associated eigenfunctions. The third question is whether the set of eigenfunctions and associated functions forms a basis in H and, if so, what are the properties of this basis. Answer to the first and second questions were given in Ramm [94], and the third question is discussed in detail in the appendix by M. S. Agranovich to the book Voitovich-Kacenelenbaum-Sivov [1]. The aim of this section is to direct the reader's attention to relatively new questions which are of interest in applications but require deep knowledge of the theory of nonselfadjoint operators. We present here a simple method for studying the first question.

2. First we remind the reader that a normal operator in H has no associated functions. A bounded linear operator B is called dissipative if $(B_J\phi,\phi) \geq 0$ for all $\phi \in H$, where $B_J \equiv (B-B^*)/2i$. In Kato [1] another terminology is used. We use terminology adopted in Gohberg-Krein [2]. The operator B is called nuclear if $\sum_{j=1}^{\infty} s_j(B) < \infty$, where the $s_j(B)$ are the singular values of B. The set of eigenfunctions and associated functions is called the root system.

<u>Theorem 3</u>. <u>Let</u> $Q \geq 0$ <u>be a compact operator on a Hilbert space</u> H, T <u>be a dissipative nuclear operator on</u> H, <u>and</u> $B = Q+T$. <u>Then the root system of the operator</u> B <u>is complete in</u> H.

<u>Proof of Theorem 3</u>: First we note that if $Q \geq 0$ is a compact operator and T is a compact selfadjoint operator, then $N_t^-(Q+T) \leq N_t^-(T)$, where $N_t^-(T) \equiv \sum_{\lambda_n^-(t) \leq -t} 1$ and the $\lambda_n^-(T)$ are the negative eigenvalues of T arranged so $|\lambda_1(T)| \geq |\lambda_2(T)| \geq \dots$. Let $T = T_R + iT_J$, where

$T_J = -\frac{1}{2}i(T-T^*)$, $T_R = \frac{1}{2}(T+T^*)$, $T_J \geq 0$, and is nuclear. Furthermore, we have

$$0 \leq \lim_{t\to 0} tN_t^-(Q + T_R) \leq \lim_{t\to 0} tN_t^-(T_R). \tag{3.3}$$

Since T is nuclear, T_R is nuclear and

$$\lim_{t\to 0} tN_t^-(T_R) = 0. \tag{3.4}$$

Hence

$$\lim_{t\to 0} tN_t^-(B_R) = 0. \tag{3.5}$$

Now we apply the following

Proposition (Gohberg-Krein [2, p. 292]). *The root system of a compact dissipative operator* B *with the nuclear imaginary part* B_J *is complete provided condition* (3.5) *holds.*

This completes the proof of Theorem 3. □

Theorem 4. *The root system of operator* (3.1) *is complete in* $H = L^2(\Gamma)$.

Proof of Theorem 4: We have

$$B = Q + T, \quad Q\phi = \int_\Gamma \frac{\phi(t)dt}{4\pi r_{st}}, \tag{3.6}$$

$$T\phi = \int_\Gamma \frac{\exp(ikr_{st}) - 1}{4\pi r_{st}} \phi(t)dt, \tag{3.7}$$

so that $Q \geq 0$ is compact in H and

$$T_J\phi = \int_\Gamma \frac{\sin(kr_{st})}{4\pi r_{st}} \phi(t)dt. \tag{3.8}$$

Since

$$\begin{aligned}\int_\Gamma\int_\Gamma &\frac{\sin(kr_{st})}{4\pi r_{st}}\phi(t)\phi^*(s)dtds \\ &= \frac{k}{4\pi}\int_{S^2}\left|\int_\Gamma \phi(t)\exp\{ik(\omega,t)\}dt\right|^2 d\omega \geq 0,\end{aligned} \tag{3.9}$$

where $\omega \in S^2$ and S^2 is the unit sphere in $\mathbb{R}^3$, we conclude that B is dissipative. It remains to prove that T_J is nuclear. That is easy to verify since its kernel is sufficiently smooth (see, for example, Dunford-Schwartz [1], Chapter 11, point 9.32). Hence we get the statement of Theorem 4 from Theorem 3. □

3. Here we answer the second question from point 1. Since a normal operator has no associated functions it is sufficient to give conditions sufficient for the operator B defined by formula (3.1) to be normal. We can easily calculate that the kernel $C(x,y)$ of the operator $BB^* - B^*B$ is

$$C(x,y) = 2i \int_\Gamma \frac{\sin\{k(r_{xt} - r_{ty})\}}{16\pi^2 r_{xt} r_{ty}} dt. \tag{3.10}$$

Hence B is normal if and only if $C(x,y) = 0$. For example, this will be the case if Γ is sphere. Another example of interest in antenna synthesis theory (see Minkovich-Jakovlev [1]) gives the equation

$$B\phi = \int_{-a}^{a} \exp(ixy)dy = \lambda\phi(x), \quad -a \leq x \leq a. \tag{3.11}$$

This equation arises in linear antenna synthesis theory. In this example,

$$C(x,y) = 2i \int_{-a}^{a} \sin\{(x-y)z\}dz = 0. \tag{3.12}$$

Hence operator (3.11) is normal in $L^2(-a,a)$ and its eigenfunctions form an orthonormal basis of $L^2(-a,a)$. This fact is used in practice.

4. Now we describe briefly what is known about expansions in eigenfunctions and associated functions of the opera-

tor (3.1). The reader can find these results in detail in Appendices 10, 11 where the necessary definitions are given, and in the book Voitovich-Kazenelenbaum-Sivov [1]. It can be proved that

$$\lambda_n \sim -\frac{1}{\sqrt{n}} \cdot \frac{\sqrt{S}}{4\sqrt{\pi}}, \quad S = \text{meas}\ \Gamma,$$

$$\frac{\text{Re}\lambda_n}{\lambda_n} \to 1 \quad \text{as} \quad n \to \infty, \quad \frac{\text{Im}\lambda_n}{\lambda_n} \to 0 \quad \text{as} \quad n \to \infty.$$

$$\left|\frac{\lambda_n - \text{Re}\lambda_n}{\lambda_n}\right| \le C_q n^{-q} \quad \text{for any} \quad q > 0.$$

We shall write $B \in \mathscr{B}(H)$ if the root system of the compact operator B forms a Bari basis with brackets in a Hilbert space H. Let $H_m = H_m(\Gamma) = W_2^m(\Gamma)$ be the Sobolev spaces. It can be proved that $B \in \mathscr{B}(H_m)$ for any $m > 0$. We end this brief account with the definition of Bari basis with brackets in a Hilbert space H. Let $\{h_j\}$ be an orthonormal basis in H, $0 < m_0 < m_1 < \dots$, and let Q_ℓ be the orthogonal projection of H onto the linear span of the vectors $\{h_{m_\ell+1},\dots,h_{m_\ell+1}\}$. Let $\{f_j\}$ be a complete minimal system in H. A system $\{f_j\}$ is called minimal if any vector f_m does no belong to the closure of the linear span of the other vectors of the system. Let $0 < m_0 < m_1 < \dots$, and let P_ℓ be the orthogonal projection of H onto the linear span of the vectors $\{f_{m_\ell+1},\dots,f_{m_\ell+1}\}$. If $\sum_{\ell=0}^{\infty} \|P_\ell - Q_\ell\| < \infty$ then the system $\{f_j\}$ is called a Bari basis with brackets in H.

4. Research Problems

1. Prove (or disprove) the conjecture in Remark 6 in Section 2.
2. Try the numerical method described in Section 1 in some practical problems.
3. Study spectral properties of the integral operator $Af = \int_{-1}^{1} e^{i(x-y)^2} f(y)dy$.
4. Investigate stability of the complex poles of Green's function under perturbations of the boundary in diffraction problems and under perturbations of the potential in scattering problems (see Appendix 10).
5. To what extent do the complex poles of Green's function determine the reflecting obstacle in diffraction problems or the potential in scattering problems?
6. Find the asymptotic distribution of the complex poles of Green's function.
7. Investigate the same questions for electromagnetic wave scattering.
8. Is it possible to represent the Green's function in diffraction and scattering theories in Mittag-Leffler form (as a sum of its principal parts)?

5. Bibliographical Note

Resonances and quantum mechanical theory of decay of particles have been discussed in the physics literature (Newton [1], Baz-Zeldovich-Perelomov [1]). There was no general approach to the problem of calculating the complex poles of Green's function which are resonances in quantum mechanics, complex eigenvalues of eigenfrequencies in diffraction theory and in the theory of open resonators. Nonselfadjoint operators as such were intensively studied (Gohberg-Krein [2], [3], Lifschitz [1], Dunford-Schwartz [1]). The author ([76]-[99]) studied spectral properties of the Schrödinger operator and proved eigenfunction expansion theorem for nonselfadjoint Schrödinger operator (Ramm [99], [93]). The results of this chapter were obtained in Ramm [54]-[56]. In Ramm [87], the domain free of resonances in the three-dimensional potential scattering problem was found. The study of the analytic continuation of the Schrödinger operator resolvent kernel was given in Ramm [78]. This was probably the first paper on the subject in multidimensional case. Lax-Phillips [1] investigated the location of resonances in the scattering problem for the Laplace operator in exterior domain.

CHAPTER V

INVESTIGATION OF SOME INTEGRAL EQUATIONS ARISING IN ANTENNA SYNTHESIS

1. A Method for Stable Solution of an Equation of the First Kind

Let A be a compact linear operator on a Hilbert space H, $N(A) = \{0\}$, $A\Phi_n = \lambda_n\Phi_n$, $|\lambda_1| \geq |\lambda_2| \geq \ldots$. We assume that the system $\{\Phi_n\}$ is an orthonormal basis of H.

Consider the equation

$$Ax = y, \quad y \in R(A). \tag{1.1}$$

The operator A^{-1} is unbounded. Suppose we know y_δ such that $||y - y_\delta|| \leq \delta$. The problem is to find $x_\delta = R_\delta y_\delta$ so that $||x_\delta - x|| \equiv ||x_\delta - A^{-1}y|| \to 0$ as $\delta \to 0$. Such problems have been studied in the literature (Tihonov-Arsenin [1]). We give an approach which is quasi-optimal in the sense defined below and an effective error estimate. Let

$$x_{h,\delta} = R_h y_\delta = \sum_{h=1}^{\infty} y_{n\delta}\lambda_n^{-1}\exp(-h|\lambda_n|^{-1}), \tag{1.2}$$

where $h > 0$, $y_{n\delta} \equiv (y_\delta,\Phi_n)$, and $(\cdot,\cdot)$ is the inner product in H. Since $y \in R(A)$ we have

$$\sum_{n=1}^{\infty} |y_n\lambda_n^{-1}|^2 < \infty, \quad y_n = (y,\Phi_n). \tag{1.3}$$

<u>Theorem 1</u>. <u>If</u> $h = h(\delta) \to 0$ <u>and</u> $\delta^2 h^{-2}(\delta) \to 0$ <u>as</u> $\delta \to 0$, <u>then</u> $\|R_{h(\delta)} y_\delta - x\| \to 0$ <u>as</u> $\delta \to 0$. <u>If</u>

$$\sum_{h=1}^{\infty} |y_n|^2 |\lambda_n|^{-2-b} \leq M, \quad b > 0, \tag{1.4}$$

$$h(\delta) = L\delta^{2/(C+2)}, \tag{1.5}$$

<u>where</u>

$$L = (4/bQe^2)^{1/(C+2)}, \quad C = \min(b,2), \quad Q = 2(\|y\|^2+M), \tag{1.6}$$

<u>then</u>

$$\|R_{n(\delta)} y_\delta - x\| \leq \delta^{C/(C+2)} (QL^C + 2(Le)^{-2})^{1/2}. \tag{1.7}$$

<u>Proof of Theorem 1</u>: We have

$$\begin{aligned} \|x_{h,\delta} - x\|^2 &= \sum_{n=1}^{\infty} |y_{n\delta}\lambda_n^{-1} \exp(-h/|\lambda_n|) - y_n\lambda_n^{-1}|^2 \\ &\leq 2 \sum_{n=1}^{\infty} |(y_{n\delta} - y_n)\lambda_n^{-1}|^2 \exp(-2h/|\lambda_n|) \\ &\quad + 2 \sum_{n=1}^{\infty} |y_n|^2 \left|\frac{1-\exp(-h/|\lambda_n|)}{\lambda_n}\right|^2 \equiv J_1 + J_2. \end{aligned} \tag{1.8}$$

We note that

$$\sup_{0<|\lambda|<\|A\|} 2|\lambda|^{-2} \exp(-2h|\lambda|^{-1}) \leq 2(eh)^{-2}. \tag{1.9}$$

Hence

$$J_1 \leq 2(eh)^{-2}\delta^2 \tag{1.10}$$

So if $h = h(\delta) \to 0$ and $\delta^2 h^{-2}(\delta) \to 0$ we obtain

$$\|x_{h(\delta),\delta} - x\| \to 0 \quad \text{as} \quad \delta \to 0. \tag{1.11}$$

We can obtain a better estimate for J_2 provided (1.4) holds. We have

$$J_2 = 2\left\{\sum_{|\lambda_n|\leq 1} + \sum_{|\lambda_n|>1}\right\} \equiv 2(J_{21} + J_{22}), \tag{1.12}$$

where

$$J_{22} \le \sum_{|\lambda_n|>1} |y_n|^2(1-\exp(-h))^2 \le \|y\|^2 h^2, \tag{1.13}$$

$$\begin{aligned} J_{21} &\le \sum_{|\lambda_n|\le 1} \frac{|y_n|^2}{|\lambda_n|^{2+b}} |\lambda_n|^b |1 - \exp(-h/|\lambda_n|)|^2 \\ &\le M \sup_{0\le|\lambda|\le 1} |\lambda|^b (1 - \exp(-h/|\lambda|))^2 \le MM_1 h^C. \end{aligned} \tag{1.14}$$

Here

$$C = \min(b,2), \quad M_1 \le 1. \tag{1.15}$$

Hence

$$J_2 \le Qh^C, \quad Q = 2(\|y\|^2 + M), \tag{1.16}$$

$$\|x_{h,\delta} - x\| \le 2(he)^{-2}\delta^2 + Qh^C. \tag{1.17}$$

Taking δ to minimize the right-hand side of this inequality, we obtain (1.7). □

Remark 1. We could have used convergence multipliers other than $\exp(-h|\lambda_n|^{-1})$. This method works for closed unbounded normal operators and for spectral operators.

2. Some Results Concerning the General Antenna Synthesis Problem

The traditional problem of antenna synthesis theory presented in the literature (Minkovich-Jakovlev [1]) can be described as follows. Given a domain in which a current flows and a vector function $f(n,k_0)$, where n is an ort, $k_0 > 0$ is a fixed wave number, find a current distribution that generates an electric field E with the asymptotics

$$E \sim \frac{\exp(ikr)}{r} f(n,k_0) \quad \text{as} \quad x \to \infty, \quad \frac{x}{|x|} = n.$$

For this statement of the antenna synthesis problem it is interesting to know conditions under which a given function $f(n,k)$, $n \in S^2$, $k > 0$, called the scattering amplitude (or the radiation pattern in electrodynamics) is the scattering amplitude of some current distribution. Another question of interest is whether a given function $f(n,k)$ can be approximated with the prescribed accuracy by a scattering amplitude. Of course one must define precisely what approximation with the prescribed accuracy means. We give here a brief summary of the conclusions obtained in Ramm [67]. A necessary condition for a vector function $f(n,k)$ to be a scattering amplitude is $f_r = 0$, where f_r is the r-component of vector f. If Maxwell's equations are

$$\text{curl } E = ikH, \quad c = \varepsilon = \mu = 1, \quad k = \frac{\omega}{c},$$

$$\text{curl } H = -ikE + j(x),$$

then

$$f = ik(I_\phi a_\phi + I_\theta a_\theta), \tag{2.1}$$

where

$$I = \int_{\mathbb{R}^3} j(y) \exp\{-ik(n,y)\} dy, \tag{2.2}$$

$n = (\theta,\phi)$ is ort; (θ,ϕ) are coordinates on S^2; a_ϕ, a_θ, a_r are the orts of the spherical coordinate system; and I_r, I_θ, I_ϕ are the projections of the vector I on the orts of the spherical coordinate system. If the domain of current flow is not all of $\mathbb{R}^3$, then given a vector function $f(n,k)$, $f_r = 0$, $f \in L^2(\mathbb{R}^3)$, we can find many current distributions $j(x)$ which generate the scattering amplitude $f(n,k)$. To obtain a unique current distribution we can fix an arbitrary scalar function $I_r(n,k) \in L^2(\mathbb{R}^3)$ in addition to the given vector function $f(n,k)$. Then there exists a unique current distribution $j(x)$ which generates the scattering amplitude $f(n,k)$ and has function (2.2) with $I_r = I_r(n,k)$. We note that only such problems in antenna synthesis theory in which amplitude defined current distribution uniquely were studied in the literature (Minkovich-Jakovlev [1]). This was usually the case because the domain current flow and the direction of current flow were fixed in advance. A typical example is the problem of linear antenna synthesis, where current flows along a line segment $(-a,a)$. Under these restrictions, the following approximation problem is interesting. We note that if $j(y) = 0$ outside a bounded domain D then from (2.2) it follows that $I = I(kn) = I(k_1,k_2,k_3)$ is an entire function of exponential type with the Fourier transform which vanishes in $\Omega = \mathbb{R}^3 \setminus D$. The problem mentioned above can be stated in the following manner: given a function $f(x)$ in $C(\Delta)$ or $L^2(\Delta)$, where Δ is some domain in $\mathbb{R}^N$, and $\varepsilon > 0$, how does one find a function $g(x) \in W_D$ such that $\|f-g\| < \varepsilon$, where $\|\cdot\|$ denotes the norm in $C(\Delta)$ or $L^2(\Delta)$, and W_D is the class of entire functions of the form

$$g(x) = \frac{1}{(2\pi)^{N/2}} \int_D \exp\{-i(x,y)\}h(y)dy, \quad h \in L^2(D). \tag{2.3}$$

This problem is of interest also for optics (apodization theory, see Ramm [14-16]). In the next section we give a solution to this problem.

3. Formula for Approximation by Entire Functions

We formulate two problems.

Problem A. Given $\varepsilon > 0$ and $f \in C(\Delta)$ how does one find $f_\varepsilon \in W_D$ such that $|f - f_\varepsilon| < \varepsilon$, $|\cdot| = \| \ \|_{C(\Delta)}$.

Problem B. Given $\varepsilon > 0$ and $f \in L^2(\Delta)$ how does one find $f_\varepsilon \in W_D$ such that $\|f - f_\varepsilon\| < \varepsilon$, $\|\cdot\| = \|\cdot\|_{L^2(\Delta)}$.

Let us set

$$g_n(y) = \left[\frac{1}{|D|}\int_D \exp\left\{-\frac{i}{2n+N}(t,y)\right\}dt\right]^{2n+N}\left(1-\frac{|y|^2}{R^2}\right)^n\left(\frac{n}{\pi R^2}\right)^{N/2} \tag{3.1}$$

Here $|D|$ = meas D, and $R > 0$ is a number such that the ball $|y| \leq R$ contains all the differences $t - y$ with $t,y \in \Delta$, the origin in R_t^N is placed at the centroid of D so that

$$\int_D t\, dt = 0. \tag{3.2}$$

We assume that $\Delta = \bar{\Delta}$ and $D = \bar{D}$ are simply connected bounded domains. Let D be convex, and

$$f_n(x) = \int_\Delta g_n(x-y)f(y)\,dy. \tag{3.3}$$

Theorem 2. There exists $n = n(\varepsilon)$ such that $f_n(x)$ is the solution to problem A for $n \geq n(\varepsilon)$ if $f(x) \in C(\Delta)$ and to problem B if $f \in L^2(\Delta)$.

Theorem 3. If $|f(x)| \leq a$ and $|\nabla f(x)| \leq b$, then

$$|f - f_n| \leq \frac{\Gamma(\frac{N+1}{2})}{\Gamma(\frac{N}{2})}\frac{bR}{\sqrt{n}} + O(\frac{1}{n}) \quad \text{as } n \to \infty, \tag{3.4}$$

where $\Gamma(x)$ is the Gamma function.

Proof of Theorem 2: Note that $g_n(y) \in W_D$. Hence $f_n(x) \in W_D$. Let us prove that

$$\left[\frac{1}{|D|}\int_D \exp\left\{-\frac{i}{2n+N}\right\}(t,y)\,dt\right]^{2n+N} \to 1 \quad \text{as} \quad n \to \infty, \tag{3.5}$$

$$\left(\frac{n}{\pi R^2}\right)^{N/2}\int_\Delta\left(1 - \frac{|y-x|^2}{R^2}\right)^n f(y)\,dy \to f(x) \quad \text{as} \quad n \to \infty \tag{3.6}$$

uniformly in Δ if $f(x) \in C(\Delta)$.

We have

$$(2n+N)\ \ln \frac{1}{|D|}\int_D \exp\left\{-\frac{i}{2n+N}(t,y)\right\}dt = (2n+N)$$
$$\cdot\ln\left[1 - \frac{i}{2n+n}\left(y, \frac{1}{|D|}\int_D t\,dt\right) + \alpha_n\right],\ \alpha_n = O(\frac{1}{n^2}). \tag{3.7}$$

From here and (3.2) we obtain (3.5). To prove (3.6) we can assume without loss of generality that $R = 1$, for otherwise we could use scaling. For any η, $0 < \eta < \frac{1}{2}$, we have

$$\lim_{n\to\infty} (\tfrac{n}{\pi})^{N/2} \int_{\eta\le|x-y|\le 1} [1 - |x-y|^2]^n dy = 0 \tag{3.8}$$

$$\lim_{n\to\infty} (\tfrac{n}{\pi})^{N/2} \int_{|x-y|<\eta} [1 - |x-y|^2]^n dy = 1. \tag{3.9}$$

Furthermore,

$$f_n - f = J_1 + J_2 \equiv (\tfrac{n}{\pi})^{N/2}\int_\Delta [1 - |x-y|^2]^n [f(y) - f(x)]\,dy$$
$$+ f(x)\left[(\tfrac{n}{\pi})^{N/2}\int_\Delta \{1 - |x-y|^2\}^n dy - 1\right], \tag{3.10}$$

$$|J_2| \le \|f\|\cdot|(\tfrac{n}{\pi})^{N/2}\int_{|u|\le 1}(1-u^2)^n du - 1| \to 0 \text{ as } n \to \infty, \tag{3.11}$$

$$|J_1| \le (\tfrac{n}{\pi})^{N/2} \int_{|u|\le\varepsilon} (1-|u|^2)^n |f(u+x)-f(x)|du$$
$$+ 2|f|(\tfrac{n}{\pi})^{N/2} \int_{\varepsilon<|u|\le 1} (1-|u|^2)^n du. \tag{3.12}$$

From (3.12) it follows that $J_1 \to 0$ as $n \to \infty$. We have proved the first statement of Theorem 2. To prove the second statement we must prove that

$$\|(\tfrac{n}{\pi R^2})^{N/2} \int_\Delta \left(1 - \frac{|x-y|^2}{R^2}\right)^n f(y)dy - f(x)\| \to 0 \quad \text{as } n \to \infty. \tag{3.13}$$

As above we assume that $R = 1$. Let us fix $\varepsilon > 0$ and find $\phi(x) \in C(\Delta)$ such that $\|f-\phi\| < \varepsilon$. Setting

$$Q_n f \equiv (\tfrac{n}{\pi})^{N/2} \int_\Delta [1 - |x-y|^2]^n f(y)dy \tag{3.14}$$

we have

$$\|Q_n f - f\| \le \|Q_n(f-\phi)\| + \|Q_n\phi - \phi\| + \|f - \phi\|. \tag{3.15}$$

Since $\|F\| \le C|F|$, $C = C(\Delta)$, we obtain

$$\|Q_n\phi - \phi\| \to 0 \quad \text{as} \quad n \to \infty. \tag{3.16}$$

It remains to prove that

$$\|Q_n\psi\| \to 0 \quad \text{as} \quad n \to \infty, \quad \psi \equiv f - \phi, \quad \|\psi\| < \varepsilon. \tag{3.17}$$

We have

$$\|Q_n\psi\|^2 = \int_\Delta dx\Big\{(\tfrac{n}{\pi})^{N/2}\int_\Delta [1-|x-y|^2]^n|\psi(y)|dy\, (\tfrac{n}{\pi})^{N/2} \cdot$$
$$\cdot \int_\Delta [1-|x-z|^2]^n|\psi(z)|dz\Big\}$$
$$\le \tfrac{1}{2}(\tfrac{n}{\pi})^N \int_{|u|\le 1}\int_{|v|\le 1} dudv(1-|u|^2)^n(1-|v|^2)^n\Big\{\int_\Delta(|\psi(x+u)|^2$$
$$+ |\psi(x+v)|^2)dx\Big\} \le \tag{3.18}$$

$$\leq \varepsilon(\tfrac{n}{\pi})^N \int_{|u|\leq 1} \int_{|v|\leq 1} (1-|u|^2)^n(1-|v|^2)^n du dv \leq \text{const } \varepsilon, \quad (3.18)$$

where const does not depend on n. □

Proof of Theorem 3: To prove this theorem we must sharpen some of the estimates given in the proof of Theorem 1. First we note that α_n in (3.7) satisfies the estimate

$$|\alpha_n| \leq \frac{C}{2|D|(2n+N)^2} \int_D |(t,y)|^2 dt \leq \frac{C_1}{(2n+N)^2}. \quad (3.19)$$

From (3.19) and (3.7) it follows that

$$\left|\left[\frac{1}{|D|}\int_D \exp\{-\frac{i}{2n+N}(t,y)\}dt\right]^{2n+N} - 1\right| = O(\tfrac{1}{n}) \quad \text{as } n \to \infty. \quad (3.20)$$

Furthermore,

$$\left(\frac{n}{\pi R^2}\right)^{N/2} \int_\Delta \left(1 - \frac{|y|^2}{R^2}\right)^n dy = 1 + O(\tfrac{1}{n}) \quad \text{as } n \to \infty. \quad (3.21)$$

Now we estimate the integral J_1 from formula (3.10) without the assumption $R = 1$. We have

$$\begin{aligned} J_1 &\leq b\left(\frac{n}{\pi R^2}\right)^{N/2} \int_{|u|\leq R} \left(1 - \frac{|u|^2}{R^2}\right)^n |u| du \\ &= bR(\tfrac{n}{\pi})^{N/2} S_N \int_0^1 (1 - v^2)^n v^N dv \\ &= \frac{bR\Gamma(\frac{N+1}{2})}{\Gamma(\frac{N}{2})\sqrt{n}} + O(\frac{1}{n^{3/2}}) \quad \text{as } n \to \infty, \end{aligned} \quad (3.22)$$

where $S_n = 2\pi^{\frac{1}{2}N}/\Gamma(\frac{1}{2}N)$ is the area of the unit sphere in $\mathbb{R}^n$. The last integral here can be evaluated as $\int_0^1 (1 - v^2)^n v^N dv = \frac{1}{2}B((N+1)/2, n+1)$, where $B(x,y)$ is the beta function. This integral can be approximated as $n \to \infty$

either by using Stirling's formula or Laplace's method for asymptotic evaluation of integrals. As a result we obtain (3.4). □

4. Nonlinear Synthesis Problems

In this section we discuss a new type of antenna synthesis theory problem, the problem of nonlinear synthesis. This problem is of interest in monoimpulse radiolocation (Hellgren [1]). Nonlinear synthesis problems have not been discussed in the literature as far as the author knows.

1. Let $0 < k_0 < \pi/2$ be a fixed number. The function

$$F^2(k+k_o) - F^2(k-k_o) = g(k) \tag{4.1}$$

is called the pelengation characteristic, where

$$F(k) = F_j(k) \equiv \int_{-\pi}^{\pi} j(x) \exp(ikx)dx. \tag{4.2}$$

$j(x)$ is the current distribution along the linear antenna. We discuss the following questions:

(a) Is $j(x)$ uniquely determined by a given pelengation characteristic $g(k)$?

(b) How does one describe the class of functions which are pelengation characteristics of current distributions $j \in L^2 = L^2(-\pi,\pi)$?

(c) How does one find explicit solutions of equation (4.1) if the equation is solvable?

(d) How does one find an approximate solution to equation (4.1) if g is not a pelengation characteristic, i.e., how does one approximate a given function $g(k)$ by a function which can be represented in the form $F^2(k+k_o) - F^2(k-k_o)$ with the prescribed accuracy?

We answer these questions below.

2. From definitions (4.1) and (4.2), one can see that $j(x)$ and $-j(x)$ generate the same pelengation characteristic so we do not consider $j(x)$ and $-j(x)$ as different current distributions. With this agreement we can formulate our first result which answers question (a).

Theorem 4. *The integral equation* (4.1) *has at most one solution* $j \in L^2$.

Proof of Theorem 4: If j,ψ are solutions to (4.1) then $F_j^2(k+k_o) - F_\psi^2(k+k_o) \equiv F_j^2(k-k_o) - F_\psi^2(k-k_o)$. We define $\Phi(k) \equiv F_j^2(k) - F_\psi^2(k)$. Then $\Phi(k+k_o) \equiv \Phi(k-k_o)$. Hence $\Phi(x)$ is a $2k_o$ -periodic function. In addition, $\Phi(k)$ is an entire function of exponential type, $\Phi(k) \in W_{2\pi}$ (for the definition of class W_D see Section 2 of this chapter), and is bounded on $I = (-\infty,\infty)$ since $|F(k)| \leq \sqrt{2\pi}\, \|j\|_{L^2} < \infty$. Now we use the following lemma due to S. Bernstein (see Ahiezer [1, p. 226]).

Lemma 1. *If* $\Phi(k)$ *is an entire function*, $|\Phi(k)| \leq A \exp(B|k|)$, *where* $A,B > 0$ *are constants*, *and* $\phi(k+\ell) = \phi(k)$ *then*

$$\phi(k) = \sum_{m=-n}^{n} C_m \exp(2\pi imk/\ell), \quad n \leq [\frac{B\ell}{2\pi}], \tag{4.3}$$

where $[x]$ *denotes the greatest integer not exceeding* x.

From this lemma it follows that

$$\Phi(k) = \sum_{m=-3}^{3} C_m \exp(i\pi mk/k_o), \tag{4.4}$$

since in our case $B = 2\pi$, $\ell = 2k_o$, $k_o < \pi/2$, so that $n \leq [\pi] = 3$. As $\phi \in L^2$, we conclude that $C_m = 0$, $|m| \leq 3$, hence $\Phi(k) \equiv 0$, $F_j = \pm F_\psi$, $j = \pm\psi$. □

3. To answer question (b) consider only even distributions $j(x) = j(-x)$, so that $F(k) = F(-k)$. In this case it is clear that a necessary condition for a function $g(k)$ to be a pelengation characteristic is

Condition A. The function $g(k)$ is an entire odd function and $g \in W_{2\pi}$.

This condition is assumed to be fulfilled. Then by the Paley-Wiener theorem (Ahiezer [1, p. 179]) there exists an odd function $\tilde{g}(x) \in L^2$ such that

$$g(k) = \int_{-2\pi}^{2\pi} \tilde{g}(x) \exp(ikx)dx. \tag{4.5}$$

By the same theorem we can write

$$F^2(k) = \int_{-2\pi}^{2\pi} \phi(x)\exp(ikx)dx, \quad \phi \in L^2. \tag{4.6}$$

Hence equation (4.1) takes the form

$$2i \int_{-2\pi}^{2\pi} \phi(x)\sin(k_o x) \exp(ikx)dx = \int_{-2\pi}^{2\pi} \tilde{g}(x) \exp(ikx)dx. \tag{4.7}$$

From here it follows that

$$\phi(x) = \frac{\tilde{g}(x)}{2i \sin(k_o x)} . \tag{4.8}$$

If $\tilde{g}(x)/(2i \sin(k_o x)) \in L^2$ then

$$F^2(k) = \int_{-2\pi}^{2\pi} \frac{\tilde{g}(x)}{2i \sin(k_o x)} \exp(ikx)dx. \tag{4.9}$$

Since $F(k)$ is an entire function, all the zeros of the function (4.9) must be of even order. If this is so, the function

$$F(x) = \left\{\int_{-2\pi}^{2\pi} \frac{\tilde{g}(x)}{2i \sin(k_o x)} \exp(ikx)dx\right\}^{1/2} \tag{4.10}$$

is entire and the corresponding current distribution

$$j(x) = \frac{1}{2\pi} \int_{-\infty}^{\infty} F(k) \exp(ikx)dk \tag{4.11}$$

is the solution of equation (4.1). We have proved

Theorem 5. *Equation* (4.1) *is uniquely solvable in the class of even current distributions* $j \in L^2$ *if and only if condition A holds, the function* (4.8) *belongs to* L^2 *and the function* (4.10) *is entire. If these conditions are satisfied the solution of equation* (4.1) *can be calculated from formulas* (4.11), (4.10), *and* (4.8).

Remark 1. The function (4.10) is entire if and only if all the zeros of function (4.9) have even order.

Example 1. Let $k_o = 1$, $g(k) = (\sin 2k\pi)/(k^2-1)$. Condition A is fulfilled but $F^2(k) = -2(\sin 2\pi k)/k$ so $g(k)$ is not a pelengation characteristic.

Example 2. If $k_o = 1$, $g(k) = -k(\sin^2 k\pi)/(k^2-1)^2$ then $g(k)$ is a pelengation characteristic and the corresponding current is

$$j(x) = \begin{cases} \dfrac{1}{2\sqrt{2}}, & |x| \le \pi, \\ 0, & |x| > \pi. \end{cases}$$

Remark 2. The same method can be applied to investigate the equation

$$\psi(F(k + k_o)) - \psi(F(k - k_o)) = g(k), \tag{4.12}$$

provided that $\psi(F(k))$ is a function of exponential type.

4. Consider the following optimization problem of finding the pelengation characteristic with maximal steepness at the origin and given level of side leaves or, what is the same, with given steepness at the origin and minimal level of side leaves. In other words, we want to find a function $g(k) \in W_{2\pi}$, satisfying the conditions of Theorem 5 and the condition

$$g'(0) = a \tag{4.13}$$

for which

$$\max_{k \in I} |g(k)| = \min_{\substack{\phi \in W_{2\pi} \\ \phi'(0)=a}} \max_{k \in I} |\phi(k)|, \quad I = (-\infty,\infty). \tag{4.14}$$

This is a nonlinear analog to the Dolph-Cebyshev distribution for linear problem (Dolph [2]). It can be proved (Ahiezer [1, p. 355]) that this problem has the unique solution $g(k) = a\,(\sin 2\pi k)/(2\pi)$ and, moreover, this function is also the solution to the problem

$$\min_{\substack{\phi \in W_{2\pi} \\ \phi'(0)=a}} \max_{|k|>d} |\phi(k)|$$

where $0 < d < \frac{1}{4}$ (see Ahiezer [1, p. 364]). The function $\sin 2\pi k$ is not a pelengation characteristic, there is no $j(x) \in L^2$ for which $\sin 2\pi k$ is a pelengation characteristic but with the help of delta-functions we can produce such $g(k)$. Namely if $g(x) = [\delta(x-\pi) - \delta(x+\pi)]/(2i)$ then $F(k) = \sin k\pi$, $g(k) = F^2(k + k_o) - F^2(k - k_o) =$ $\sin 2k\pi \sin 2k_o$.

5. Suppose $f(k) = -f(-k) \in L^2(-k_b,k_b)$, where k_b is some arbitrarily large fixed number. Is it possible to approximate with the prescribed accuracy $\varepsilon > 0$ in the norm

$L^2(-k_b,k_b)$ the given function $f(k)$ by a pelengation characteristic, i.e., by an odd function $f_\varepsilon(k) \in W_{2\pi}$ such that function (4.10) constructed with respect to f_ε instead of $\tilde{g}$ is entire? If the answer is affirmative we consider the corresponding function $j_\varepsilon(x)$ as an approximate solution to the synthesis problem. It can be proved that such an approximation is possible provided that $f(k)$ is odd and for large n the function

$$\left\{\int_{-2\pi}^{2\pi} \frac{f_n(x)}{2i\,\sin(k_o x)} \exp(ikx)\,dx\right\}^{1/2} \neq 0 \quad \text{on} \quad (-k_b,k_b)$$

or is continuous in some neighborhood of this segment on the complex plane k. Here $f_n(x)$ is the Fourier transform of $f_n(k)$ and $f_n(k) \in W_{2\pi}$ is the function, which approximates $f(k)$, constructed by the formula described in Section 3 (formula (3.3) of this section with $\Delta = (-k_b,k_b)$, $x = k$) (see Ramm [71]).

6. The methods given in this section can be applied to the synthesis problem for a plane aperture. Instead of the Paley-Wiener theorem one must apply its generalization, the Plancherel-Pólya, for entire functions of exponential type in several variables (see, for example, Ronkin [1]), and instead of Lemma 1 its obvious generalization for entire periodic functions in several variables (see Ramm [72]).

5. Inverse Diffraction Problems

In the literature the inverse problem for scattering theory has been intensely studied (Chadan-Sabatier [1]). Most of the results were obtained for the potential scattering for one-dimensional problems.

For the inverse problem in three dimensions a uniqueness theorem was obtained for potential scattering but there is no complete solution of the inverse problem. Roughly speaking, the problem is to find the scattering potential and the surface of the scattering body from scattering data, for example, from the scattering amplitude. In this section we state some inverse problems and give some results due to the author Ramm [61], [64], [73].

1. First we consider the following model diffraction problem

$$\begin{aligned} \Delta\psi + k_1^2\psi &= 0 \quad \text{in} \quad D \\ \Delta\psi + k^2\psi &= 0 \quad \text{in} \quad \Omega \end{aligned} \tag{5.1}$$

$$\psi^+|_\Gamma = \psi^-|_\Gamma, \quad \frac{\partial\psi^+}{\partial N}\bigg|_\Gamma = \frac{\partial\psi^-}{\partial N}\bigg|_\Gamma, \tag{5.2}$$

where D is a bounded domain in $\mathbb{R}^3$ with smooth boundary Γ, $+(-)$ denote limiting values from inside (outside) Γ,

$$\psi = \exp\{ik(\nu,x)\} + v, \tag{5.3}$$

where ν is an ort, and v satisfies the radiation condition. The solution to problem (5.1)-(5.3) satisfies the equation

$$\psi(x,k) = \exp\{ik(\nu,x)\} - (k^2-k_1^2)\int_D \frac{\exp(ikr_{xy})}{4\pi r_{xy}}\psi(y,k)dy. \tag{5.4}$$

The scattering amplitude is defined as

$$f(k,k_1,n,\nu) \tag{5.5}$$
$$= \lim_{\substack{|x|\to\infty,\\ x/|x|=n}} |x|\exp(ik|x|)v = \frac{(k_1^2-k^2)}{4\pi}\int_D \exp\{-ik(n,y)\}\psi(y,k)dy.$$

From (5.4) and (5.5) we obtain

$$f(k,k_1,n,\nu) = \frac{k_1^2-k^2}{4\pi}\int_D \exp\{ik(\nu-n,y)\}dy \tag{5.6}$$
$$+ \frac{(k_1^2-k^2)^2}{(4\pi)^2}\int_D \exp\{-ik(n,y)\}\int_D \frac{\exp(ikr_{yz})}{r_{yz}}\psi(z,k)dz.$$

We fix k_1^2 and show that if the function $\partial f(k,k_1,n,\nu)/\partial k^2\Big|_{k^2=k_1^2}$ known for $n,\nu \in S^2$, where S^2 is the unit sphere in $\mathbb{R}^3$, then the shape of D can be uniquely determined. From (5.6) we find

$$\left.\frac{\partial f}{\partial k^2}\right|_{k^2=k_1^2} = \int_D \exp\{ik_1(\nu-n,y)\}dy. \tag{5.7}$$

It is not difficult to justify the passage from (5.6) to (5.7) using, for example, theorems about the dependence of solutions of linear operator equations on a parameter (see, for example, Ramm [89]). If the function (5.7) is known we know the entire function

$$\int_D \exp\{i(z,y)\}dy \equiv \phi(z) = \phi(z_1,z_2,z_3) \tag{5.8}$$

in a neighborhood of the origin in $\mathbb{R}_z^3$. By the uniqueness theorem for analytic functions this means that the Fourier transform (5.8) of the characteristic function η of domain D,

$$\eta = \begin{cases} 1, & x \in D \\ 0, & x \in \Omega, \end{cases}$$

is known. Hence we know the shape of D. To prove that

$\phi(z)$ is known in some ball $|z| < \varepsilon$, we note that $k_1(\nu-n) = k_1|\nu-n|\cdot\ell$, where $\ell = (\nu-n)/|\nu-n|$, $0 \le |\nu-n| \le 2$. Hence $0 \le k_1|\nu-n| \le 2k_1$ so that $\phi(z)$ is known in the ball $x_1^2 + x_2^2 + x_3^2 \le 4k_1^2$, $x_j = \operatorname{Re} z_j$. By the uniqueness theorem, the analytic function $\phi(z)$ is uniquely determined by the data. So we have proved that the shape of D is uniquely determined if function (5.7) is known for $n,\nu \in S^2$.

Remark 1. By similar arguments it can be proved that if we replace the operator Δ in (5.1) and (5.2) by $\Delta + p(x)$ then the scattering amplitude for $n,\nu \in S^2$, $k = k_1$, $0 < k < \infty$ determines D and $p(x)$ uniquely (see Ramm [64]).

2. Consider a convex centrally symmetric reflecting body D with boundary Γ. Assume that the Gaussian curvature of Γ is positive and continuous. Let

$$(\Delta+k^2)u = 0 \quad \text{in} \quad \Omega, \quad u|_\Gamma = 0, \quad u = \exp\{ik(\nu,x)\}+v, \tag{5.9}$$

where v satisfies the radiation condition. The scattering amplitude for the problem of scattering of the plane wave $\exp ik(\nu,x)$ from the body D is

$$f(n,\nu,k) = \frac{-1}{4\pi}\int_\Gamma \exp\{-ik(n,s)\}\frac{\partial u}{\partial N_s}\, ds. \tag{5.10}$$

We prove that a knowledge of $f(n,\nu,k)$ for $n,\nu \in S^2$ and $k \to \infty$ uniquely determines the shape of Γ and give some formulas for calculating this shape.

Let Γ_+ be the illuminated part of Γ and let Γ_- be the part of Γ which is in shadow. We set, in the short-wave approximation $k \to \infty$,

$$\left.\frac{\partial u}{\partial N}\right|_{\Gamma_-} = 0, \quad \left.\frac{\partial u}{\partial N}\right|_{\Gamma_+} = 2\,\frac{\partial \exp\{ik(\nu,s)\}}{\partial N}$$

and obtain

$$f(n,\nu,k) = \frac{k}{2\pi i}\int_{\Gamma^+} \exp\{-ik|\nu-n|(\ell,s)\}(\nu,N_s)ds, \quad (5.11)$$

where $\ell = (n-\nu)/|n-\nu|$. Evaluating integral (5.11) by the method of stationary phase (see, for example, Fedorjuk [1]) we obtain

$$f(n,\nu,k) = -\frac{1}{2}\sqrt{R_1R_2}\exp\{-2ika(\ell)\cos n\ell\}, \quad n \neq \nu \quad (5.12)$$

where the origin of the coordinate system is placed at the center of symmetry of D, M is the point on Γ at which the normal to Γ is directed along ℓ, R_1 and R_2 are the principal radii of curvature at M, and $a(\ell)$ is the semi-width of D in the direction ℓ. In the coordinate system with origin at M and X_3-axis directed along ℓ, we obtain

$$f(n,\nu,k) = -\frac{1}{2}\sqrt{R_1R_2} = -\frac{1}{2\sqrt{K}}, \quad (5.13)$$

where $K = (R_1R_2)^{-1}$ is the Gaussian curvature at M. Knowing K on Γ we can determine Γ uniquely by Minkowski's theorem (see Blaschke [1, p. 182]). Knowing $a(\ell)$ for all $\ell \in S^2$, we can calculate the shape of Γ by the formula (Blaschke [1], p. 168):

$$x_j = \frac{\partial a(\ell)}{\partial \alpha_j}, \quad 1 \leq j \leq 3, \quad (5.14)$$

where $a(\ell) = a(\alpha_1,\alpha_2,\alpha_3)$ and α_j are the Cartesian coordinates of ℓ. For example if $a(\ell) = R = \text{const}$, then $x_j = R\alpha_j$, $1 \leq j \leq 3$. This is the parametric equation of the sphere with radius R.

For $n = \nu$, it is easy to calculate the integral (5.11),

$$f(n,\nu,k) = -\frac{k}{2\pi i}S(\nu), \quad (5.15)$$

where $S(\nu)$ is the area of the projection of Γ onto the plane orthogonal to ν. According to the optical theorem (Newton [1]) for the cross-section $\sigma = \sigma(\nu)$ we have $\sigma = (4\pi/k)\operatorname{Im} f(\nu,\nu,k)$. Hence $\sigma = 2S(\nu)$, which is a well-known result of geometrical optics. We can find the area S of Γ by the formula (Blaschke [1, p. 176])

$$S = \int_{S^2} K^{-1}(\omega)\,d\omega \tag{5.16}$$

and the area of the projection F_t of D in the direction t by the formula (Blaschke [1, p. 76])

$$|F_t| = \frac{1}{2}\int_{S^2} |\cos t\omega| K^{-1}(\omega)\,d\omega. \tag{5.17}$$

Another way to calculate the shape of Γ can be described as follows. From (5.11) we obtain

$$\begin{aligned} g(n,\nu,k) &\equiv f(n,\nu,k) + f^*(-n,-\nu,k) \\ &= \frac{k}{2\pi i}\int_\Gamma \exp\{-ik(n-\nu,s)\}(N,\nu)\,ds \\ &= \frac{k\nu}{2\pi i}\int_D \nabla_j \exp\{-ik(n-\nu,y)\}\,dy \\ &= \frac{1-(\nu,n)}{2\pi} k^2 \int_D \exp\{-ik|\nu-n|(\ell,y)\}\,dy, \\ &\qquad \ell = \frac{n-\nu}{|n-\nu|}. \end{aligned} \tag{5.18}$$

Using the asymptotic formula for the Fourier transform as $k \to \infty$ of the characteristic function of a convex domain (John [1]) we obtain

$$g(n,\nu,k) \sim -\frac{\cos\{k_1 a(\ell)\}}{\sqrt{K(\ell)}} \quad \text{as} \quad k \to \infty, \tag{5.19}$$

where $k_1 = k|n-\nu|$. Hence we know $a(\ell)$ and can calculate the shape of Γ from formula (5.14).

6. Optimal Solution to the Antenna Synthesis Problem

1. The problem of linear antenna synthesis can be reduced to solving the equation

$$Aj \equiv \int_{-\ell}^{\ell} j(z)\exp(ikz)dz = f(k), \quad -k_o \le k \le k_o, \tag{6.1}$$

where $f(k)$ is the given radiation pattern, $j(z) \in L^2 = L^2(-\ell,\ell)$, $j(z)$ is the desired current distribution, and $j(z)$, $f(k)$ are scalar functions. It is quite clear that equation (6.1) has at most one solution and is solvable if and only if $f(k) \in W_\ell = W_{(-\ell,\ell)}$, where the class W_D was defined near the end of Section 2 of this chapter. If the given function $f(k) \notin W_\ell$ but $f_\varepsilon(k) \in W_\ell$, $\|f_\varepsilon - f\| < \varepsilon$, $\|\cdot\| = \|\cdot\|_{L^2(-k_o,k_o)}$, then we consider the j_ε corresponding to f_ε as an approximate solution to the synthesis problem. Since the operator A is compact, small perturbations of f can cause large variations of j. So j_ε can change greatly when f_ε is changed slightly. This phenomenon has been discussed in the literature (Minkovich-Jakovlev [1]) in connection with the superdirectivity of antennas. From a practical point of view, we should find a stable current distribution which generates a radiation pattern close to the desired pattern $f(k)$.

So we require that

$$\int_{-\ell}^{\ell} |j|^2 dz \le M_o, \quad \int_{-\ell}^{\ell} |j'(z)|^2 dz \le M_1, \tag{6.2}$$

where M_o, M_1 are some constants.

We denote the set of functions satisfying conditions (6.2) by $\Omega(M_o,M_1)$. It is convex and compact in L^2. So we arrive at the problem of solving equation (6.1) under conditions (6.2) in the following sense: given a function $g(k) \in L^2(-k_o,k_o) = L^2$ we want to find $f(k) \in W_\ell$ such that

$$J = \int_{-\infty}^{\infty} |g(k)-f(k)|^2 dk = \min, \quad g(k) = 0 \text{ for } |k| > k_o, \tag{6.3}$$

$$f(k) = Aj \quad (\text{see } (6.1)), \tag{6.4}$$

$$j \in \Omega(M_o,M_1). \tag{6.5}$$

This optimization problem can, in principle, be solved by methods of nonlinear programming, by direct methods of the calculus of variations, and by methods of calculus of variations based on the Euler equation. We consider these possibilities.

2. Let

$$\tilde{g}(z) = \frac{1}{2\pi} \int_{-\infty}^{\infty} g(k) \exp(-ikz) dk. \tag{6.6}$$

From Parseval's equality we obtain

$$\begin{aligned} \frac{J}{2\pi} &= \int_{-\infty}^{\infty} |g(z)-j(z)|^2 \\ &= \int_{|z|>\ell} |\tilde{g}(z)|^2 dt + \int_{-\ell}^{\ell} |\tilde{g}(z)-j(z)|^2 dz \\ &= \delta(g,\ell) + J_1, \end{aligned} \tag{6.7}$$

where $\delta(g,\ell)$ does not depend on $j(z)$. Hence J and J_1 attain their minimum at the same function and the problem (6.3)-(6.5) is equivalent to the problem

$$\begin{cases} J_1 = \min \\ j \in \Omega(M_o,M_1). \end{cases} \tag{6.8}$$

We assume at first, for simplicity, that

$$j(\ell) = j(-\ell), \tag{6.9}$$

but later we eliminate this assumption. Let

$$j(z) = \sum_{n=-\infty}^{\infty} j_n \frac{\exp(in\pi z/\ell)}{\sqrt{2\ell}}, \quad \tilde{g} = \sum_{n=-\infty}^{\infty} \tilde{g}_n \frac{\exp(in\pi z/\ell)}{\sqrt{2\ell}},$$
$$j_n \equiv \int_{-\ell}^{\ell} j(z) \frac{\exp(-in\pi z(\ell))}{\sqrt{2\ell}} \, dz. \tag{6.10}$$

We note that assumption (6.9) will be used only once, to justify term by term differentiation of the Fourier series of $j(z)$. Substituting (6.10) into (6.8) we obtain

$$\begin{cases} J_1 = \sum_{h=-\infty}^{\infty} |j_n - \tilde{g}_n|^2 = \min \\ \sum_{n=-\infty}^{\infty} |j_n|^2 \le M_o, \quad \sum_{n=-\infty}^{\infty} n^2 |j_n|^2 \le M_1 \frac{\ell^2}{\pi^2}. \end{cases} \tag{6.11}$$

This is a convex programming problem. Since $\Omega(M_o, M_1)$ is uniformly convex and compact in ℓ^2 and the functional $J_1 = \|\tilde{g} - j\|^2$ is strictly convex, problem (6.11) has a unique solution. A numerical solution to problem (6.11) can be obtained in the following manner. We fix N and set

$$\begin{cases} J_{1N} = \sum_{n=-N}^{N} |j_n - \tilde{g}_n|^2 \\ \sum_{n=-N}^{N} |j_n|^2 \le M_o, \quad \sum_{n=-N}^{N} n^2 |j_n|^2 \le M_1 \frac{\ell^2}{\pi^2}. \end{cases} \tag{6.12}$$

This is a finite-dimensional convex programming problem. In Zuhovickij-Avdeeva [1], Polak [1], there are algorithms for solving this problem. We denote the solution to problem (6.12) by $j^{(N)}$ and show that $j^{(N)} \to j$ as $N \to \infty$. Here j is the solution to problem (6.8) and $\to$ denotes convergence in ℓ^2. We denote $J_{1N}(j^{(N)}) = \delta_N$,

$$J_1(j) = d, \quad \sum_{n=-N}^{N} |j_n - \tilde{g}_n|^2 = d_N, \quad \sum_{|n|>N} |\tilde{g}_n|^2 = \varepsilon_N.$$

Since $d \le d_N + \varepsilon_N$, $\delta_N \le d_N$ we obtain $\lim_{N\to\infty} \delta_N \le d$. It is clear that $\delta_N < \delta_{N+1}$. Let $\delta = \lim_{N\to\infty} \delta_N$, $\delta < d$, $\lim_{N\to\infty} j^{(N)} = j^{(\infty)} \in \Omega(M_o,M_1)$, $J_1(j^{(\infty)}) = \delta < d$. Hence by the definition of d we obtain $\delta = d$ and by the uniqueness of solution to problem (6.8) we obtain $j^{(\infty)} = j$. Hence $j^{(N)} \to j$ as $N \to \infty$. If $\tilde{g} \in \Omega(M_o,M_1)$ then the solution $j_n = \tilde{g}_n$ to problem (6.11) is trivial, but is uninteresting since in practice $\tilde{g} \notin \Omega(M_o,M_1)$. If $\tilde{g} \notin \Omega(M_o,M_1)$ then it is clear for geometrical reasons that the solution to problem (6.11) lies on the boundary of the set $\Omega(M_o,M_1)$ and hence equality should hold in at least one of the inequalities (6.11). If one of the inequalities (6.11) is strict and equality holds in the other, we can omit the strict inequality and consider the problem (6.11) with the equality as the only restriction. We do not want to discuss this relatively simple problem and pass to the more difficult case in which equality holds in both inequalities (6.11). In that case we obtain the problem

$$\begin{cases} J_1 = \sum_{n=-\infty}^{\infty} |j_n - \tilde{g}_n|^2 = \min \\ \sum_{n=-\infty}^{\infty} |j_n|^2 = M_o, \quad \sum_{n=-\infty}^{\infty} n^2 |j_n|^2 = M_2 = M_1 \dfrac{\ell^2}{\pi^2}. \end{cases} \tag{6.13}$$

The same is valid for problem (6.12), for which we have

$$\begin{cases} J_{1N} = \sum_{h=-N}^{N} |j_n - \tilde{g}_n|^2 = \min \\ \sum_{n=-N}^{N} |j_n|^2 = M_o, \quad \sum_{n=-N}^{N} n^2 |j_n|^2 = M_2. \end{cases} \tag{6.14}$$

The latter problem can be easily solved with the help of the Lagrange multipliers. We set

$$\Phi = J_{1N} + \lambda\left(\sum_{n=-N}^{N} |j_n|^2 - M_o\right) + \mu\left(\sum_{n=-N}^{N} n^2|j_n|^2 - M_2\right) \tag{6.15}$$

and write the Euler equations

$$\frac{\partial \Phi}{\partial j_n} = 0, \quad \frac{\partial \phi}{\partial j_n^*} = 0, \quad -N \le n \le N, \tag{6.16}$$

whose solution

$$j_n = \frac{\tilde{g}_n}{1+\lambda+\mu n^2}, \quad -N \le n \le N, \tag{6.17}$$

can easily be found. The parameters λ, μ should be found from the equations

$$\sum_{n=-N}^{N} \frac{|\tilde{g}_n|^2}{(1+\lambda+\mu n^2)^2} = M_o, \quad \sum_{n=-N}^{N} \frac{n^2|g_n|^2}{(1+\lambda+\mu n^2)^2} = M_2. \tag{6.18}$$

Equations (6.18) can be easily solved by numerical methods.

3. Problem (6.13) can be written as

$$\begin{cases} J_1 = \int_{-\ell}^{\ell} |j(z) - \tilde{g}(z)|^2 dz = \min, \\ \int_{-\ell}^{\ell} |j|^2 dz = M_o, \quad \int_{-\ell}^{\ell} |j'(z)|^2 dz = M_1. \end{cases} \tag{6.19}$$

Using the calculus of variations, we obtain the Euler equation and the natural boundary conditions

$$\begin{aligned} &(1+\lambda)j(z) - \mu j''(z) = \tilde{g}(z), \quad -\ell \le z \le \ell, \\ &j'(-\ell) = j'(\ell) = 0. \end{aligned} \tag{6.20}$$

This equation can be solved explicitly and then the parameters λ, μ can, in principle, be found from equalities (6.19). Problem (6.19) can be also solved by direct methods of the calculus of variations, for example, by the Ritz method

(Mihlin [1]). If we set

$$j^{(N)} = \sum_{n=-N}^{N} j_n \frac{\exp(in\pi z/\ell)}{\sqrt{2\ell}},$$

we obtain the system (6.16) for the unknown coefficients j_n with restrictions (6.14).

4. Here we eliminate assumption (6.9). If $j(\ell) \neq j(-\ell)$ then even a function smooth in the closed segment $[-\ell,\ell]$ may have Fourier coefficients which are $O(\frac{1}{n})$, and so the Fourier coefficients of $j_n'(z)$ are not equal to $(i\pi n/\ell)j_n$. This is true because the 2ℓ-periodic continuation of the function $j(z)$ may be not smooth. To avoid this difficulty we can use the orthonormal basis in $L^2(-\ell,\ell)$ given by

$$\phi_n = \frac{1}{\sqrt{\ell}} \cos(\frac{n\pi z}{2\ell} + \frac{\pi n}{2}), \quad n = 0,1,2,\ldots . \tag{6.21}$$

Note that the ϕ_n are the eigenfunctions of the following problem

$$y'' + \lambda y = 0, \quad -\ell \le z \le \ell; \quad y'(-\ell) = y'(\ell) = 0. \tag{6.22}$$

The Fourier coefficients

$$C_n = \int_{-\ell}^{\ell} j(z)\phi_n(z)dz \tag{6.23}$$

are $O(n^{-2})$ as $n \to \infty$ and we can differentiate the Fourier series of $j(z)$ with respect to the system $\{\phi_n\}$ term by term.

5. One can solve other antenna synthesis problems by the method given in this section, for example, the problem of synthesizing the spherical antenna and the directional antenna (Ramm [70]).

7. Research Problems

1. Investigate the following inverse problem. Let

$$\begin{cases} (\Delta + k_1^2 + p(x))u = 0 \quad \text{in} \quad D \\ (\Delta + k^2 + p(x))u = 0 \quad \text{in} \quad \Omega. \\ \left.\frac{\partial u^+}{\partial N}\right|_\Gamma = a(x) \left.\frac{\partial u^-}{\partial N}\right|_\Gamma, \quad u^+|_\Gamma = u^-|_\Gamma. \\ u = \exp\{ik(\nu,x)\} + v \\ v \quad \text{satisfies the radiation condition.} \end{cases}$$

Is it possible to find $a(x)$, $p(x)$, and Γ from the scattering amplitude?

2. Let

$$\begin{cases} (\Delta + k^2 + p(x))u = 0 \quad \text{in} \quad \Omega, \\ \frac{\partial u^-}{\partial N} - h(x)u^-|_\Gamma = 0, \\ u = \exp\{ik(\nu,x)\} + v. \end{cases}$$

Is it possible to find $h(x)$, $p(x)$, and Γ from the scattering amplitude?

3. Investigate the stability of solutions to inverse problems 1 and 2.

4. Is it possible to find a basis $\{\phi_j\}$ in $L^2[-\ell,\ell]$ such that

$$f^{(n)}(x) = \sum_{j=1}^{\infty} f_j \phi_j^{(n)}(x)$$

whenever $f \in C^n[a,b]$, where $f_j = (f,\phi_j) = \int_a^b f(x)\phi_j^*(x)dx$? For any fixed n this is possible by the method given in point 4 of Section 6 of this chapter for $n = 1$.

5. Suppose we know the integral $\int_a^b g(u(x))dx$ for any $g(x) \in C(I)$, $I = (-\infty,\infty)$. When is it possible to find the

unknown $u(x)$ from this data? Give an algorithm for finding $u(x)$.

6. Construct in detail an iterative process for solving equation (16) from Appendix 1 and give an estimate for N so that for an operator similar to operator K (see formula (5) and (7) of Appendix 1) the estimate $||K|| < 1$ holds.

8. Bibliographical Note

In this chapter we present some of the results from papers by Ramm [59]-[75]. There are many papers on improperly posed problems (see Lattes-Lions [1], Tihonov-Arsenin [1]) but very few contain any effective error estimates with explicitly given constants. In Ramm [65], [66] there are examples of such estimates. In Section 1 we present the contents of the paper Ramm [65]. A survey of antenna synthesis theory is given in Minkovich-Jakovlev [1]. Nonlinear problems of antenna synthesis theory have not been investigated in the literature. In Section 4 we present some results due to the author ([68], [71], [73]). R. Kuhn [1] is a monograph on antennas.

In Section 3 a result from Ramm [58] is given. One of the first papers in which optimal solutions to antenna synthesis problems were obtained was the paper presented by the author to the international URSI Symposium (Stresa, Italy, 1968) (see Ramm [69], [70]).

Inverse problems in diffraction theory are interesting for applications but their theory is not sufficiently developed. Some other inverse problems of interest for seismology are studied in Gerver [1] and Lavrentjev-Vasiljev-Romanov [1].

The approximation formula from Section 3 is of use in apodization theory (Ramm [14]-[16]). In Appendix 1 we consider an integral equation of the type which is of interest in potential theory and for the problem of numerical analytical continuation and give a stable method to solve this equation (Ramm [74]).

In Ramm [100] some stable methods to sum Fourier series with perturbed coefficients are given and numerical examples are presented. In Ramm [115] other ill-posed problems are studied.

There was much activity in diffraction theory in recent years. In particular, the high frequency asymptotics of the scattering amplitude discussed in Section 5 was studied for non-convex bodies in Petkov [1] and for convex bodies by Majda and Taylor [1]. Additional references one can find in Petkov [1].

APPENDIX 1

STABLE SOLUTION OF THE INTEGRAL EQUATION OF THE INVERSE PROBLEM OF POTENTIAL THEORY

1. Consider the equation

$$\int_{-1}^{1} \frac{\phi(t)}{x-t}\, dt = f(x), \quad x > 1, \tag{1}$$

where the integral operator maps some space of functions defined on $[-1,1]$ into some space of functions defined on the semiaxis $x > 1$. It is clear that if equation (1) is solvable, then $f(x)$ can be analytically continued into the complex plane with cut $[-1,1]$ to a function which vanishes at infinity. It is also clear that equation (1) has no more than one solution $\phi \in C[-1,1]$ or $\phi \in L^p[-1,1]$, $p > 1$. We assume that equation (1) has a solution and construct an iterative process for solving this equation.

We set

$$x = N+y, \quad -1 \leq y \leq 1, \quad f(N+y) = g(y), \tag{2}$$

where $N > 2$ will be chosen later, and rewrite equation (1) as

$$A\phi \equiv \int_{-1}^{1} \frac{\phi(t)dt}{N+y-t} = g(y), \quad -1 \leq y \leq 1. \tag{3}$$

Equation (3) is equivalent to the equation

$$K\phi = \int_{-1}^{1} k(z,t)\phi(t)dt = \psi(z), \quad -1 \leq z \leq 1, \tag{4}$$

where

$$\begin{cases} k(z,t) = \int_{-1}^{1} \frac{dy}{(N+y-t)(N+y-z)} = \frac{1}{t-z} \ln \frac{N+1-t}{N+1-z} \frac{N-1-z}{N-1-t} . \\ \psi = A^*g. \end{cases} \tag{5}$$

Here we need the known Lemma 1.

Lemma 1. Let $A: H \to H$ *be a closed linear operator,* $f \in D(A^*)$, *and* $f \in R(A)$. *Then the equation* $A^*A\phi = A^*f$ *is equivalent to the equation* $A\phi = f$.

Remark 1. If A is bounded, then the condition $f \in D(A^*)$ can be omitted.

Proof of Lemma 1: It is obvious that every solution of the equation $A\phi = f$ is a solution of the equation $A^*A\phi = A^*f$. Conversely, suppose $A^*A\phi = A^*f$. Since $f \in R(A)$ there is an element ψ such that $A\psi = f$. Hence $A^*A(\phi-\psi) = 0$. Taking the inner product of this equality with $\phi - \psi$, we obtain $A(\phi-\psi) = 0$. Therefore $A\phi = A\psi = f$. □

The operator K with kernel (5) is selfadjoint and positive definite in $L^2(-1,1]$. Equation (4) is equivalent to the equation

$$\phi = (I-K)\phi + \psi. \tag{6}$$

It can be verified that the kernel (5) satisfies the following asymptotic equality

$$k(z,t) \sim \frac{2}{N^2} \quad \text{as} \quad N \to \infty. \tag{7}$$

Hence the operator

$$B = I - K \tag{8}$$

is positive definite as $N \to \infty$ and $||B|| \leq 1$. Since $K = K^*$ is compact we conclude that $||B|| = ||I-K|| = 1$. Indeed, $||B|| = \sup_{||\phi|| \leq 1} ||\phi - K\phi|| \geq ||\phi_n - K\phi_n|| \geq 1-\lambda_n \to 1$ as $n \to \infty$. Here $\{\phi_n\}$, $\{\lambda_n\}$ are the systems of eigenfunctions and eigenvalues of K. Hence $||B|| = 1$. Now we make use of the following theorem due to M. Krasnoselskij (Krasnoselskij, Vainikko, et al. [1, p. 71]): Let B be a selfadjoint linear operator on a Hilbert space H such that $||B|| = 1$ and -1 is not an eigenvalue of B. Let $f \in R(I-B)$. Then the iterative process $\phi_{n+1} = B\phi_n + f$, with $\phi_o \in H$ arbitrary, converges in H to a solution of the equation $\phi = B\phi + f$.

In our case the operator (8) is selfadjoint, has norm 1, and is nonnegative definite so that -1 is not an eigenvalue. Hence we prove

Theorem 1. _If equation_ (1) _has a solution, then this solution can be calculated by the iterative process_

$$\phi_{n+1} = (I-K)\phi_n + \psi, \quad \psi \in L^2[-1,1], \tag{9}$$

with ϕ_o _is arbitrary_. $\lim_{n\to\infty} \phi_n = \phi$ _is the solution of equation_ (1), _provided that_ N _is large enough_ (_for example_, $N > 10$).

2. The iterative process (9) is a stable method for solving the equation (6). Indeed, assume that in the right-hand side of equation (1) is substituted by f_δ, $||f - f_\delta||_{L^2[1,\infty)} \leq \delta$, so that in equation (6) ψ is substituted by ψ_α, $||\psi - \psi_\alpha||_{L^2[-1,1]} \leq \alpha$, $\alpha = \alpha(\delta) \to 0$ as $\delta \to 0$. Let $\phi_{n,\alpha}$ denote the sequence (9) for $\psi = \psi_\alpha$ and ϕ denote the solution of equation (6).

Theorem 2. There exists $n = n(\alpha)$ such that

$$\|\phi_{n(\alpha),\alpha} - \phi\|_{L^2[-1,1]} \to 0 \quad \text{as} \quad \alpha \to 0. \tag{10}$$

Proof of Theorem 2: The proof we give is also a proof that if we construct an iterative process to solve an operator equation of the first kind we are able to construct a stable method to solve this equation. Let $\|\cdot\| = \|\ \|_{L^2[-1,1]}$. We start with the inequality

$$\|\phi - \phi_{n,\alpha}\| \leq \|\phi - \phi_n\| + \|\phi_n - \phi_{n,\alpha}\| . \tag{11}$$

Theorem 1 shows that

$$\varepsilon(n) \equiv \|\phi - \phi_n\| \to 0 \quad \text{as} \quad n \to \infty. \tag{12}$$

It is clear that

$$\|\phi_n - \phi_{n,\alpha}\| \leq C(n)\,\|\psi - \psi_\alpha\| = C(n)\alpha, \tag{13}$$

where $C(n)$ does not depend on α, ψ, ψ_α. Hence

$$\|\phi - \phi_{n,\alpha}\| \leq \varepsilon(n) + \alpha C(n). \tag{14}$$

We take $n(\alpha)$ so that $n(\alpha) \to \infty$ as $\alpha \to 0$ and $\alpha C(n(\alpha)) \to$ as $n \to \infty$. Then

$$\|\phi - \phi_{n(\alpha),\alpha}\| \to 0 \quad \text{as} \quad \alpha \to 0 \tag{15}$$

The proof is completed. □

Remark 2. In our case $\|B\| \leq 1$. If $\phi_o = \psi$ then

$$\|\phi_n - \phi_{n,\alpha}\| \leq \sum_{j=0}^{n} \|B^j\|\,\|\psi - \psi_\alpha\| \leq (n+1)\|\psi - \psi_\alpha\| \leq (n+1)\alpha.$$

In general, $C(n) \leq \sum_{j=0}^{n} \|B^j\|$.

Remark 3. Without any essential alterations the given method can be applied to the equation

$$\int_D \frac{\phi(t)dt}{r_{st}} = f(x), \quad x \in \Delta, \tag{16}$$

where D is a bounded domain in $\mathbb{R}^N$, $r_{st} = |x-t|$, $x,t \in \mathbb{R}^N$, and Δ is a domain which contains all vectors $N+t$, $t \in D$, with sufficiently large N.

APPENDIX 2

ITERATIVE PROCESSES FOR SOLVING BOUNDARY VALUE PROBLEMS

Here we formulate iterative processes for solving the interior and exterior boundary value problems

$$\Delta u = 0 \quad \text{in} \quad D, \quad u|_\Gamma = \phi, \tag{1}$$

$$\Delta v = 0 \quad \text{in} \quad \Omega, \quad \partial v/\partial N_e|_\Gamma = \psi, \quad v(\infty) = 0, \tag{2}$$

$$\Delta u = 0 \quad \text{in} \quad \Omega, \quad u|_\Gamma = \phi, \quad u(\infty) = 0, \tag{3}$$

$$\Delta v = 0 \quad \text{in} \quad D, \quad \partial v/\partial N_i|_\Gamma = \psi, \quad \int_\Gamma \psi dt = 0, \tag{4}$$

where D is a bounded domain in $\mathbb{R}^3$ with smooth boundary Γ. We set

$$v = \int_\Gamma \frac{\sigma(t)dt}{4\pi r_{st}}, \quad u = \int_\Gamma \mu(t) \frac{\partial}{\partial N_t} \frac{1}{4\pi r_{xt}} dt. \tag{5}$$

Taking into account the formulas

$$u_{\substack{i\\e}} = \frac{A^*\mu \mp \mu}{2}, \quad A^*\mu = \int_\Gamma \mu(t) \frac{\partial}{\partial N_t} \frac{1}{2\pi r_{st}} dt, \tag{6}$$

$$\frac{\partial v}{\partial N_{\substack{i\\e}}} = \frac{A\sigma \pm \sigma}{2}, \quad A\sigma = \int_\Gamma \sigma(t) \frac{\partial}{\partial N_s} \frac{1}{2\pi r_{st}} dt, \tag{7}$$

we obtain from (1) and (2) the integral equations

$$\mu = A^*\mu - 2\phi, \tag{1'}$$

$$\sigma = A\sigma - 2\psi. \tag{2'}$$

These equations have a unique solution. It is well-known that the spectral radius of A and A^* is equal to 1 and that -1 is the only characteristic value of A and A^* on the circle $|\lambda| = 1$ (see Theorem 3 in Chapter 2).

Theorem 1. _Let_ A _be a bounded linear operator in a Hilbert space_ H _with spectral radius_ $r_A = 1$, 1 _does not belong to the spectrum of_ A. _Then for any_ $a > 0$, _the equation_

$$\phi = A\phi + f \tag{8}$$

can be solved by the iterative process

$$\phi_{n+1} = B\phi_n + \phi_o, \quad \phi_o = \frac{f}{1+a}, \quad B = \frac{A+aI}{1+a}, \ a > 0 \tag{9}$$

which converges at the rate of a convergent geometric series.

Remark 1. The iterative process (9) can be applied to solve equations (1') and (2'). Equation (2') can also be solved by the iterative process

$$\sigma_{n+1} = A\sigma_n - 2\psi, \quad \sigma_o = -2\psi. \tag{10}$$

Proof of Theorem 1: It is clear that equation (8) is equivalent to the equation

$$\phi = B\phi + \phi_o, \tag{11}$$

where B and ϕ_o are defined in formula (9). Consider the equation

$$\phi = \lambda B\phi + \phi_o. \tag{12}$$

For $\lambda = 1$, equation (12) coincides with equation (11). For $|\lambda| \leq \delta$, $\delta > 0$ sufficiently small, equation (12) has the solution

$$\phi = \sum_{n=0}^{\infty} \lambda^n B^n \phi_o . \tag{13}$$

Here $\phi = \lim \phi_n$ where the ϕ_n are obtained from the iterative process $\phi_{n+1} = \lambda B \phi_n + \phi_o$, and ϕ is analytic in λ for $|\lambda| \leq \delta$. If ϕ is analytic in a disk $|\lambda| \leq R$, $R > 1$, then the iterative process (9) converges and $q = R^{-1}$. Let u prove that the function (13) is analytic in the disk $|\lambda| \leq R$ $R > 1$. Equation (12) can be rewritten in the form

$$\phi = zA\phi + f, \quad z = \frac{\lambda}{1+a-\lambda a}, \quad b = \frac{1}{1+a-\lambda a} . \tag{14}$$

The solution of equation (14) is analytic in a domain Δ which includes the disk $|z| < 1$ and a neighborhood of the point $z = 1$. For any $a > 0$ we can find $R > 1$ such that the disk $|\lambda| \leq R$ is mapped by the function $z = \lambda/(1+a-\lambda a)$ onto a disk $K_r \subset \Delta$. From here we get the statement of Theorem 1. Indeed the function $z = \lambda/(1+a-\lambda a)$ is analytic in the disk $|\lambda| \leq R$, $R > 1$, and maps this disk onto $K_r \subset \Delta$. The solution $\phi(z)$ of equation (14) is analytic in Δ. Hence the solution $\phi(\lambda)$ of equation (12) is analytic in the disk $|\lambda| \leq R$. It remains to prove that the disk $|\lambda| \leq R$, for some $R > 1$, is mapped by the function $z(\lambda) = \lambda/(1+a-\lambda a)$ into Δ. It is clear that $z(\lambda^*) = \{z(\lambda)\}$ and that $z(\lambda)$ maps circles in the λ-plane into circles in the z-plane. Hence the circle $|\lambda| = R$ is mapped onto the circle K_r with the diameter $[z(-R), z(R)]$ lying on the real axis, radius $r = \frac{1}{2}(z(R) + z(-R))$ and the center

$\frac{1}{2}(z(R) + z(-R))$. Hence $K_r \subset \Delta$ provided that $z(-R) > -1$, $|z(R)-1| < \alpha$, $\alpha > 0$, are sufficiently small. We set $R = 1 + \varepsilon$, $\varepsilon > 0$ sufficiently small and note that

$$z(R) = \frac{1+\varepsilon}{1+a-a(1+\varepsilon)} = 1 + \varepsilon(1+a) + O(\varepsilon^2);$$

$$z(-R) = \frac{-1}{1+2a}\left[1 + \varepsilon \frac{1+a}{1+2a} + O(\varepsilon^2)\right].$$

Therefore $z(-R) > -1$ and $|z(R)-1| < \alpha$ provided that $\varepsilon > 0$ is sufficiently small. Hence $K_r \subset \Delta$. We can choose a so that R will be maximal, that $q = R^{-1}$ will be minimal, and the process (9) will converge at the maximal rate. □

<u>Remark 2</u>. Setting $a = 1$ in (9), we obtain the known Neumann process of solving the interior Dirichlet problem (see Gunter [1]).

For problem (4) we obtain the equation

$$\sigma = -A\sigma - 2\psi. \tag{4'}$$

The number $\lambda = -1$ is a characteristic value of operator A. Since $\lambda = -1$ is a simple characteristic value and ψ is orthogonal to the subspace $N(I+A^*)$ in $H = L^2(\Gamma)$ the restriction of the operator A to the invariant subspace $R(I+A)$ of this operator has no characteristic values in the disk $|\lambda| < |\lambda_2|$, where λ_2 is the second characteristic value of A ($\lambda_1 = -1$ is the first). So the iterative process

$$\sigma_{n+1} = -A\sigma_n - 2\psi, \quad \sigma_o = -2\psi \tag{15}$$

converges to the unique solution of equation (4') in the subspace $R(I+A)$ at the rate of a geometric series with ratio

$q = |\lambda_2|^{-1} < 1$. The process (15) can be changed so that it will be stable under small perturbations of the right-hand side of equality (15) as was done in Theorem 1 in Chapter 2.

Let us construct an iterative process to solve problem (3). We look for a solution of this problem of the form

$$u = \frac{\alpha}{|x|} + \int_\Gamma \mu(t)\frac{\partial}{\partial N_t}\frac{1}{4\pi r_{xt}}\,dt, \quad \alpha = \text{const} > 0, \qquad |x|^2 = x_1^2 + x_2^2 + x_3^2. \tag{16}$$

Substituting (16) into the boundary condition (3) we obtain

$$\mu = -A^*\mu + 2\left(\phi - \frac{\alpha}{|s|}\right), \quad s = x|_\Gamma. \tag{17}$$

Consider the equation

$$\nu = M\nu + 2\left(\phi - \frac{\alpha}{|s|}\right), \quad M\nu \equiv -A^*\nu + \int_\Gamma \nu\,dt. \tag{18}$$

We prove that the operator M has no characteristic value in the disk $|\lambda| \leq 1$ so that the iterative process

$$\nu_{n+1} = M\nu_n + 2\left(\phi - \frac{\alpha}{|s|}\right), \quad \nu_0 = 2\left(\phi - \frac{\alpha}{|s|}\right) \tag{19}$$

converges to the solution of equation (18). Moreover, we can choose α so that equations (18) and (17) are equivalent, i.e., so that

$$\int_\Gamma \nu(t)\,dt = 0. \tag{20}$$

This will be true if

$$\alpha = \frac{\int_\Gamma Q\phi\,dt}{\int_\Gamma Q\left(\frac{1}{|t|}\right)dt}, \quad Q \equiv (I - M)^{-1}. \tag{21}$$

It remains to prove that M has no characteristic value in the disk $|\lambda| \leq 1$. Let

$$\nu = \lambda M\nu = -\lambda A^*\nu + \lambda\int_\Gamma \nu dt. \tag{22}$$

Putting

$$u = \int_\Gamma \nu(t) \frac{\partial}{\partial N_t} \frac{1}{4\pi r_{xt}} dt, \tag{23}$$

we obtain from (22) and (6) the equality

$$(1+\lambda)u_e = (1-\lambda)u_i + \lambda\int_\Gamma (u_e - u_i)dt. \tag{24}$$

Multiplying (24) by $\frac{\partial u}{\partial N}$, taking into account that

$$\frac{\partial u}{\partial N_e} = \frac{\partial u}{\partial N_i} \tag{25}$$

by Lyapunov's theorem about the normal derivatives of the double layer potential and integrating over Γ, we obtain

$$\begin{aligned} \frac{1+\lambda}{1-\lambda} \int_\Gamma u_e \frac{\partial u}{\partial N_e} dt &= \int_\Gamma u_i \frac{\partial u}{\partial N_i} dt \\ &+ \frac{\lambda}{1-\lambda} \int_\Gamma (u_e - u_i) dt \int_\Gamma \frac{\partial u}{\partial N} dt. \end{aligned} \tag{26}$$

By Green's formula we have

$$\int_\Gamma u_e \frac{\partial u}{\partial N_e} dt \le 0, \quad \int_\Gamma u_i \frac{\partial u}{\partial N_i} dt \ge 0, \quad \int_\Gamma \frac{\partial u}{\partial N} dt = 0. \tag{27}$$

From (26) and (27) it follows that $(1+\lambda)/(1-\lambda) \le 0$. Hence λ is real and $|\lambda| \ge 1$. It remains to prove that $\lambda = \pm 1$ are not characteristic values of M. If $\lambda = -1$ then from equality (24) we obtain

$$2 \int_\Gamma u_i \frac{\partial u}{\partial N_i} dt = 0. \tag{28}$$

Hence

$$\int_D |\nabla u|^2 dx = 0, \quad u = \text{const} \quad \text{in } D, \quad \frac{\partial u}{\partial N_i} = 0 \tag{29}$$

Therefore $\partial u/\partial N_e = 0$ and hence $u = 0$ in Ω, $\nu = u_e - u_i =$ const. We can assume that $\nu = 1$. But $\nu = 1$ is not a solution of equation (22) for $\lambda = -1$. Indeed, let

$$1 = A^*1 - S, \quad S = \text{meas}\ \Gamma, \tag{30}$$

and let ν_o be the electrostatic density:

$$\nu_o = -A\nu_o, \quad \int_\Gamma \nu_o\, dt = 1. \tag{31}$$

Taking the inner product of equality (30) with ν_o in $H = L^2(\Gamma)$, we obtain $1 = (\nu_o, A^*1) - S = -(\nu_o, 1) - S = -1-S$. Hence we obtain a contradiction. Therefore $\lambda = -1$ is not a characteristic value of M.

If $\lambda = 1$ then $\nu = -A^*\nu + \int_\Gamma \nu dt$. Hence the right-hand side of the last equation is orthogonal to ν_o, $\int_\Gamma \nu dt \int_\Gamma \nu_o dt = 0$. Therefore

$$\int_\Gamma \nu dt = 0. \tag{32}$$

From here it follows that $\nu = -A^*\nu$ so that $\nu = \text{const} \neq 0$. This contradicts equality (32). We have proved

<u>Theorem 2</u>. <u>The solution to problem</u> (3) <u>can be found from formula</u> (16) <u>with</u> $\mu(t) = \lim_{n\to\infty} \nu_n(t)$, α <u>given by formula</u> (21), <u>and</u> ν_n <u>given by formula</u> (19).

<u>Remark 3</u>. We can find α using an iterative process similar to (19). For example,

$$Q\phi = \lim_{n\to\infty} h_n, \tag{33}$$

where

$$h_{n+1} = Mh_n + \phi, \quad h_o = \phi. \tag{34}$$

APPENDIX 3

ELECTROMAGNETIC WAVE SCATTERING BY SMALL BODIES

1. Let D be a body with boundary Γ and characteristic dimension a; ε, μ, σ are the dielectric permeability, magnetic permeability, and conductivity of the body; ε_o, μ_o, $\sigma_o = 0$ are the corresponding parameters of the medium in which the body is placed, $\varepsilon' = \varepsilon + i\sigma\omega^{-1}$, λ_o and ω are the wavelength and frequency of the initial wave field, $\lambda = \lambda_o/\sqrt{|\varepsilon'\mu|}$ is the wave length in the body, and $\delta = \sqrt{2/|\varepsilon'\mu\omega^2|}$ is the depth of skin layer. We consider electromagnetic wave scattering by the body D under the following assumptions which will be treated separately:

$$|\varepsilon'| >> 1, \quad \delta >> a; \quad \lambda_o >> a, \tag{1}$$

$$|\varepsilon'| >> 1, \quad \delta << a; \quad \lambda_o >> a, \tag{2}$$

$$\left|\frac{\varepsilon' - \varepsilon_o}{\varepsilon_o}\right| << 1, \quad \left|\frac{\mu - \mu_o}{\mu_o}\right| << 1. \tag{3}$$

Our aim is to give a new proof of formulas (1.37) in Chapter 2. We start from the equations

$$\text{curl } E = i\omega\mu H, \quad \text{curl } H = -i\omega\varepsilon' E \quad \text{in } D, \tag{4}$$

$$\operatorname{curl} E = i\omega\mu_o H, \quad \operatorname{curl} H = -i\omega\varepsilon_o E + j_o \quad \text{in} \quad \Omega, \tag{5}$$

with the known boundary conditions: $[N,E]$ and $(\mu H,N)$ are continuous when crossing interface Γ, where N is the outward pointing normal to Γ. If $\sigma = \infty$ then $[N,E]|_\Gamma = 0$. This case is possible only under assumption (2). In (5), j_o is the initial current source. Let

$$A_o = \int G(x,y)j_o(y)dy, \quad G(x,y) = \frac{\exp(ik_o|x-y|)}{4\pi|x-y|} \tag{6}$$

$$k_o^2 = \omega^2\varepsilon_o\mu_o, \quad \int = \int_{\mathbb{R}^3}, \tag{7}$$

where E_o, H_o are the electromagnetic fields generated by the initial current in free space, i.e., in the space outside the body D,

$$E_o = \frac{1}{-i\omega\varepsilon_o}(\operatorname{curl}\operatorname{curl} A_o - j_o), \quad H_o = \operatorname{curl} A_o. \tag{8}$$

We can find E, H from the formulas

$$E = E_o = E_1, \quad H = H_o + H_1, \tag{9}$$

$$E_1 = \frac{1}{-i\omega\varepsilon_o}\operatorname{curl}\operatorname{curl} A - \operatorname{curl} F, \tag{10}$$

$$H_1 = \frac{1}{-i\omega\varepsilon_o}\operatorname{curl}\operatorname{curl} F + \operatorname{curl} A, \tag{11}$$

where

$$A = \int_\Gamma G(x,s)[N,H_1]ds, \quad F = -\int_\Gamma G(x,s)[N,E_1]ds. \tag{12}$$

2. To prove formulas (1.37) in Chapter 2 we start with the asymptotic expansion of the vector potential as $|x| \to \infty$, $n = x/|x|$:

$$A \equiv \int G(x,y)j(y)dy \sim \frac{\exp(ik_o|x|)}{4\pi|x|}\left\{\int j(y)dy - ik_o \int (n,y)j(y)dy\right\}$$
$$= \frac{\exp(ik_o|x|)}{4\pi|x|}(-i\omega P - ik_o[M,n]), \tag{13}$$

where

$$P = \int y\rho(y)dy, \quad M = \frac{1}{2}\int [y,j]dy, \tag{14}$$

$$\rho(y) = \frac{\text{div } j}{i\omega}, \tag{15}$$

P, M are the electric and magnetic dipole moments. Indeed, using Gauss' formula and taking into account that $j_N|_\Gamma = 0$ and $j = 0$ in Ω in our scattering problem, we obtain

$$-i\omega P = -i\omega \int_D \rho y(y)dy = -\int_D y \text{div } j dy$$
$$= -\int_\Gamma y j_N ds + \int_D j(y)dy = \int j(y)dy, \tag{16}$$

$$\int_D (n,y)jdy = \frac{1}{2}\int_D \{[[y,j],n] + j(n,y) + y(n,j)\}dy$$
$$= [M,n] + K = [M,n], \tag{17}$$

where M is defined in formula (14) and

$$K = \frac{1}{2}\int_D \{j(n,y) + y(n,j)\}dy$$
$$= \frac{1}{2}\int_\Gamma j_N s(s,n)ds - \frac{1}{2}\int_D y(y,n)\text{div } jdy = 0. \tag{18}$$

Here we took into account that $j_N|_\Gamma = 0$ and $\text{div } j \approx 0$ if $k_o a << 1$. If the current j is distributed on Γ, for example, $j = [N,H_1]$, we obtain from formula (14) the following formulas

$$P = \int_\Gamma t\sigma(t)dt, \quad M = \frac{1}{2}\int_\Gamma [t,j]dt, \tag{19}$$

where $\sigma(t)$ is the surface density of charge, $j(t)$ is the surface density of current (cf. formula (1.29) in Chapter 2). From (13) and the following formula, which is valid in the domain free of currents, in particular in the far distance zone,

$$E = -\frac{1}{i\omega\varepsilon_o} \text{ curl curl } A \sim \frac{k_o^2}{i\omega\varepsilon_o}[n[n,A]] \quad \text{as } |x| \to \infty, \tag{20}$$

we obtain the first formula of (1.37) in Chapter 2. The second formula in (1.37) is an immediate consequence of the first.

3. Under assumptions (3) we can rewrite equations (4) and (5) as

$$\text{curl } E = i\omega\mu_o H + i\omega(\mu-\mu_o)\eta(x)H, \tag{21}$$

$$\text{curl } H = -i\omega\varepsilon_o E + j_o - i\omega(\varepsilon'-\varepsilon_o)\eta(x)E, \tag{22}$$

where

$$\eta(x) = \begin{cases} 1, & x \in D, \\ 0, & x \in \Omega. \end{cases} \tag{23}$$

Let us set

$$j_m = -i\omega(\mu-\mu_o)\eta H, \quad j_e = -i\omega(\varepsilon'-\varepsilon_o)\eta E, \tag{24}$$

$$A = \int_D G(x,y)j_e(y)dy, \quad F = \int_D G(x,y)j_m(y)dy. \tag{25}$$

Then the vectors E_1, H_1 defined in formula (9) can be calculated from the formulas

$$E_1 = \frac{1}{-i\omega\varepsilon_o}(\text{curl curl } A-j_e) - \text{curl } F, \tag{26}$$

$$H_1 = \frac{1}{-i\omega\mu_o}(\text{curl curl } F-j_m) + \text{curl } A. \tag{27}$$

From (9) and (24)-(27) we obtain the integro-differential equations

$$E(x) = E_o(x) + \frac{\varepsilon' - \varepsilon_o}{\varepsilon_o} \operatorname{curl} \operatorname{curl} \int_D G(x,y)E(y)dy$$
$$- \frac{\varepsilon' - \varepsilon_o}{\varepsilon_o} \eta(x)E(x) + i\omega(\mu - \mu_o) \operatorname{curl} \int_D G(x,y)H(y)dy, \tag{28}$$

$$H(x) = H_o(x) + \frac{\mu - \mu_o}{\mu_o} \operatorname{curl} \operatorname{curl} \int_D G(x,y)H(y)dy$$
$$- \frac{\mu - \mu_o}{\mu_o} \eta(x)H(x) - i\omega(\varepsilon' - \varepsilon_o) \operatorname{curl} \int_D G(x,y)E(y)dy. \tag{29}$$

The system (28)-(29) can be solved by a simple iterative method if

$$\left(\left|\frac{\mu - \mu_o}{\mu_o}\right| + \left|\frac{\varepsilon' - \varepsilon_o}{\varepsilon_o}\right| \right) (1 + (k_o a)^2) << 1, \tag{30}$$

since the norm in $L^2(D)$ of the integro-differential operator of system (28)-(29) is less than one provided condition (30) holds. Let us show that

$$\left\|\frac{\varepsilon' - \varepsilon_o}{\varepsilon_o} \eta E\right\| \leq \left|\frac{\varepsilon' - \varepsilon_o}{\varepsilon_o}\right| \|E\|, \tag{31}$$

$$\left\|\operatorname{curl} \int_D G(x,y)E(y)dy\right\| \leq C(1 + k_o a) \|E\|, \tag{32}$$

$$\left\|\operatorname{curl} \operatorname{curl} \int_D \psi(x,y)E(y)dy\right\| \leq C(1 + k_o^2 a^2) \|E\|, \tag{33}$$

where $C = C(D)$ and $\|\cdot\| = \|\cdot\|_{L^2(D)}$. The estimate (31) is obvious. To prove (32) and (33) we note that

$$4\pi G(x,y) = \frac{1}{|x-y|} + \frac{\exp(ik_o|x-y|) - 1}{|x-y|}$$
$$= \frac{1}{|x-y|} + ik_o - \frac{k_o^2|x-y|}{2} + O(k_o^3|x-y|^2)$$

if $k_o|x-y| << 1$. Hence

$$D^2G = D^2 \frac{1}{4\pi|x-y|} + O\left(\frac{k_o^2}{|x-y|}\right). \tag{34}$$

We have

$$\left\|\int_D \frac{|E|}{|x-y|}\,dy\right\| \le \|E\|\left(\int_D\int_D \frac{dxdy}{|x-y|^2}\right)^{1/2} = \|E\|\,O(a^2), \tag{35}$$

$$\left\|\int_D \frac{Edy}{|x-y|}\right\|_{W_2^2(D)} \le C\,\|E\|. \tag{36}$$

Inequality (36) can be found, for example, in Kantorovich-Akilov [1]. From (34)-(36) we obtain inequalities (32) and (33). One can see that condition (30) can be valid for large bodies $k_oa > 1$ if $|(\varepsilon'-\varepsilon_o)/\varepsilon_o| + |(\mu-\mu_o)\mu_o|$ is small enough. If we put

$$g(n) \equiv \int_D \exp\{-ik_o(n,y)\}dy, \tag{37}$$

and use the first iteration of the system (28)-(29), we obtain the formula

$$f_E = -\frac{\varepsilon'-\varepsilon_o}{4\pi\varepsilon_o} k_o^2 g(n)[n,[n,E_o]] - \frac{\omega(\mu-\mu_o)k_o g(n)}{4\pi}[n,H_o] \tag{38}$$

for the scattering amplitude. If D is a ball with radius a,

$$g(n) = 4\pi a^3 \frac{\sin(k_oa) - k_oa\cos k_oa}{(k_oa)^3} \tag{39}$$

so that $g(n)$ does not actually depend on n. If D is a cylinder with radius a and length $2L$ then

$$g(n) = 2L\frac{\sin(k_oL\cos\theta)}{k_o L\cos\theta}\,\frac{J_1(k_oa\sin\theta)}{k_o a\sin\theta}, \tag{40}$$

where θ is the angle between the axis of the cylinder and ort n and $J_1(x)$ is the Bessel function.

APPENDIX 4

TWO-SIDED ESTIMATES OF THE SCATTERING AMPLITUDE FOR LOW ENERGIES

1. Introduction.

Consider the problem

$$(H-k^2)\psi = (-\nabla^2+V(x)-k^2)\psi = 0, \quad x \in \mathbb{R}^3, \qquad \psi = \exp\{ik(n,x)\} + v, \tag{1}$$

$$v \sim \frac{\exp(ik|x|)}{|x|} f(n,\nu,k) \quad \text{as} \quad |x| \to \infty, \quad \nu = \frac{x}{|x|}. \tag{2}$$

Our main assumptions are $(\int \equiv \int_{\mathbb{R}^3})$:

$$V(x) = 0 \quad \text{for} \quad |x| > a, \quad \int |V(x)|dx < \infty, \quad ka \ll 1, \qquad H > 0, \quad V(x) \in L^2_{loc}. \tag{3}$$

The problem is to estimate the scattering amplitude $f(n,\nu,k)$. The contents of this appendix can be summarized as follows:

(1) Two-sided estimates for f are obtained;

(2) An iterative process to calculate f is given; the process converges at the rate of a geometric series;

(3) Hard core potentials are considered;

(4) Some qualitative properties of f are described;

(5) An explicit formula for one-dimensional scattering is given.

2. Preliminaries. We start with the equation

$$\psi(x) = \exp\{ik(n,x)\} - \int \frac{\exp(ik|x-y|)}{4\pi|x-y|} V(y)\psi(y)dy. \tag{4}$$

If $ka \ll 1$ we can write this equation as

$$\psi(x) = 1 - \int \frac{V(y)\psi(y)}{4\pi|x-y|} dy, \tag{5}$$

with error $O(ka)$. With the same accuracy we obtain

$$\begin{aligned} f(n,\nu,k) \equiv f &= -\frac{1}{4\pi}\int \exp\{-ik(\nu,y)\}V(y)\psi(y)dy \\ &= -\frac{1}{4\pi}\int V\psi dy. \end{aligned} \tag{6}$$

Equation (5) is equivalent to the problem

$$H\psi = -\nabla^2\psi + V(x)\psi = 0 \quad \text{in } \mathbb{R}^3, \quad \psi(\infty) = 1. \tag{7}$$

Let $\psi = 1 + \phi$. Then

$$H\phi = -\nabla^2\phi + V\phi = -V, \quad \phi(\infty) = 0, \tag{8}$$

$$f = -\frac{1}{4\pi}\int Vdx - \frac{1}{4\pi}\int V\phi dy. \tag{9}$$

In what follows we make use of the following theorem which was formulated and proved in Chapter 2, Section 1, Theorem 5.

Theorem 1. *Let* A *be a selfadjoint linear operator on a Hilbert space* H, $D(A)$ = dom A, $R(A)$ = range A, $f \in R(A)$, *and* $A\phi = f$. *Then the representation*

$$(A\phi,\phi) = \max_{g\in D(A)} \frac{|(g,f)|^2}{(Ag,g)} \tag{10}$$

holds if and only if $A \geq 0$, i.e., $(Ag,g) \geq 0$ *for all* $g \in H$. *If* $(Ag,g) = 0$ *we take the expression under the sign max as zero*.

3. <u>Positive Potential</u>. If $V \geq 0$ we derive from equations (5) the equation

$$(I+B)h = V(x)^{1/2}, \quad Bh = \int \frac{V(x)^{1/2}V(y)^{1/2}}{4\pi|x-y|} h(y)dy, \tag{11}$$

where $B \geq 0$ in $H = L^2(\mathbb{R}^3)$. From (11) and Theorem 1 it follows, that

$$-4\pi f = \int V^{1/2}(x)h(x)dx = \max \frac{|(V^{1/2},g)|^2}{(g+Bg,g)}, \tag{12}$$

where (h,g) denotes the inner product in H. Hence

$$f \leq -\frac{1}{4\pi} \frac{|(V^{1/2},g)|^2}{(g+Bg,g)}, \quad g \in H. \tag{13}$$

To obtain a lower bound for f we apply (10) to equation (8) and use the inequality $H > 0$. As a result we obtain

$$(-V,\phi) = \max_{g \in D(H)} \frac{|(g,V)|^2}{(Hg,g)}. \tag{14}$$

From (14) it follows that

$$f \geq -\frac{1}{4\pi} \int V dx + \frac{|(g,V)|^2}{4\pi(Hg,g)}, \quad g \in D(H), \quad H = -\nabla^2 + V(x). \tag{15}$$

<u>Remark 1</u>. The bound (15) was obtained without the assumption $V(x) \geq 0$. We used only the assumption $H > 0$.

Another lower bound can be obtained as follows. Consider the functional

$$E(g) = \int \{|\nabla g|^2 + Vg^2 + 2Vg\}dx \tag{16}$$

defined on $W_2^1(\mathbb{R}^3)$. Equation (8) is the necessary condition for g to be a minimum of functional (16). Since functional (16) is a quadratic equation, (8) is also a sufficient condition for g to be a minimum of this functional. Hence

$$E(g) \geq E(\phi) = \int \phi V dx = -4\pi f - \int V dx. \tag{17}$$

Here we took into account formula (9) and the identity

$$\int\{|\nabla\phi|^2 + V\phi^2 + V\phi\}dx = 0, \quad (\text{see } (8)). \tag{18}$$

From (17) we obtain

$$f \geq -\frac{1}{4\pi}\int V dx - \frac{1}{4\pi}\int\{|\nabla g|^2 + Vg^2 + 2Vg\}dx, \qquad g \in W_2^1(\mathbb{R}^3). \tag{19}$$

This inequality is similar to the estimate of electrical capacitance in Pólya-Szegö [1]. An equality similar to (13) can be found in Blankenbecler-Sugar [1].

4. Iterative Process to Calculate f.

Theorem 2. *Equation (11) can be solved by the iterative process*

$$h_{n+1} = \gamma h_n - qBh_n + qV^{1/2}(x), \quad h_o = qV^{1/2}(x), \tag{20}$$

where

$$\gamma = \frac{\|B\|}{2 + \|B\|}, \quad q = \frac{2}{2 + \|B\|}, \tag{21}$$

$\|B\|$ *is the norm of the operator* $B: L^2(D) \to L^2(D)$, *and* $D = \operatorname{supp} V(x)$. *If* $h = \lim_{n\to\infty} h_n$ *then* $\|h - h_n\| = O(\gamma^n)$.

Corollary. We can calculate f from the formula

$$f = -\frac{1}{4\pi}\lim_{n\to\infty}\int V^{1/2}(x)h_n(x)dx. \tag{22}$$

Remark 2. It is clear that $\|B\|^2 \leq \iint \frac{V(x)V(y)dxdy}{16\pi^2|x-y|^2}$, so $\|B\| \leq \|V\|_R/4\pi$, where $\|V\|_R$ is the Rollnik norm of $V(x)$.

To prove Theorem 2 we let $I + B = A$, $V^{1/2}(x) = \omega(x)$, so that $Ah = \omega$, $h = (I - qA)h + q\omega$. If we take q as in (21) and set $\gamma = 1-q$, we can see that

$\|1 - qA\| \leq \max_{1 \leq \lambda \leq 1+\|B\|} |1-q\lambda| \leq \|B\|/(2+\|B\|) = \gamma$. Hence process (20) converges at the rate of a geometric series with ratio γ. Such an iterative process can be found, for example, in Krasnoselskij et al. [1].

5. <u>One-Dimensional Scattering Problem</u>. If $x \in \mathbb{R}^1$ we get instead of (5) and (6) the formulas

$$\psi = 1 + \int \frac{V(y)\psi(y)dy}{2ik}, \quad f = \frac{1}{2ik} \int V\psi dy. \tag{23}$$

Multiplying by V and integrating we find

$$f = \frac{\int V dx}{2ik - \int V dx}, \quad x \in \mathbb{R}^1. \tag{24}$$

6. <u>Hard Core Potential</u>. If $V(x) = +\infty$ in a bounded domain $D \in \mathbb{R}^3$ with smooth boundary Γ, $a = \operatorname{diam} D$, $ka \ll 1$, then the solution to problem (1)-(2) can be found in the form

$$V = \int_\Gamma \frac{\exp(ik|x-s|)}{4\pi|x-s|} \sigma(s)ds, \quad v|_\Gamma = -\exp\{ik(n,s)\}|_{ka\ll 1} = -1 \tag{25}$$

$$f = \frac{1}{4\pi} \int_\Gamma \exp\{-ik(\nu,s)\}\sigma(s)ds|_{ka\ll 1} = \frac{1}{4\pi} \int_\Gamma \sigma(s)ds. \tag{26}$$

From (25) we obtain

$$\int_\Gamma \frac{\sigma(s)ds}{4\pi|t-s|} = -1. \tag{27}$$

Hence $\int \sigma(s)ds = -C$, where C is the electrical capacitance of the conductor D. Therefore $f = -\frac{1}{4}C/\pi$. In Chapter 2, Section 4 two-sided estimates for C are given, and in Section 1 an iterative process and approximate analytical formulas for calculating C with a prescribed accuracy were given. In particular, $C \geq 4\pi S^2 J^{-1}$ where $S = \operatorname{meas} \Gamma$ and

$J = \int_\Gamma \int_\Gamma |s-t|^{-1} ds dt$. In many cases the formula $C \simeq 4\pi S^2 J^{-1}$ gives an error of ~3%, (e.g., in the calculation of the capacitance of a parallelepiped of arbitrary shape, or a circular cylinder of an arbitrary shape). If R, r are the radii of the minimal ball containing D contained in D, then $r < C < R$. So

$$- \frac{R}{4\pi} \leq f \leq - \frac{r}{4\pi}. \tag{28}$$

7. Monotonicity of Ψ and f. If $0 \leq V_1 \leq V_2$ then from (5) and maximum modulus principle it follows that $\Psi_1 \geq \Psi_2 \geq 0$ where Ψ_j corresponds to V_j, $j = 1,2$. If $V_1 \leq V \leq V_2$, V_j = const, then $\Psi_1 \geq \Psi \geq \Psi_2 \geq 0$, where for Ψ_j we can give an explicit formula since V_j = const. If $0 \leq V_1 \leq V_2$ then $f_1 \geq f_2$. Indeed $-4\pi f = \int h\, dx$ where $h \equiv V\psi$, $(V^{-1}+G)h = 1$, $Gh = \int \frac{h(y)dy}{4\pi|x-y|}$. From (10) we get $-4\pi f = \max_{g \in D(T)} |(g,1)|^2/(Tg,g)$, $T \equiv V^{-1} + G$. If $0 < V_1 \leq V_2$ then $V_1^{-1} + G \geq V_2^{-1} + G$ and $f_1 \geq f_2$.

8. Alternating Potential. Because of Remark 2 we can obtain only an upper bound for f. We will not get the best estimates, but rather will describe a simple method of obtaining an upper bound and give a simple example. First we note that if $V(x) = V_+(x) - V_-(x)$, we first define $V_+(x) \equiv \max(V(x),0)$ and $V_-(x) \equiv V - V_+(x) = -\min(V(x),0)$. We also define $H \equiv -\nabla^2 - V_-(x)$ and observe $(Hg,g) \geq (H_-g,g)$. To simplify matters, we make the additional assumption

$$H_- > 0. \tag{29}$$

Then from (9) and (14) it follows that

$$f \leq -\frac{1}{4\pi}\int V(x)dx + \max_{g\in D(H_-)} \frac{|(g,V)|^2}{4\pi(H_-g,g)}. \tag{30}$$

To obtain a simple upper bound for f we must estimate (H_-g,g) from below. In Glaser et al. [1], the following inequality was proved ($\|\ \| = \|\ \|_{H=L^2(\mathbb{R}^3)}$):

$$(H_-g,g) \geq \|\nabla g\|^2(1 - \mu_p^{-1}N_p), \tag{31}$$

where $H_- = H_o - V_-$, $N_p^p = \int|y-x|^{2p-3}V_-^p(x)dx$, $p' = \frac{p}{p-1}$, $p > 1$, $\mu_p \equiv \frac{p}{p-1}[\frac{4\pi(p-1)\Gamma^2(p)}{\Gamma(2p)}]^{1/p}$, $\Gamma(p)$ is the Gamma function, and $y \in \mathbb{R}^3$ is an arbitrary point. Our assumption concerning V_- can now be formulated as:ththere exist a point $y \in \mathbb{R}^3$ and a number $p > 1$ such that

$$N_p < \mu_p. \tag{32}$$

We note that inequality (32) implies (29).

Theorem 3. *If inequality* (32) *holds, then*

$$f \leq -\frac{1}{4\pi}\int V(x)dx + \frac{1}{4\pi}\int |V(x)|dx \frac{\mu_p}{\mu_p - N_p}. \tag{33}$$

Proof: We have

$$\max \frac{|(g,V)|^2}{(Hg,g)} \leq \frac{\int|V|dx \int|V(x)||g|^2dx}{\|\nabla g\|^2(1-\mu_p^{-1}N_p)+(V_+g,g)}. \tag{34}$$

Here we used inequality (31). Since $H_o - V_- \geq 0$ we have

$$\|\nabla g\|^2 \geq \int V_-|g|^2dx. \tag{35}$$

Hence the right-hand side of (34) is less than

$$\int |V|dx \frac{A + B}{cA + B}, \quad A \equiv (V_-g,g), \quad B \equiv (V_+g,g), \quad c \equiv 1 - \mu_p^{-1}N_p. \tag{36}$$

From (32) we conclude that $0 < c < 1$. Therefore

$$\frac{A + B}{cA + B} \leq \frac{1}{c}. \tag{37}$$

From (37), (36) and (34) we obtain (33).

Remark 3. We could obtain the upper bound in different ways. Here are two examples. If for some $\sigma(x) > 0$, $\int |V|^2 \sigma^{-1}(x)\,dx < \infty$ and $(Hg,g) \geq \int \sigma |g|^2 dx$, then $\max(|(g,V)|^2/(Hg,g)) \leq \int |V|^2\, dx$. This is the first way to get an upper bound for f. In Grosse [1], the inequality $(H_o g,g) \geq u_q \|(q-3)/r^{2q} g\|^2_{L^{2q}(\mathbb{R}^3)}$ was used, connections with Pade approximation were indicated, and a very good upper bound for f was obtained. In general, any norm $|||g|||$ such that $|(V,g)| \leq C_1(V)\,|||g|||$ and $(Hg,g) \geq C_2(V)\,|||g|||^2$ can be used for obtaining the upper bound:

$$f \leq -\frac{1}{4\pi}\int V dx + \frac{C_1^2(V)}{4\pi C_2(V)}\,. \tag{38}$$

Remark 4. If the inequality

$$\|H_o^{-1}V\|_{C(\mathbb{R}^3)} \leq b < 1 \tag{39}$$

holds, then equation (5) can be solved by means of the iterative process $\psi = \sum_{j=0}^{\infty} (-1)^j (H_o^{-1}V)^j 1$. In this case we also obtain a simple upper bound for f:

$$\begin{aligned} f &= -\frac{1}{4\pi}\int V dx + \frac{1}{(4\pi)^2}\iint \frac{V(x)V(y)}{|x-y|}\,dxdy + \cdots \\ &\leq -\frac{1}{4\pi}\int V dx + \frac{1}{16\pi^2}\iint \frac{V(x)V(y)}{|x-y|}\,dxdy + \frac{1}{4\pi}\int |V|\,dx\,\frac{b^2}{1-b}\,. \end{aligned} \tag{40}$$

APPENDIX 5

VARIATIONAL PRINCIPLES FOR EIGENVALUES OF COMPACT NONSELFADJOINT OPERATORS

1. Let T be a compact linear operator in a Hilbert space H with eigenvalues λ_j, $|\lambda_1| \geq |\lambda_2| \geq \ldots$, and let the r_j be the moduli of the real parts of the eigenvalues ordered so that $r_1 \geq r_2 \ldots$. Note that r_j is not necessarily equal to $|\mathrm{Re}\, \lambda_j|$. Let L_j be the eigenspace of T corresponding to λ_j, let M_j be the eigensubspace of T corresponding to r_j, $\tilde{L}_j = \sum_{k=1}^{j} \dot{+} L_k$, and $\tilde{M}_j = \sum_{k=1}^{j} \dot{+} M_k$. Let the t_j be the moduli of the imaginary parts of the eigenvalues, $t_1 \geq t_2 \geq, \ldots$, let $\tilde{N}_j = \sum_{k=1}^{j} \dot{+} N_k$, N_j be the eigenspace of T corresponding to t_j, and $\tilde{L}_j^{\Pi} \dot{+} \tilde{L}_j = H$, where the sign $\dot{+}$ denotes the direct sum and Π denotes the direct complement in H.

<u>Theorem 1.</u> <u>Under the above assumptions the following formulas hold:</u>

$$|\lambda_j| = \max_{x \in \tilde{L}_{j-1}^{\Pi}} \min_{\substack{y \in H \\ (x,y)=1}} |(Tx,y)|, \tag{1}$$

$$r_j = \max_{x \in \tilde{M}_{j-1}^{\Pi}} \min_{\substack{y \in H \\ (x,y)=1}} |\mathrm{Re}(Tx,y)|, \tag{2}$$

$$t_j = \max_{x \in \tilde{N}^{II}_{j-1}} \min_{\substack{y \in H \\ (x,y)=1}} |\mathrm{Im}(Tx,y)|. \tag{2'}$$

Proof of Theorem 1: The proof of (2') is similar to the proof of (2). So let us prove (1) and (2). First let us prove formulas (1) and (2) for $j = 1$, i.e.,

$$|\lambda_1| = \max_x \min_{\substack{y \\ (x,y)=1}} |(Tx,y)|, \tag{3}$$

$$r_1 = \max_x \min_{\substack{y \\ (x,y)=1}} |\mathrm{Re}(Tx,y)|. \tag{4}$$

For a fixed x it is possible to write $Tx = \lambda x + z$, where $z \in x^{\perp}$, $x^{\perp}$ is the subspace of all vectors orthogonal to x, and λ is a number. Thus $(Tx,y) = \lambda + (z,y)$. Let us represent y in the form $y = |x|^{-2}x + u$, $u \in x^{\perp}$. Here the condition $(x,y) = 1$ was taken into account. Finally, one obtains $(Tx,y) = \lambda + (z,u)$. From here it follows that

$$\min_{\substack{y \\ (x,y)=1}} |(Tx,y)| = \min_u |\lambda+(z,u)| = \begin{cases} |\lambda|, & z = 0, \\ 0, & z \neq 0, \end{cases} \tag{5}$$

$$\max_x \min_{\substack{y \\ (x,y)=1}} |(Tx,y)| = |\lambda_1|, \tag{6}$$

$$\min_{\substack{y \\ (x,y)=1}} |\mathrm{Re}(Tx,y)| = \min_u |\{\mathrm{Re}\,\lambda + \mathrm{Re}(z,u)\}| = \begin{cases} |\mathrm{Re}\,\lambda|, & z = 0, \\ 0, & z \neq 0. \end{cases} \tag{7}$$

Formula (4) follows from (7). Suppose that formulas (1) and (2) are proved for $j \leq n$. Then we can follow the same line of reasoning and take into account that all the eigenvalues of T in the subspace $\tilde{L}_n^{II}$ have moduli no greater than $|\lambda_{n+1}|$ and in the subspace $\tilde{M}_n^{II}$, the operator T has $\max|\text{Re }\lambda_j| = r_{n+1}$. For example

$$\max_{x\in\tilde{L}_{j-1}^{II}} \min_{\substack{y\in H\\(x,y)=1}} |T(x,y)| = \max_{\substack{x\in\tilde{L}_{j-1}^{II}\\Tx=\lambda x}} |\lambda| = |\lambda_j|.$$

Remark 1. There is a one-to-one correspondence between $\{M_j\}$ and $\{L_j\}$. Namely, take $M_i = L_{j(i)}$ where $j(i)$ is so chosen that the eigenvalue $\lambda_{j(i)}$ has $|\text{Re }\lambda_{j(i)}| = r_i$.

Remark 2. If T_n is compact and $||T_n - T|| \to 0$ then $\lambda_j(T_n) \to \lambda_j(T)$ for all j. Thus $|\lambda_j(T_n)| \to |\lambda_j(T)|$ and $\text{Re }\lambda_j(T_n) \to \text{Re }\lambda_j(T)$. This fact permits an approximate calculation of the spectrum of T using in (1) and (2) with the operator T_n instead of T. One can take T_n, for example, to have rank n.

Remark 3. Principles similar to (1) and (2) were announced by P. Popov [1] for the case $\text{Re }\lambda_j > 0$, but $\tilde{L}_{n-1}^{II}$ appears in (2) instead of $\tilde{M}_{n-1}^{II}$, which seems to be erroneous. Also minimization in P. Popov [1] is taken over the different set and his arguments are quite different from those given above.

2. The same arguments lead to variational principles for the spectrum of an unbounded linear operator with discrete spectrum (i.e., a spectrum consisting of isolated eigenvalues of finite algebraic multiplicity). Let A be an unbounded closed densely defined linear operator in H with its eigenvalues $\lambda_j = \lambda_j(A)$ and spectrum $\sigma(A) = \{\lambda_j\}$. Each λ_j is an isolated eigenvalue of finite algebraic multiplicity. The eigenvalues are ordered so that

$|\lambda_1| \leq |\lambda_2| \leq \ldots$. Let $r_1 \leq r_2 \leq \ldots$ be the moduli of real parts of the eigenvalues of A. Again we emphasize the fact that r_j is not necessarily equal to $|\mathrm{Re}\ \lambda_j|$, but it is possible to establish a one-to-one correspondence between $\{\lambda_j\}$ and $\{r_j\}$ by setting $|\mathrm{Re}\ \lambda_{j(i)}| = r_i$, as above. The variational principles read

$$|\lambda_j| = \min_{x \in \tilde{L}^{\text{II}}_{j-1}} \min_{\substack{y \in H \\ (x,y)=1}} |(Tx,y)|, \tag{8}$$

$$r_j = \min_{x \in \tilde{M}^{\text{II}}_{j-1}} \min_{\substack{y \in H \\ (x,y)=1}} |\mathrm{Re}(Tx,y)|. \tag{9}$$

Here, as in Section 1, $\tilde{L}_j = \sum_{k=1}^{j} \dot{+}\, L_k$ and L_k is the eigen-space corresponding to λ_k. A similar meaning is ascribed to M_j. Principles (8) and (9) can be proved similarly to (1) and (2).

For the moduli of the imaginary parts of the eigenvalues, $t_1 \leq t_2 \leq \ldots$, the following formula holds

$$t_j = \min_{x \in \tilde{N}^{\text{II}}_{j-1}} \min_{\substack{y \in H \\ (x,y)=1}} |\mathrm{Im}(Tx,y)|, \tag{9'}$$

and it can be proved similarly.

APPENDIX 6

BOUNDARY-VALUE PROBLEMS WITH DISCONTINUOUS BOUNDARY CONDITIONS

The following boundary-value problem is of interest in applications:

$$\Delta u + k^2 u = 0, \quad u\Big|_{S^+} = f, \quad \frac{\partial u}{\partial z}\Big|_{S^-} = h, \tag{1}$$

$$r\left(\frac{\partial u}{\partial r} - iku\right) \to 0 \quad \text{as} \quad r \to \infty. \tag{2}$$

Here S is a bounded plane domain with boundary L, S^+ is its upper part, and S^- is its lower part. It is assumed that S lies on the plane $z = 0$, $z = +0$ $(z = -0)$ corresponds its upper (lower) part and the edge condition is satisfied.

Applying Green's formula one obtains

$$u = -\int_S \left(g \frac{\partial u}{\partial N} - u \frac{\partial g}{\partial N}\right) ds, \quad g = \frac{\exp(ik|x-y|)}{4\pi|x-y|}, \tag{3}$$

where N is the outward pointing normal to S.

Let us look for a solution of problem (1)-(2) of the form

$$u = \int_{S^+} \left(-\mu \frac{\partial g}{\partial z'} - \sigma g\right) ds, \tag{4}$$

where μ, σ are unknown functions, $z = X_3$, and

$$\mu = u|_{S^+}, \quad \sigma = \frac{\partial u}{\partial Z}\Big|_{S^+}. \tag{5}$$

The integral representation (4) has an immediate physical interpretation (5). From (4) and (1) one obtains

$$f = -\int_{S^+} \sigma g ds + \frac{\mu}{2}, \tag{6}$$

$$h = \frac{\partial^2}{\partial z^2}\int_{S^+} \mu g ds - \frac{\sigma}{2}. \tag{7}$$

Let us note that

$$\frac{\partial^2}{\partial z^2}\int_{S^+}\mu g ds = (\Delta-\hat{\Delta})\int_{S^+}\mu g ds = -k^2\int_{S^+}\mu g ds - \hat{\Delta}\int_{S^+}\mu g ds, \tag{8}$$

where

$$\hat{\Delta} = \frac{\partial^2}{\partial X_1^2} + \frac{\partial^2}{\partial X_2^2}. \tag{9}$$

Thus

$$\mu(t) = 2\int_{S^+} g(t-s)\sigma(s)ds + 2f \tag{10}$$

$$\sigma(t) = -2(\hat{\Delta}+k^2)\int_{S^+} g(t-s)\mu(s)ds - 2h. \tag{11}$$

Hence

$$\begin{aligned}\sigma = -4(\hat{\Delta}+k^2)\int_{S^+} g(t-s)ds\int_{S^+} g(s-v)\sigma(v)dv \\ - 4(\hat{\Delta}+k^2)\int_{S^+} g(t-s)f(s)ds - 2h.\end{aligned} \tag{12}$$

Finally, we have

$$\sigma = T\sigma + v, \tag{13}$$

where

$$v = -2h - 4(\hat{\Delta}+k^2)\int_{S^+} g(t-s)f(s)ds, \tag{14}$$

$$T\sigma = -4(\hat{\Delta}+k^2)\int_{S^+} dv\sigma(v)\int_{S^+} dsg(t-s)g(s-v). \tag{15}$$

Before proceeding, recall that

$$u|_L = 0, \quad |\nabla u| \sim \rho^{-3/4}, \tag{16}$$

where ρ is the distance between the point x and the edge L of S. This can be verified, taking into account that the exact solution of the problem

$$\Delta u = 0, \quad u\big|_{z=+0,y>0} = 0, \quad \frac{\partial u}{\partial z}\Big|_{z=-0,y>0} = 0,$$

has this kind of behavior near the edge, $u \sim \rho^{1/4}\sin(\frac{1}{4}\phi)$ if $\rho \to 0$. The boundary condition in the boundary-value problem (1) should be satisfied at every point of $S \setminus L$. Near L the edge condition should be satisfied.

Equation (13) can be solved numerically, e.g., by the method of moments.

APPENDIX 7

POLES OF GREEN'S FUNCTION

Theorem 1. The set of complex poles of Green's function of the exterior Dirichlet problem coincides with the set of zeros of the functions $\rho_n(k)$ defined by formula (1.23) of Chapter 4.

Proof: Green's function satisfies the equation

$$G(x,y,k) = g(x,y,k) - \int_\Gamma g(x,t,k)\mu(t,y,k)dt, \qquad \mu \equiv \frac{\partial G(t,y,k)}{\partial N_t}, \quad x,y \in \Omega \tag{1}$$

and the boundary condition

$$G(x,y,k)|_\Gamma = 0. \tag{2}$$

Suppose that z is a complex pole of G,

$$G(x,y,k) = \frac{R(x,y)}{(k-z)^r} + \dots . \tag{3}$$

Multiplying (3) by $(k-z)^r$ and letting $k \to z$, we obtain

$$\int_\Gamma g(x,t,z) \frac{\partial R(t,y)}{\partial N_t} dt = 0, \quad x \in \Gamma. \tag{4}$$

The kernel $R(x,y)$ is degenerate, i.e., a function $\phi \not\equiv 0$ exists such that

$$\int_\Gamma g(x,t,z)\phi(t)dt = 0, \quad x \in \Gamma. \tag{5}$$

Hence z is a zero of $\rho_n(k)$ for some n. Conversely, let $\phi \not\equiv 0$ be a solution of equation (5), i.e., z is a zero of $\rho_n(k)$ for some n. The function

$$u(x) = \int_\Gamma g(x,t,z)\phi(t)dt, \quad x \in \Omega \tag{6}$$

is a solution of the exterior Dirichlet problem:

$$(\Delta + z^2)u = 0 \quad \text{in} \quad \Omega, \quad u|_\Gamma = 0 \tag{7}$$

satisfying the radiation condition.

The function $u(x) \equiv 0$ in D because z^2 is a complex number and $u(x)$ is a solution of the interior Dirichlet problem. Since $\phi \not\equiv 0$ we conclude that $u(x) \not\equiv 0$ in Ω. If z is not a complex pole of G then the exterior Dirichlet problem (7) has no nontrivial solutions. This contradiction proves that z is a complex pole of G.

Remark 1. Let us prove that if a complex number z is not a complex pole of G, then the problem (7) has only the trivial solution satisfying the radiation condition. For a complex z we say that a function u satisfies the radiation condition, if this function can be represented in the form

$$u(x) = \int_\Gamma \left(u \frac{\partial g}{\partial N} - g \frac{\partial u}{\partial N}\right) dt, \tag{8}$$

where N is the outward pointing normal to Γ. In other words the radiation condition for a function $u(x)$ is satisfied if and only if

$$\lim_{R\to\infty} \int_{S_R} \left(g \frac{\partial u}{\partial |x|} - u \frac{\partial g}{\partial |x|} \right) dS_R = 0, \tag{9}$$

where $S_R = \{x: |x| = R\}$.

It is not difficult to prove that G satisfies the radiation condition. If z is not a pole of G one can apply Green's formula to obtain

$$u(x) = \int_\Gamma \left(u \frac{\partial G}{\partial N_t} - G \frac{\partial u}{\partial N_t} \right) dt + \int_{S_R} \left(G \frac{\partial u}{\partial |y|} - u \frac{\partial G}{\partial |y|} \right) dy. \tag{10}$$

The integral over Γ is equal to zero because of the boundary conditions for u and G. The proof will be completed if we show that the integral over S_R approaches zero as $R \to \infty$. To this end, we substitute (8) into (10).

We obtain the formula

$$u(x) = \lim_{R\to\infty} \left\{ \int_\Gamma dt u(t) \frac{\partial u}{\partial N_t} \int_{S_R} \left[G(x,y) \frac{\partial g(y,t)}{\partial |y|} - g(y,t) \frac{\partial G(x,y)}{\partial |y|} \right] d\ldots \right.$$

$$- \int_\Gamma dt \frac{\partial u}{\partial N_t} \int_{S_R} \left[G(x,y) \frac{\partial g(y,t)}{\partial |y|} - g(y,t) \frac{\partial G(x,y)}{\partial |y|} \right] dy. \tag{11}$$

From (1) it follows that

$$\int_{S_r} \left[G(x,y) \frac{\partial g(y,t)}{\partial |y|} - g(y,t) \frac{\partial G(x,y)}{\partial |y|} \right] dy \to 0 \quad \text{as } R \to \infty. \tag{12}$$

It follows that $u(x) \equiv 0$ from (11) and (12). □

APPENDIX 8

A UNIQUENESS THEOREM FOR SCHRÖDINGER EQUATION

1. Consider the problem

$$\Delta u + c(x)u = 0 \quad \text{in} \quad D, \quad u = 0 \quad \text{on} \quad \Gamma, \tag{1}$$

where D is a bounded domain with a sufficiently smooth boundary Γ (e.g., $\Gamma \in C^{1,\alpha}$, $\alpha > 0$), $c(x) \in L^2(D)$. It is well-known that if $\sup_{x \in D}|c(x)|$ is sufficiently small, then $u \equiv 0$. Here we use integral equations and variational inequalities to prove that if $c(x)$ is sufficiently small in L^2 then $u \equiv 0$.

<u>Theorem 1</u>. <u>Let</u> $D \subset \mathbb{R}^2$ <u>and assume</u>

$$|D|\int |c(x)|^2 dx < \pi^2 (5/3)^{1/2}, \quad |D| = \text{meas } D, \quad \int \equiv \int_D . \tag{2}$$

<u>Let</u> u <u>be a solution of</u> (1) <u>such that</u> $u \in W_2^1(D)$. <u>Then</u> $u \equiv 0$.

<u>Proof</u>: We have

$$\int |u|^2 dx = \int c(x)|u|^2 dx \leq \left(\int |c(x)|^2 dx\right)^{1/2} \left(\int |u|^4 dx\right)^{1/2}$$

Thus

$$\left(\int |\nabla u|^2 dx\right)^2 \left(\int |u|^4 dx\right)^{-1} \leq \int |c(x)|^2 dx. \tag{3}$$

Consider the problem

$$I = \int |\nabla u|^2 dx = \inf, \quad u \in \overset{\circ}{W}{}^1_2(D), \quad \int |u|^4 dx = 1. \tag{4}$$

Let $\inf I = d$. Then it follows from (3) that the inequality

$$\int |c(x)|^2 dx < d^2 \tag{5}$$

implies uniqueness of the solution of (1) in $\overset{\circ}{W}{}^1_2(D)$. It remains to estimate d. The solution of (4) satisfies the Euler equation

$$-\Delta u = du^3, \quad u = 0 \quad \text{on} \quad \Gamma, \quad \int |u|^4 dx = 1. \tag{6}$$

Let $G(x,y)$ be Green's function for the Dirichlet problem for the Laplace operator. Then $u = d\int G(x,y)u^3 dy$. From this and Hölder's inequality it follows that

$$\begin{aligned} 1 = \int |u|^4 dx \le d^4 \int dx \int |G(x,y)|^4 dy \left(\int |u|^4 dy\right)^3 \\ = d^4 \iint |G|^4 dxdy. \end{aligned} \tag{7}$$

From (7) we obtain

$$\left(\iint |G|^4 dxdy\right)^{-1/4} \le d. \tag{8}$$

In order to estimate the left-hand side in (8) we use the Steiner symmetrization (see Pólya-Szegö [1]) and conclude that among all figures with fixed $|D|$ the cirlce has minimal d. If D is the circle with area $|D|$ then, taking into account that $0 < G(x,y) \le (2\pi)^{-1}\ln(ar_{xy}^{-1})$, where $r_{xy} = |x-y|$, $a = \operatorname{diam} D$, we obtain

$$\iint |G|^4 dxdy \leq \iint (2\pi)^{-4} \ln^4(ar_{xy}^{-1})dxdy$$
$$\leq 2^{-1}(2\pi)^{-4} \int_{D_1}\int_{D_1} \ln^4(a|t|^{-1})dtdz$$
$$\leq 2^{-1}(2\pi)^{-4}\, 4^{-1}\pi(2a)^2\, 2\pi \int_0^a \ln^4(ar^{-1})rdr \tag{9}$$
$$= (16\pi^2)^{-1}a^4 \int_0^1 \ln^4(r^{-1})rdr = 3(5\pi^4)^{-1}|D|^2.$$

Here D_1 is the circle with diameter $2a$. From (9) it follows that $d \geq \pi(5/3)^{1/4}|D|^{-1/2}$. Therefore inequality (2) is a sufficient condition for the uniqueness of solution of (1) in $\overset{\circ}{W}{}_2^1(D)$.

2. Consider now the case in which $D \subset \mathbb{R}^3$ and $0 < G \leq (4\pi r_{xy})^{-1}$. The integral $\iint r_{xy}^{-4}dxdy$ diverges and we must change the arguments.

Theorem 2. *Let $D \subset \mathbb{R}^3$, and $a = \operatorname{diam} D$. Let $u(x) \in \overset{\circ}{W}{}_2^1(D)$ be a solution of (1) and assume the inequality*

$$\int |c(x)|^2 dx \leq 4\pi a^{-1} \tag{10}$$

holds. Then $u \equiv 0$.

To prove Theorem 2 we need the following theorem (see Kantorovich-Akilov [1]).

Theorem 3. *Let $Bv = \int K(x,y)v(y)dy$, $B: L^p(D) \to L^q(D')$, $1 \leq p,q \leq \infty$, $D \subset \mathbb{R}^m$, $D' \subset \mathbb{R}^n$, $|D| = 1$, $|D'| < \infty$. Let*

$$\left(\int |K(x,y)|^2 dy\right)^{1/r} \leq c_1, \quad \left(\int_{D'} |K(x,y)|^s dx\right)^{1/s} \leq c_2, \quad r>0,\ s>0,$$

$q \geq p, q \geq s$, $(1 - sq^{-1})p' \leq r$, $p' = p(p-1)^{-1}$. *Then*
$\|B\|_{L^p(D) \to L^q(D')} \leq c_1^{1-sq^{-1}} c_2^{sq^{-1}}$.

Remark 1. If $|D| \neq 1$ then $||B||_{L^p(D) \to L^2(D')} \leq c_1^{1-sq^{-1}} c_2^{sq^{-1}} \cdot |D|^{p'^{-1}-(1-sq^{-1})r^{-1}}$.

Proof of Theorem 2: We have

$$1 = \int |u|^4 dx \leq d^4 \int \left(\int |u|^3 (4\pi r_{xy})^{-1} dy \right)^4 \leq d^4 b^4, \quad (11)$$

where $b = b(D)$ is the norm of the operator $Bv = \int v(y)(4\pi r_{xy})^{-1} dy$, $B: L^{4/3}(D) \to L^4(D)$, $v = |u|^3$. It is known that $b < \infty$. Therefore

$$\int |c|^2 dx \leq b^{-2} \quad (12)$$

is a sufficient condition for the uniqueness of the solution of (1) in $\overset{o}{W}{}^1_2(D)$. In order to estimate b we apply Theorem 3. In our case $D = D'$, $m = n = 3$, $K(x,y) = r_{xy}^{-1}$, $p = 4/3$, and $p' = q = 4$. Let $r = s = 2$. Then $p'^{-1}(1 - sq^{-1})r^{-1} = sq^{-1} = 1/2$, $c_1 = c_2$. Thus $||B||_{L^{4/3}(D) \to L^4(D)} \leq b$, where $b = c_1(4\pi)^{-1}$ and $c_1 = \max_{x \in D} (\int r_{xy}^{-2} dy)^{1/2} \leq (4\pi a)^{1/2}$, $a = \text{diam } D$. Hence $b^{-2} = 4\pi a^{-1}$.

Remark 2. If $n > 3$ then $0 < G(x,y) \leq \{(n-2)S_n r^{n-2}\}^{-1}$, where S_n is the area of the unit sphere in $\mathbb{R}^n$. The operator $Bv = \int r_{xy}^{2-n} v dy$ is bounded as an operator from $L^{4/3}(D)$ into $L^q(D)$ for $q < 4n(3n - 8)^{-1}$. If $u \in L^4(D)$ then $q \geq 4$. Therefore $4n(3n - 8)^{-1} > 4$, $n < 4$. This means that the arguments used in the proof of Theorem 2 are not valid for $n > 3$. The results presented in Theorems 1 and 2 are due to Ramm [116].

APPENDIX 9

STABLE SOLUTION OF INTEGRAL EQUATIONS OF THE FIRST KIND WITH LOGARITHMIC KERNELS

1. Introduction. Consider the equation

$$Af \equiv \int \ln(r_{xy}^{-1}) f(y)dy = g(x), \quad x \in D, \quad \int \equiv \int_{\Gamma}, \tag{1}$$

where $\Gamma \subset \mathbb{R}^m$ is a closed surface. Because $\ln(r_{xy}^{-1})$ changes sign, it is possible that for some Γ, the homogeneous equation (1) (equation (1_0)) will have a nontrivial solution. Let us assume throughout this appendix that: Equation (1) is solvable in $H = L^2(\Gamma)$ and equation (1_0) has a nontrivial solution in H. The purpose of this appendix is to give an iterative process for calculating solutions of equations (1) and (1_0). Our method holds for any A semibounded below with the kernel $A(x,y) > -k$, $k = \text{const}$.

2. Preliminaries. Let us take a number $d > \operatorname{diam} D$. Then $\ln(dr_{xy}^{-1}) > 0$ for $x,y \in \Gamma$. Equation (1) is equivalent to the equation

$$Bf \equiv \int \ln(dr_{xy}^{-1}) f(y)dy = h(x), \tag{2}$$

where

$$h(x) \equiv g(x) + \ln d \int f(y)dy \equiv g + C(f). \tag{3}$$

Note that B is positive in H because of the choice of d.

Thus

$$Bf = 0 \text{ implies that } f = 0. \tag{4}$$

Lemma 1. $\dim N(A) = 1$.

Proof: Suppose that $f_o(x)$ and $f_1(x)$ are two linearly independent solutions of equation (1_o) in H. Then

$$\int f_j \, dx \neq 0, \quad j = 0,1. \tag{5}$$

Indeed, equation (1_o) is equivalent to the equation

$$Bf = C(f), \tag{6}$$

where $C(f)$ is defined by formula (3). If $C(f) = 0$ then it follows from (4) that $f \equiv 0$. Thus (5) holds. From (5) it follows that there exists a number b such that

$$\int (f_o - bf_1) dx = 0. \tag{7}$$

Thus

$$B(f_o - bf_1) = 0. \tag{8}$$

From (8) and (4) it follows that

$$f_o = bf_1, \quad b = \text{const}. \tag{9}$$

This completes the proof.

Remark 1. We have also proved that if $N(A) \neq \{0\}$, then the equation

$$Bf = C = \text{const} \tag{10}$$

is solvable in H.

3. Iterative process for solution of equations (1_o), (2)

Denote by f_o the solution of the equation $Af_o = 0$, $\int f_o dx =$
If f_1 is a solution of equation (1) and $a = \text{const}$, then

$f_1 - af_o$ is a solution of equation (1). Thus, let us look for a solution of equation (1) which satisfies the condition $\int f dx = 0$. Such a solution must be unique and exist if equation (1) is solvable. For this solution, equation (1) is equivalent to the equation

$$Bf = g \tag{11}$$

Equation (11) has a positive kernel. In the Lemma 2 below an iterative process to solve equation (11) is given.

Lemma 2. _If equation_ (1) _is solvable in_ H _then the equation_

$$Bf = g \tag{12}$$

has the solution and $f = a(x)\psi$ _in_ H, _where_

$$a(x) = \left\{\int \ln(dr_{xy}^{-1})dy\right\}^{-1}, \quad 0 < m \leq a(x) \leq M, \quad x \in D, \tag{13}$$

and $\psi = \lim_{n\to\infty} \psi_n$. _Here_

$$\psi_{n+1} = (I - B_1)\psi_n + g, \quad \psi_o = g, \tag{14}$$

$$B_1\psi \equiv B(a\psi), \tag{15}$$

and the sequence ψ_n _converges in_ H.

Remark 1. Because $B > 0$ in H it is possible to construct other iterative processes which converge in H to a solution of (12), e.g.,

$$\phi_{n+1} = (I - \alpha B)\phi_n + \alpha g, \quad 0 < \alpha < 2\,||B||^{-1}. \tag{16}$$

The statement of Lemma 2 follows from Theorem 6, Section 1, Chapter 2.

APPENDIX 10

NONSELFADJOINT OPERATORS IN DIFFRACTION AND SCATTERING

1. <u>Introduction</u>. Consider the following problem

$$(\Delta + k^2)u = 0 \quad \text{in} \quad \Omega, \tag{1}$$

$$\partial u | \partial N = f \quad \text{on} \quad \Gamma, \tag{2}$$

$$|x|(\partial u/\partial |x| - iku) \to 0 \quad \text{as} \quad |x| \to \infty, \tag{3}$$

where Ω is an unbounded domain with a smooth closed compact surface Γ, $\Gamma \in C^2$.

If we look for a solution of the form

$$u = \int_\Gamma \frac{\exp(ikr_{xy})}{4\pi r_{xy}} g(y)dy, \quad r_{xy} = |x-y|, \tag{4}$$

then

$$g = Ag - 2f, \tag{5}$$

where

$$Ag = \int_\Gamma \frac{\partial}{\partial N_t} \frac{\exp(ikr_{ty})}{2\pi r_{ty}} g(y)dy. \tag{6}$$

If the boundary condition is of the form

$$u = f \quad \text{on} \quad \Gamma, \tag{7}$$

then the integral equation for g takes the form

$$Tg = f, \tag{8}$$

where

$$Tg = \int_\Gamma \frac{\exp(ikr_{ty})}{4\pi r_{ty}} g(y)dy. \tag{9}$$

If one wishes to solve equations (8), (5) by means of expansions in root vectors, one must prove that the root vectors of operators A and T form a basis of $H = L^2(\Gamma)$. Both operators are compact and nonselfadjoint. A priori it is not clear why these operators have eigenvectors: e.g., the Volterra operator has no eigenvectors. In applications it is more convenient to use only eigenvectors, because calculations with the root vectors are more complicated. This leads to the following question: when does a nonselfadjoint operator have no root vectors? Here and below we use the phrase root vector to mean an associated root vector. The definition is: if $Ag = \lambda g$, $g \neq 0$, then g is an eigenvector; if the equation $Ah_1 - \lambda h_1 = g$ is solvable, then h_1 is an associated vector (or root vector); the set $(g, h_1, \ldots, h_s)$ is called a Jordan chain of length $s + 1$ if $(A - \lambda)g = 0$, $(A - \lambda)h_1 = g$, $(A - \lambda)h_k = h_{k-1}$, $2 \leq k \leq 2$; vectors $h_1, \ldots, h_s$ are called root vectors. An isolated eigenvalue λ is called a normal eigenvalue if its algebraic multiplicity is finite and the Hilbert space H can be decomposed into the direct sum of subspaces $H = L_\lambda \dot{+} R_\lambda$, where L_λ is the root subspace of A and R_λ is an invariant subspace for A in which $(A - \lambda I)^{-1}$ exists. The root subspace L_λ is the linear span of all eigenvectors and root vectors of A corresponding to λ. It is well known that λ is a normal eigenvalue if the projector $P = -(2\pi i)^{-1} \int_{|z-\lambda|=\varepsilon} (A - zI)^{-1} dz$ is finite-dimensional (Gohberg-Krein [2]). If λ is a normal eigenvalue of A, then $(A - zI)^{-1} \equiv R_z$ has a simple pole at λ iff the length of the Jordan chain is equal to 1. This means that the eigensubspace of A corresponding to λ coincides with the root subspace of A corresponding to λ. From the definition given above it follows that the pole λ is simple iff $(A - \lambda I)^2 f = 0 \Rightarrow (A - \lambda I)f = 0$. In the physics literature there is a great interest in equations of type (6), (8) and in their counterparts in electromagnetic wave scattering theory (Dolph-Scott [1]). Engineers used the singularity and eigenmode expansion methods for solution of exterior boundary-value problems (Voitovich et al. [1], Baum [1]). What they call the eigenmode expansion method (EEM) is actually Picard's method for solution of selfadjoint integral equations of the first kind. They suppose that the operator T defined by (9) has eigenvectors

$$Tf_j = \lambda_j f_j, \quad j = 1,2,\ldots, \quad |\lambda_1| \geq |\lambda_2| > \ldots, \tag{10}$$

has no root vectors, and the set of these eigenvectors $\{f_j\}$ forms a Riesz basis of $H = L^2(\Gamma)$. We remind the reader that $\{f_j\}$ is a Riesz basis of H (or basis equivalent to an orthonormal basis $\{h_j\}$ of H) if a bounded invertible linear operator B exists such that $Bh_j = f_j$. We call an operator B invertible if B^{-1} is bounded and defined on H. Under such an assumption engineers solve equation (8) using the formula

$$g = \sum_{j=1}^{\infty} \lambda_j^{-1}(k)(f,f_j)f_j. \tag{11}$$

The following questions are open and of interest to mathematicians:

(1) When do the eigenvectors of T and A form a basis of H?

(2) When do there not exist root vectors of T and A?

These questions are far from trivial. In fact, nothing is known about the existence and properties of eigenfunctions for the basic equation of the theory of lasers,

$$\int_{-1}^{1} \exp\{i(x-y)^2\}f(y)dy = \lambda f(x). \tag{12}$$

Fortunately the situation is much better for the operators A and T and later we give some reasons for this statement.

The singularity expansion method (SEM) consists in the following. Given the nonstationary problem:

$$\begin{cases} u_{tt} = \Delta u, \quad t \geq 0, \quad x \in \Omega \in \mathbb{R}^3, \\ \partial u/\partial N = 0 \quad \text{on} \quad \Gamma, \\ u(0,x) = 0, \quad u_t(0,x) = f(x), \end{cases} \tag{13}$$

and assuming

$$v(x,k) = \int_0^\infty \exp(ikt)u(x,t)dt, \tag{14}$$

we obtain:

$$\begin{cases} \Delta v + k^2 v = -f, \quad \partial v/\partial N = 0 \quad \text{on} \quad \Gamma, \\ \partial v/\partial |x| - ikv = o(|x|^{-1}). \end{cases} \tag{15}$$

If $G(x,y,k)$ is Green's function for this problem, then

$$v = \int_\Omega G(x,y,k)fdy. \tag{16}$$

From (14), (16) we obtain

$$u(x,t) = \frac{1}{2\pi}\int_{-\infty}^{\infty} v \exp(-ikt)dk. \tag{17}$$

For the sake of simplicity we assume that $f \in C_0^\infty(\Omega)$. The function $G(x,y,k)$ can be continued analytically on the whole complex plane k. It is analytic in the upper half-plane $\mathrm{Im}\, k \geq 0$ and meremorphic in the lower half-plane $\mathrm{Im}\, k < 0$. For details see Ramm [78] and [90]. Suppose that

$$|v| \leq \frac{C}{1 + |k|^a}, \quad \mathrm{Im}\, k > -b, \quad b > 0, \quad a > \frac{1}{2}. \tag{18}$$

Then we can move down the contour of integration in (17)

$$u(x,t) = \frac{1}{2\pi} \int_{-ic-\infty}^{-ic+\infty} v(x,k) \exp(ikt) dk, \quad 0 < c < b. \tag{19}$$

From this it follows that

$$u(x,t) = \exp(-ct) w(x,t), \quad 0 < c < b, \tag{20}$$

where

$$w(x,t) = \frac{1}{2\pi} \int_{-\infty}^{\infty} v(x,-ic+y) \exp(-iyt) dy. \tag{21}$$

From (18) it follows that $v \in L^2(\mathbb{R})$ and $w \in L^2(\mathbb{R})$. Suppose that the poles k_j of $v(x,k)$ satisfy the inequality

$$\mathrm{Im}\, k_j < -F(|\mathrm{Re}\, k_j|), \tag{22}$$

where $F(x)$ is a continuous positive function,

$$F(x) > 0, \quad F(-x) = F(x), \quad F(x) \to +\infty \quad \text{as } x \to \infty. \tag{23}$$

If (18) holds in the domain

$$\mathrm{Im}\, k_j > -F(|\mathrm{Re}\, k_j|), \tag{24}$$

then by moving the contour of integration in (17) we get the asymptotic expansion (singularity expansion):

$$u(x,t) = \sum_{j=1}^{n} e^{-ik_j t} v_j(x,t) + o(e^{-|\mathrm{Im}\, k_n| t}). \tag{25}$$

where $|v_j| = O(t^{\nu_j - 1})$ as $t \to +\infty$, ν_j is the multiplicity of the pole k_j. This leads to the following questions:

(3) What can be said about location of the poles k_j? When does (18) hold? When does (18) hold in the domain (22)?

(4) What can be said about the properties of the poles $\{k_j\}$? How can these poles be calculated? Do these poles depend continuously on the boundary?

(5) To what extent does the set of poles $\{k_j\}$, $\mathrm{Im}\, k_j < 0$ determine the shape of the obstacle?

These questions are discussed in this appendix. They are of interest in applications and difficult from the mathematical point of view.

All of the results concerning operators A and T can be obtained for the analogous integral operators in electromagnetic wave scattering theory.

2. When do the Eigenvectors of T and A Form a Basis of H?

1. Bases with brackets and Tests for completeness and basisness. Let $\{h_j\}$ be an orthonormal basis of H, $m_0 < m_1 \ldots$ a sequence of integers, $m_\ell \to \infty$ as $\ell \to \infty$, and let H_ℓ be the linear span of the vectors h_{m_ℓ}, $h_{m_\ell + 1}, \ldots h_{m_{\ell+1} - 1}$. Let $\{f_j\}$ be a complete minimal system in H, and $\mathscr{F}_\ell$ be the linear span of vectors $f_{m_\ell}, \ldots f_{m_{\ell+1} - 1}$. By basisness we mean the property of a system of vectors or subspaces to form a basis of H.

Definition 1. If a linear, bounded, invertible operator B exists such that $BH_\ell = \mathscr{F}_\ell$, then the system $\{f_j\}$ is called a Riesz basis of H with brackets (notation: $\{f_j\} \in R_b(H)$). If $m_j = j$ then $\{f_j\}$ is called a Riesz basis of H ($\{f_j\} \in R(H)$).

Remark 1. It is known [1], that $\{f_j\} \in R_b(H)$ if and only if $C_1|f|^2 \leq \sum_{\ell=0}^{\infty} |P_\ell f|^2 \leq C_2|f|^2$, where $|\cdot|$ is the norm in H, $C_2 \geq C_1 > 0$ are constants, P_ℓ is the projector on F_ℓ, and $f \in H$ is an arbitrary element of H. The projector P_ℓ is defined by the direct decomposition $H = F_\ell \dot{+} G_\ell$, where G_ℓ is the union of the subspaces F_j for $j \neq \ell$.

Definition 2. Denote by Q_ℓ the orthoprojector on H_ℓ. If $\sum_{\ell=0}^{\infty} |P_\ell - Q_\ell|^2 < \infty$, then the system $\{f_j\}$ is called a Bari basis with brackets (notation: $\{f_j\} \in B_b(H)$).

Definition 3. A linear closed densely defined operator L on a Hilbert space H is called an operator with discrete spectrum if and only if its spectrum $\sigma(L)$ consists only of normal eigenvalues λ_j, $|\lambda_1| \leq |\lambda_2| \leq \cdots |\lambda_j| \leq \cdots$, $|\lambda_j| \to \infty$ as $j \to \infty$.

Remark 2. If L is a normal operator with discrete spectrum, $0 \notin \sigma(L)$, then L^{-1} is compact.

In what follows we assume for the sake of simplicity that L is a selfadjoint operator with a discrete spectrum $\{\lambda_j\}$, $0 \notin \sigma(L)$, and

$$\lambda_j = cj^p + O(j^{p_1}), \text{ as } j \to \infty, \quad p > 0,\ c > 0,\ p_1 < p. \tag{26}$$

Consider the operator

$$A = L + Q \tag{27}$$

where Q is a (nonselfadjoint) linear operator

$$|L^{-a}Qf| \leq C_a|f|, \text{ for all } f,\ a < 1,\ D(Q) \supset D(L). \tag{28}$$

Since

$$(L+Q-\lambda I)^{-1} = \{I + (L-\lambda I)^{-1}L^{a}L^{-a}Q\}^{-1}(L-\lambda I)^{-1} \quad \text{for} \quad \lambda \notin \sigma(L) \tag{29}$$

it is clear that

$$\lambda \notin \sigma(A) \quad \text{if} \quad |(L - \lambda I)^{-1}L^{a}| < C_a^{-1}. \tag{30}$$

It is clear that

$$|(L - \lambda I)^{-1}L^{a}| \leq \sup_j |\lambda_j - \lambda|^{-1}|\lambda_j|^{a}. \tag{31}$$

If $|\lambda_j - \lambda| \geq |\lambda_j|^{a}C_a q$, where $q > 1$ is arbitrary, then (30) holds. Hence we have proved the main part of the following lemma.

<u>Lemma 1</u>. <u>Suppose that</u> L <u>is a selfadjoint operator with a discrete spectrum</u>, Q <u>is a linear operator</u>, $A = L + Q$, <u>and</u> (28) <u>holds</u>. <u>Then</u> $\sigma(A) \subset K$, <u>where</u>

$$K = \bigcup_{j=1}^{\infty} \{\lambda: |\lambda - \lambda_j| < |\lambda_j|^{a}C_a q,\ q > 1\}, \tag{32}$$

<u>and</u> $\sigma(A)$ <u>is discrete</u>.

<u>Proof</u>: It remains to prove the last statement of Lemma 1. The statement follows immediately from the compactness of $(L - \lambda I)^{-1}$ and boundedness of the operator $\{I + (L - \lambda I)^{-1}L^{a}L^{-a}Q\}^{-1}$ in (29).

<u>Remark 3</u>. Estimates of type (31) were used earlier by Kacnelson [1] and Voitovich et al. [1]. We made no use of assumption (26) so far.

The following theorem is due to Kacnelson (Kacnelson [1]).

Theorem 1. <u>Under the assumptions</u> (26), (28), $A \in R_b(H)$ <u>if</u> $p(1-a) = 1$, <u>and</u> $A \in B_b(H)$ <u>if</u> $p(1 - a) > 1$.

Remark 4. We write $A \in R_b(H)$ $(B_b(H))$ if the root system $\{f_j\}$ of A form a Riesz (Bari) basis of H with brackets and $A \in R(H)$ if $\{f_j\} \in R(H)$.

Remark 5. Actually for Theorem 1 to be true it is sufficient to use the following estimate instead of (26): $\lambda_j \geq cj^p$ (see Kacnelson [1]).

Remark 6. Under some additional assumptions M. S. Agranovich proved that the series in root vectors of A converges rapidly (see the Appendix in Voitovich et al. [1]).

Remark 7. Completeness of the root system of a linear operator A in a Hilbert space H can be proved by means of the following theorems.

Theorem 2. (Gohberg-Krein [2]) <u>If</u> L <u>is a selfadjoint operator on a Hilbert space</u> H <u>with a discrete spectrum</u>, $0 \notin \sigma(L)$, Q <u>is a linear operator</u> $D(Q) \supset D(L)$, $L^{-1}Q$ <u>is compact and</u> $p(L^{-1}QL^{-1}) < \infty$, <u>then the system of root vectors of</u> $A = L + Q$ <u>is complete in</u> H.

Remark 8. The symbol $p(A) < \infty$ means that A is compact and $\Sigma_1^\infty s_n^p < \infty$, where $s_n = \lambda_n\{(A^*A)^{1/2}\}$ are the s-values of A.

Theorem 3. (Gohberg-Krein [2]) <u>The system of root vectors of a compact dissipative operator</u> A <u>with nuclear imaginary component is complete in</u> H <u>if</u> $\liminf_{n \to \infty} ns_n(A) = 0$.

Remark 9. A linear operator A is called dissipative if $\text{Im}\,(Af,f) \geq 0$ for all $f \in D(A)$. A compact linear operator is called nuclear if $\Sigma_1^\infty s_n(A) < \infty$.

<u>Theorem 4</u>. (Ramm [94]). <u>If</u> $A \geq 0$ <u>is compact</u>, B <u>is dissipative and nuclear</u>, <u>then the root system of</u> $A + B$ <u>is complete in</u> H.

<u>Example</u>. (Ramm [94]). Operator (9) can be split into the sum $T = T_o + T_1$, where $T_o g = \int_\Gamma (4\pi r_{ty})^{-1} g(y)dy$, $T_o > 0$, and $T_1 = T - T_o$ is nuclear and dissipative. The last statements is easy to verify (see Ramm [94] for details). Thus from Theorem 4 it follows that the root system of operator (9) is complete in $H = L^2(\Gamma)$. Actually this system forms a Riesz basis as we shall prove later.

2. <u>Elliptic pseudo-differential operators (PDO) on Γ</u>

In order to explain how to prove that the root system of operators A (formula (6)) and T (formula (9)) form a Riesz basis of H we start with the operator T. It is clear that

$$T = T_o + T_1,$$

where T_o, T_1 are defined in Example 1. It is easy to verify that T_o is an elliptic pseudo-differential operator on Γ of order -1 and T_1 is a PDO of order $\gamma < -1$ (in fact $\gamma = -3$). Suppose that $N(T_o) = \{0\}$. Then $L = T_o^{-1}$ exists, L is a selfadjoint operator with discrete spectrum. If $N(T) = \{0\}$, then $(T_o+T_1)^{-1} = (I+LT_1)^{-1}L = L + Q$, where $Q = -(I+LT_1)^{-1}LT_1L$, $|L^{-a}Q| \leq C$ for

$$a = 2 + \gamma < 1 \tag{33}$$

because $\text{ord}\, LT_1L = 2 + \gamma < 1$. Condition (26) is valid for PDO under very general assumptions (Agmon [2]). Therefore one can apply Theorem 1 and obtain

<u>Proposition 1</u>. <u>The root system of operator</u> T <u>defined by formula</u> (9) <u>forms a Riesz basis of</u> $H = L^2(\Gamma)$ <u>with brackets</u>.

<u>Remark 10</u>. It is easy to verify that $N(T_o) = \{0\}$ and $N(T) = \{0\}$ if k^2 is not an eigenvalue of the interior Dirichlet problem.

<u>Remark 11</u>. One can find, e.g. in Seeley [1], how to calculate the order of an elliptic PDO.

<u>Remark 12</u>. It is possible (and in a way more reasonable) to choose $T_o = \frac{1}{2}(T + T^*)$, because in this case T_1 will be of the order $-\infty$ for real $k > 0$ since the kernel of T_1 is $\sin kr_{ty}/r_{ty} \in C^\infty$ and $\text{ord } T_1 = -3$ for complex k.

Consider now the operator A defined by formula (6). It is easy to verify that A is a pseudo-differential elliptic operator, and $\text{ord } A = -1$. If $A_o = \frac{1}{2}(A+A^*)$, $A_1 = A - A_o$, then $\text{ord } A_o = -1$, $\text{ord } A_1 < -1$. If $N(A_o) = \{0\}$, and $N(A) = \{0\}$ one can use the arguments similar to ones used above and obtain the analogue of Proposition 1 for the operator A. If $N(A_o) \neq \{0\}$ then $\dim N(A_o) < \infty$ and $N(A_o) \subset C^\infty$. This statement follows from the a priori estimates for an elliptic PDO (Seeley [1]). Thus, one can add a finite-dimensional operator P to A_o and subtract this operator from A_1. Since $N(A_o) \subset C^\infty$, operator P can be chosen so that $\text{ord } (A_o + P) = \text{ord } A_o = -1 (\text{ord } P = -\infty)$, and $N(A_o + P) = \{0\}$. Hence, one can assume that $N(A_o) = \{0\}$. If $N(A_o) = \{0\}$, then A_1^{-1} exists and has a discrete spectrum. Since $\text{ord } A_1 < \text{ord } A_o$ the operator $A_o^{-1}A_1$ is compact in H. From this argument and the formula $A = A_o(I + A_o^{-1}A_1)$ it follows that the root subspace N of A corresponding to $\lambda = 0$ is finite-dimensional. Therefore one can split H into a direct sum $H = N \dot{+} M$, where N and M are invariant subspaces of A and $N(A|_M) = \{0\}$, $A|_M$ denoting the restriction of A to M. Hence, one can assume that $N(A) = \{0\}$. This completes the proof of the following proposition.

Proposition 2. *The root system of operator* A *defined by formula* (6) *forms a Riesz basis of* H *with brackets*.

3. When do T and A Have no Root Vectors?

1. A simple sufficient condition was given in Chapter 4, §3: in order that T (or A) has no root vectors it is sufficient that T is normal. This condition $T^*T = TT^*$ can be written explicitly (see 4.3.10)) and it is a condition concerning the surface Γ. In §4.3 it was verified that for an operator T this condition is satisfied if Γ is sphere. For a linear antenna this condition is also satisfied (see (4.3.12)). Of course, this condition is not necessary. In a finite-dimensional Hilbert space H every linear operator A without root vectors is similar to a normal operator. Indeed, if A has no root vectors, then its eigenvectors $\{f_j\}$ form a basis of H. If $\{h_j\}$ is an orthonormal basis of H, $Af_j = \lambda_j f_j$ and $f_j = Ch_j$, then $C^{-1}ACh_j = \lambda_j h_j$. This means that operator $C^{-1}AC$ is normal.

In an infinite-dimensional Hilbert space H this is not true: there exist compact operators whose eigenvectors span H but these operators are not similar to normal operators (an example is given in Decuard et al. [1]).

2. In Ramm [32] the following observation was formulated: the eigensubspace and the root subspace of a compact operator T, corresponding to the number λ, coincide if and only if (1) λ is a simple pole of the resolvent $(T - \lambda I)^{-1}$, or if

(2) $(T - \lambda I)^2 f = 0 \Rightarrow (T - \lambda I)f = 0$, or iff (3) the operator $T - \lambda I$ does not have zeros in the subspace $R(T - \lambda I)$, where $R(A)$ denotes the range of A.

4. What can Be Said about the Location and Properties of the Complex Poles?

1. Consider Green's function $G(x,y,k)$ of the exterior Dirichlet problem:

$$(\Delta + k^2)G = -\delta(x-y) \quad \text{in} \quad \Omega \tag{34}$$

$$G|_\Gamma = 0 \tag{35}$$

$$|x|(\partial G/\partial |x| - ikG) \to 0 \quad \text{as} \quad |x| \to \infty,\ k > 0. \tag{36}$$

Let $G_o = (4\pi r_{xy})^{-1}\exp(ikr_{xy})$. Then

$$G(x,y,k) = G_o(x,y,k) - \int_\Gamma G_o(x,t,k)\mu(t,y,k)dt, \qquad \mu = \frac{\partial G}{\partial N_t}, \tag{37}$$

where N is the unit of the outer normal to Γ at the point t, and μ satisfies the equation

$$\mu + A\mu = 2\frac{\partial G_o}{\partial N}, \tag{38}$$

where A is defined by formula (6). Operator $A = A(k)$ is an entire function of k and $A(k)$ is compact in $H = L^2(\Gamma)$ for any k since Γ is smooth. It is invertible for $\operatorname{Im} k > 0$. Hence, $(I + A(k))^{-1}$ is meromorphic and is defined on the whole complex plane k. Since $\partial G_o/\partial N$ for $y \notin \Gamma$ is an element of H which is an entire function of k, one can see from (38), that $\mu = 2(I + A(k))^{-1}\partial G_o/\partial N$ is meromorphic. From this argument and formula (37) it follows that $G(x,y,k)$ is meromorphic in k.

In §1 we emphasized that the location and properties of the complex poles of G are of interest in applications. By the properties of the poles we mean mostly whether the poles are simple or not.

Proposition 3. The set of poles of G coincide with the set of zeros of the functions $\lambda_n(k)$, $n = 1,2,3,\ldots$, where $\lambda_n(k)$ are the eigenvalues of the operator $T(k)$ defined by formula (9).

Proof: Let z be a pole of G,

$$G = \frac{R(x,y)}{(k-z)^r} + \cdots . \tag{39}$$

From (37), (39), (35) after multiplying (37) by $(k-z)^r$ and taking $k = z$ we obtain

$$\int_\Gamma G_o(s,t,z) \frac{\partial R(t,y)}{\partial N_t} dt = 0, \quad s \in \Gamma. \tag{40}$$

Since $R(x,y)$ is a degenerate kernel it follows from (40) that a function $f(t) \not\equiv 0$ exists such that

$$\int_\Gamma G_o(s,t,z) f(t) dt = 0, \quad s \in \Gamma. \tag{41}$$

This means that $\lambda_n(z) = 0$ for some n.

Conversely, suppose equation (41) has a nontrivial solution. The function

$$u(x) = \int_\Gamma G_o(x,t,z) f(t) dt \tag{42}$$

is a solution of the exterior Dirichlet problem

$$(\Delta + z^2)u = 0 \quad \text{in} \quad \Omega, \; u|_\Gamma = 0, \tag{43}$$

and

u satisfies the asymptotic condition at infinity. (44)

If z is not a pole of G, $\operatorname{Im} z \neq 0$, then $u \equiv 0$ in Ω and in D. This means that $f \equiv 0$ according to the jump relation. This is a contradiction. If z is not a pole of G and $\operatorname{Im} z = 0$, then $u = 0$ in Ω and $u \neq 0$ in D only if z^2 is an eigenvalue of the interior Dirichlet problem for the Laplace operator. But such an eigenvalue is a (real) pole of $G(x,y,k)$. Again, we obtain a contradiction. This completes the proof. More details are given in Appendix 7.

Remark 13. It is possible to find other functions whose zeros are poles of G (Ramm [55]).

Not much is known about the location of the complex poles of G:

1. It is proved in Lax-Phillips [2], [3] that the complex poles k_j of G (only the Dirichlet boundary condition was considered) satisfy the following inequality:

$$\operatorname{Im} k_j < a - b \ln |k_j|, \quad b > 0 \tag{45}$$

2. In Ramm [96] it was proved that a strip $-\varepsilon < \operatorname{Im} k < 0$, $\varepsilon > 0$ is free of the poles of the resolvent kernel of the Schrödinger operator with a finite potential $q(x) \in C_0^1$ for the exterior Dirichlet problem. This result shows that there exists a function $F(x)$ with the properties (23) such that the complex poles of the resolvent kernel of the Schrödinger operator with $q(x) \in C_0^1$ satisfy inequality (22) for the exterior Dirichlet problem.

3. In Lax-Phillips [3] a study of the poles $k_j = i\sigma_j$, $\sigma_j < 0$ was carried out. It was proved that there exist infinitely many of such poles, and the number of poles with $|\sigma_j| < \sigma$ was estimated asymptotically for $\sigma \to \infty$.

4. The resolvent kernel of the Laplace operator of the exterior boundary-value problem with the third boundary condition can have a pole $k = 0$. In this case the solution of the corresponding nonstationary problem for the wave equation does not necessarily decay as $t \to \infty$. An example is given in Asakura [1] where the problem

$$u_t = \Delta u \quad \text{in} \quad \Omega = \{|x| \geq R,\ t \geq 0\} \tag{46}$$

$$u(x,0) = 0, \quad u_t(x,0) = f(r), \tag{47}$$

$$\partial u/\partial r + R^{-1}u = 0, \quad \text{for} \quad r = |x| = R, \quad t \geq 0 \tag{48}$$

was considered. The solution can be found in the form

$$u = \sum_{n,m} u_{nm}(r,t)Y_{nm}(w), \tag{49}$$

where Y_{nm} are the spherical harmonics. From the explicit formula for u_{nm} it can be seen that $u_{00}(r,t)$ does not decay as $t \to \infty$ is $f(r) \geq 0$ and is finite. Another example is given in Dolph [3].

5. In Howland [1] a criterion is given for an operator function $[I + A(k)]^{-1}$ to have only simple poles. If z is a pole of this function, $I + A(k) = I + A(z) + (k-z)A_1 + \ldots$, then z is a simple pole iff

$$H = R(I + A(z)) \dotplus A_1 \ker\{I + A(z)\}. \tag{50}$$

Unfortunately in order to apply this criterion in practice it is necessary to have such information about $I + A(z)$ and A_1, which is usually unavailable.

5. <u>How to Calculate the Poles of Green's Function</u>. <u>Do the Poles Depend Continuously on the Boundary of the Obstacle?</u>

1. A general method for calculating the poles of Green's functions in diffraction and scattering was given in Chapter IV. The poles coincide with the numbers k_j for which $I + A(k)$ is not invertible (see equation (38)). Let $\{f_j\}$ be an orthonormal basis in $H = L^2(\Gamma)$,

$$\mu_n = \sum_1^n c_j f_j. \tag{51}$$

Substituting (51) in (38) and multiplying in H by f_i one obtains the system for unknown c_j:

$$\sum_{i=1}^n b_{ij}(k)c_j = 0, \quad b_{ij} \equiv ([I + A(k)]f_j, f_i). \tag{52}$$

Here $(.,.)$ denotes the scalar product in H. System (52) has a nontrivial solution if and only if

$$\det [b_{ij}(k)] = 0. \tag{53}$$

The left-hand side of this equation is an entire function of k. Let $k_m^{(n)}$, $m = 1,2,3,\ldots$ be its roots. In Chapter IV the following proposition is proved (see §4.1).

<u>Proposition 4</u>. <u>The limits</u> $\lim_{n\to\infty} k_m^{(n)} = k_m$ <u>exist and are the poles of Green's function</u> $G(x,y,k)$ <u>of the exterior Dirichlet problem</u>. <u>Every pole of</u> $G(x,y,k)$ <u>can be obtained in such a way</u>.

<u>Remark 14</u>. The same approach is valid for various boundary conditions (Neumann and third boundary conditions included), and for potential scattering by a finite potential (Chapter IV).

<u>Remark 15</u>. This approach is a variant of the general projection method.

Sketch of the Proof: First we show that $k_m^{(n)} \to k_m$ as $n \to \infty$. In the complex plane we choose a circle K_R of arbitrary radius R. Suppose that the points $k_1,\dots,k_s$ for which $I + A(k)$ is not invertible lie inside K_R and the remaining points k_m lie outside K_R. Denote by $\varepsilon > 0$ a small number, by $D_{\varepsilon,R} = \{k: |k - k_j| \geq \varepsilon, |k| \leq R\}$. We assume that the circles $|k - k_j| \leq \varepsilon$, $1 \leq j \leq s$ do not overlap. The operator $[I + A(k)]^{-1}$ is uniformly bounded on $D_{\varepsilon,R}$:

$$\| [I + A(k)]^{-1} \| \leq M, \quad k \in D_{\varepsilon,R}, \quad M = M_{\varepsilon,R}. \tag{54}$$

Equation (52) can be written as

$$\mu_n + P_n A(k) \mu_n = 0, \tag{55}$$

where P_n is the projector on the span of $f_1,\dots,f_n$. Since $P_n \to I$, where $\to$ denotes strong convergence of the operators on H, and $A(k)$ is compact, we conclude that A $\|A(k) - P_n A(k)\| \to 0$ as $n \to \infty$. Therefore $\|I + A(k) - [I + P_n A(k)]\| \to 0$ as $n \to \infty$. This means that for n sufficiently large, operators $I + P_n A(k)$ are invertible in $D_{\varepsilon,R}$, because $I + A(k)$ is invertible in $D_{\varepsilon,R}$. Therefore all roots of equation (53) for n sufficiently large lie in the union of the circles

$$|k - k_j| \leq \varepsilon, \quad |k| \leq R. \tag{56}$$

Since $\varepsilon > 0$ is arbitrarily small, this means that uniformly in the domain $|k| \leq R$ the limits exist:

$$\lim_{n \to \infty} k_j^{(n)} = k_j. \tag{57}$$

Conversely, let k_j, $|k_j| < R$ be an arbitrary pole of $G(x,y,k)$. Then operator $I + A(k_j)$ is not invertible. Suppose that in the circle $|k_j - k| < \varepsilon$ there are no numbers $k_m^{(n)}$ and no points k_i for $i \neq j$. Then $\|[I + A(k)]^{-1}\| \leq M$ for $|k - k_j| = \varepsilon$ and for n sufficiently large $\|[I + P_nA(k)]^{-1}\| \leq M_1$. Since there are no numbers $k_m^{(n)}$ inside the circles $|k - k_j| < \varepsilon$, the operator $I + P_nA(k)$ is invertible for $|k - k_j| \leq \varepsilon$ and $[I + P_nA(k)]^{-1}$ is an analytic operator function for $|k - k_j| \leq \varepsilon$. From the maximum modulus principle we obtain a uniform (with respect to n) estimate $\|[I + P_nA(k)]^{-1}\| \leq M_1$ for $|k - k_j| \leq \varepsilon$. But from this estimate we conclude that the operator $[I + A(k)]^{-1}$ exists for $|k - k_j| \leq \varepsilon$, which is a contradiction. This completes the proof.

Remark 16. The method gives a uniform approximation to the complex poles in any compact domain of the complex plane k.

2. In this section we show that in any compact domain of the complex plane the complex poles depend continuously on the boundary in the following sense. Consider a parametrized equation of the boundary

$$x_j = x_j(t_1,t_2), \quad 1 \leq j \leq 3, \quad 0 \leq t_1,t_2 \leq 1, \tag{58}$$

where $x_j \in C^2$.

Assume that a boundary Γ_ε obeys the following equation

$$x_j(\varepsilon) = x_j(t_1,t_2) + \varepsilon y_j(t_1,t_2), \quad 1 \leq j \leq 3, \tag{59}$$

where $y_j \in C^2$. Let $G(G_\varepsilon)$ be Green's function of the exterior Dirichlet problem in Ω, $\partial\Omega = \Gamma(\Omega_\varepsilon, \partial\Omega_\varepsilon = \Gamma_\varepsilon)$. Let $k_j(k_j(\varepsilon))$ be the poles of $G(G_\varepsilon)$.

<u>Proposition 5</u>. <u>If</u> $\varepsilon \to 0$, <u>then</u> $k_j(\varepsilon) \to k_j$ <u>uniformly for</u> $|k_j| \le R$, <u>where</u> $R > 0$ <u>is an arbitrary large fixed number.</u>

<u>Proof</u>: Let $\Delta = \{0 \le t_1, t_2 \le 1\}$. Then k_j, $|k_j| \le R$ are the points of the complex plane k at which the operator $I + A(k)$ defined by formula (38) is not invertible. Operator $I + A(k,\varepsilon)$ is not invertible at the points $k_j(\varepsilon)$. Here the operator $A(k,\varepsilon)$ is the counterpart of $A(k)$ for Γ_ε. Both operators can be written in the form

$$A(k,\varepsilon) = \int_\Delta \frac{\partial G_o}{\partial N}\mu\, J(t,\varepsilon)dt_1dt_2, \tag{60}$$

where $J(t,\varepsilon)dt_1dt_2$ is the element dt of the area of Γ_ε; for $\varepsilon = 0$ we obtain the operator $A(k)$. Since $x_j, y_j \in C^2$ the function $J(t,\varepsilon)$ is continuous (actually $J(t,\varepsilon) \in C^1$)

$$\lim J(t,\varepsilon) = J(t) \quad \text{as} \quad \varepsilon \to 0. \tag{61}$$

Thus,

$$||A(k,\varepsilon) - A(k)|| \to 0 \quad \text{as} \quad \varepsilon \to 0,\ |k| \le R. \tag{62}$$

Now we can use the arguments given in the proof of Proposition 4. The role of n is played by ε. Consider the union K_δ of the circles $|k - k_j| \le \delta$, where $\delta > 0$ is an arbitrary small fixed number, $|k_j| < R$, $1 \le j \le s$ and the circles do not overlap. By $D_{R,\delta}$ we denote $K_R \setminus K_\delta$, $K_R = \{k: |k| \le R\}$.

In $D_{R,\delta}$ operator $I + A(k)$ is invertible. Because of condition (62) for ε sufficiently small the operator $I + A(k,\varepsilon)$ is also invertible in $D_{R,\delta}$. This means that $k_j(\varepsilon) \in K_\delta$ for an ε sufficiently small. Since $\delta > 0$ is arbitrarily small the proof of Proposition 5 is complete.

<u>Remark 17</u>. It is possible to estimate $k_j(\varepsilon) - k_j$. In a general setting this type of perturbation theory was studied in Vainberg-Trenogin [1] and in Kato [1].

6. An Example on Complex Scaling. The complex scaling technique has recently attracted much attention in connection with spectral properties of the Schrödinger operator (see Complex Scaling in the Spectral Theory of the Hamiltonian Int. Journ. of Quantum Chemistry, 14, 1978). The main idea is to consider solutions of the Schrödinger equation for complex values of $r = |x|$.

This idea was used by the author as early as 1963 in order to prove the absense of a positive discrete spectrum of the Laplace operator of the Dirichlet problem in some infinite domains with infinite boundaries (Ramm [76]). The arguments given in Ramm [76] are not elementary. Here we use the same idea as in that paper and give a very simple proof of the following (known) proposition.

Proposition 6. Let $D \subset R^3$ be a bounded domain with a smooth closed connected boundary Γ, $\Omega = R^3 \setminus D$,

$$(\Delta + k^2)u = 0 \quad \text{in} \quad \Omega, \quad k^2 > 0, \tag{63}$$

$$u \in L^2(\Omega), \tag{64}$$

$$u|_\Gamma = 0. \tag{65}$$

Then $u(x) \equiv 0$ in Ω.

Proof: By Green's formula we have

$$u(x) = -\int_\Gamma g^+ \mu dt, \quad \mu = \frac{\partial u}{\partial N}, \quad g^+ = \frac{\exp(ikr_{xy})}{4\pi r_{xy}}. \tag{66}$$

(From (64) it follows that $\nabla u \in L^2(\Omega)$ and hence a sequence $r_n \to \infty$ exists such that

$$\int_{|x|=r_n} \{|u|^2 + |\partial u/\partial N|^2\} ds \to 0 \quad \text{as} \quad n \to \infty. \tag{67}$$

Therefore the integral over the large sphere in Green's formula tends to zero.) Let $x = rw$, where w is a unit vector, and let $z = r\exp(i\theta)$. The function $u(x) = u(rw)$ is considered as a function of the complex variable z. Since

$$g^{+} = \exp \frac{\{ik\sqrt{r^2 - 2r|t|\cos\alpha + |t|^2}\}}{\sqrt{r^2 - 2|t|r\cos\alpha + |t|^2}}, \quad \alpha = wt, \tag{68}$$

it is clear that G_0 is analytic in $z = r\exp(i\theta)$ for $|z| \geq R$, where R is sufficiently large, such that if $r > R$ then the inequality

$$r^2 > 2rd + d^2, \quad d = \max_{t\in\Gamma} |t| \tag{69}$$

holds. Thus for $|z| > R$ the function $\sqrt{z^2 - 2z\, t\cos\alpha + |t|^2}$ is analytic if we fix some branch of the radical. From (66) it follows that

$$u = \frac{\exp(ikz)}{z} f_1(z), \tag{70}$$

where f_1 is analytic in $|z| > R$ and

$$f_1 = 0(1) \quad \text{for} \quad |z| > R. \tag{71}$$

Exactly the same arguments lead to the formulas:

$$u = -\int_{\Gamma} g^{-}\mu dt, \quad g^{-} = \frac{\exp(-ikr_{xy})}{4\pi r_{xy}}, \tag{72}$$

$$u = \frac{\exp(-ikz)}{z} f_2(z), \tag{73}$$

where $f_2(z)$ is analytic in $|z| > R$ and

$$f_2 = 0(1) \quad \text{for} \quad |z| > R. \tag{74}$$

Hence

$$u(z) = \frac{e^{ikz}}{z} f_1 = \frac{e^{-ikz}}{z} f_2(z) \quad \text{for} \quad |z| > R. \tag{75}$$

Formula (75) is contradictory unless $u \equiv 0$. To prove $u \equiv 0$ we use a known uniqueness lemma for analytic functions.

Lemma. *Let D be a domain on the complex plane z, C be its boundary. Let D contain the half-plane $\operatorname{Re} z > a$. Let $f(z)$ be analytic in D, continuous in $D + C$ and*

$$\ln|f(z)| \leq A|z| \quad \text{for} \quad |z| > R, \quad z \in D, \tag{76}$$

where $A = \text{const} > 0$, and R is an arbitrary large fixed number,

$$\ln|f(z)| \leq -h(|z|), \quad z \in C, \tag{77}$$

where $h(t) > 0$ is a continuous function such that

$$\int_1^\infty t^{-2} h(t)dt = \infty. \tag{78}$$

Then $f(z) \equiv 0$ in D.

In our case $f(z) = u(z)$, D can be chosen so that C coincides outside of some large circle with the rays $\arg z = 3\pi/4$, $\arg z = 5\pi/4$, $h(t) = \text{const} + t/\sqrt{2}$, so that (78) is satisfied. We have

$$\ln|u(z)| \leq k|z| - \ln|z| + \ln|f_1| \leq A|z|, \quad z \in D,$$

(since $|f_1| \leq C_1$ we have $\ln|f_1| \leq C_2$). $\ln|u(z)| \leq -k|z|/\sqrt{2} + \text{cost}$, for $|z| > R$, $z = |z|\exp(i3\pi/4)$. A similar estimate holds for the ray $\arg z = 5\pi/4$. From the preceding lemma it follows that $u(z) \equiv 0$. Thus $u(r,w) = 0$ for $r > R$. By the unique continuation theorem we conclude that $u \equiv 0$ in Ω.

Bibliographical Notes

Section 1. The questions discussed here are of interest to engineers and physicists (Dolph-Scott [1], Voitovich et al. [1], Baum [1], Dolph [3]). They have also attracted considerable attention from mathematicians in recent years (Gohberg-Krein [2], Lax-Phillips [3], and the appendix of Voitovich et al. [1]). Our knowledge of the spectral structure of nonselfadjoint operators is very limited. For example, it is not known how to investigate this structure for equation (12). If a nonselfadjoint operator is a weak perturbation (in the sense defined in section 2) of a selfadjoint operator, some information is available (see Marcus [1], Kacnelson [1], and the appendix of Voitovich et al. [1]). There exist some theorems about completeness of root systems for dissipative operators (Gohberg-Krein [2] and Ramm [94]). No answer to question (5) is known.

Section 2. Properties of the bases of a Hilbert space are described in Gohberg-Krein [2] in a form convenient for our purpose. A rigorous study of the spectral properties of the integral operators arising in diffraction theory was initiated in Ramm [54], [55], [94]. Questions put forward by B. Kacnelenbaum were stimulating for these studies. M. S. Agranovich (the appendix to Voitovich et al. [1]) has made further contributions to this theory. Essential to his results were the results due to Markus [1] and Kacnelson [1]. The theory of pseudo-differential operators is now well developed. A summary of main results of this theory is given in Agmon [2], Seeley [1], Voitovich et al. [1] and Shubin [1].

M. S. Agranovich in the appendix to Voitovich et al. [1] applied the theory of pseudo-differential operators to the integral equations of diffraction theory.

Section 3. References are given in the section.

Section 4. The questions discussed here are of interest in applications. Proposition 3 was proved in Appendix 7 of this book. A part of it was proved in Ramm [55]. The scheme for the study of analytic continuation of the resolvent kernel of the Schrödinger operator was given in Ramm [78], [79], [81], [90]. Analytic properties of the scattering matrix for acoustic wave scattering by an obstacle were studied in Lax-Phillips [1]. Eigenfunction expansion theorems for nonselfadjoint Schrödinger operators are proved in Ramm [93], [99] and the properties of the resolvent in the complex plane of the spectral parameter k were used in the proofs. In Lax-Phillips [3] there is a study of the purely imaginary poles of Green's functions of the exterior Dirichlet and the Neumann problem. The known criteria for a pole of an operator valued function to be simple, including criterion (50) unfortunately are difficult to apply: so far no applications of these criteria appear to be known.

In Babich-Grigorjeva [1] it is proved that for the complex poles of Green's function of the exterior Neumann problem for a convex domain in $\mathbb{R}^3$ with a smooth boundary which has a positive Gaussian curvature, the function $F(x)$ in formula (22) can be taken as $F(x) = \varepsilon|x|^{1/3}$, for some small $\varepsilon > 0$. In B. R. Vainberg [1] it was shown how to pose correctly the problem of finding root vectors corresponding to the complex poles of Green's functions.

In B. R. Vainberg [2] analytic continuation of the resolvents of some general differential operators is studied.

There is an example in Il'in [1] which shows that a root system of a nonselfadjoint operator may form a basis of H, but some other root system of the same operator may not form a basis of H.

In the literature, the radiation condition in the form $u \sim (\exp(ikr)/r)(f(k,n)+O(1/r))$ as $r \to \infty$ is often used for $\operatorname{Im} k < 0$, i.e., for exponentially increasing solutions of the problem (43). It is assumed in such cases that the solution of the boundary-value problem satisfying the radiation condition is unique. This is false. A simple example is the function $u = g^+ * f - g^- * f$, where g^+, g^- are defined by formulas (68), (72), $f \in C_0^\infty$ is arbitrary, and $*$ denotes convolution. It is clear that $(\Delta + k^2)u = 0$ in $\mathbb{R}^3$ and u satisfies the radiation condition for $\operatorname{Im} k < 0$, but $u \not\equiv 0$. The right asymptotic condition for exponentially increasing solutions is given in B. R. Vainberg [1], where it is proved that for $\operatorname{Im} z < 0$ the solution of the problem (43) has, in a neighborhood of infinity, the form $u = r^{-1}\exp(izr) \sum_{j=0}^{\infty} f_j(\alpha) r^{-j}$, $r = |x|$, $\alpha = x|x|^{-1}$, and the series converges absolutely and uniformly for sufficiently large r.

Section 5. The simple method for calculation of the complex poles is given in Ramm [54], [55]. It is essentially a variant of the projection method and the arguments show that the complex poles depend continuously on the boundary. The same arguments prove the continuous dependence of these poles on the parameters if the kernel depends continuously on these parameters.

In Ramm [83], [91] it was shown rigorously that the solution of the exterior Dirichlet boundary-value problem is the limit of the solutions of the potential scattering problem when the potential goes to infinity in D and is equal to 0 in Ω. Here, as usual, $D = \mathbb{R}^3 \setminus \Omega$, Ω being the exterior domain. In Ramm [85], [97] behavior as $t \to \infty$ of the solution of the wave equation in the exterior domain was studied in case when the resolvent kernel of the corresponding stationary problem cannot be analytically continued through the continuous spectrum.

It is possible to conclude from formula (75) in Appendix 2 that $u(z) \equiv 0$ without making use of the lemma following equation (75). Indeed, since f_1, f_2 are analytic and bounded in some neighborhood of infinity they behave asymptot cally as $C_n z^{-n}$, $n \geq 0$. If $z = iy$ in formula (75) of Appendix 2 and $y \to +\infty$, then the left-hand side of this formula goes to zero, while the right-hand side goes to infinity unless $f_1 = f_2 = 0$. This simple argument was pointed out by B. A. Taylor. In Ramm [76], where the boundary of the domai was infinite it was necessary to use the Lemma following equa tion (75). It is interesting to mention that exactly the sa arguments prove the following proposition.

<u>Proposition 7</u>. <u>Let</u> u <u>be a solution of problem</u> (63)-(64). <u>Then</u> $u \equiv 0$.

Note that no assumptions about boundary values of u a made in this proposition.

<u>Unsolved Problems</u>

1. To what extent do the complex poles of Green's function determine the obstacle?

2. Is it true that the complex pole of Green's function for the exterior Dirichlet problem are simple?
3. Does the order of a complex pole coincide with the order of zero of the corresponding eigenvalue? (see Proposition 2).

APPENDIX 11

ON THE BASIS PROPERTY FOR THE ROOT VECTORS OF SOME NONSELFADJOINT OPERATORS

1. Introduction. Let A be a densely defined linear operator on a Hilbert space H of the form $A = L + T$, where L is a selfadjoint operator with discrete spectrum $\{\lambda_n\}$, $\lambda_1 \leq \lambda_2 \leq \ldots$, $D(A) = D(L)$, $D(A) \equiv \text{dom } A$. We assume that

$$\lambda_n = cn^p(1 + o(n^{-1})), \quad c = \text{const} > 0, \quad p > 0. \tag{1}$$

This assumption is satisfied by some elliptic differential and pseudo-differential operators (PDO). An operator T is said to be subordinate to L if:

$$|Tf| \leq M|L^a f|, \quad a < 1, \quad f \in D(L^a). \tag{2}$$

Here and below M denotes various constants, and $|T|$ denotes the norm of operator T in H.

Under the assumptions (1), (2) the operator $A = L + T$ has a discrete spectrum, that is, every point of its spectrum is an eigenvalue of finite algebraic multiplicity. If λ is an eigenvalue of A, then the linear span of the corresponding eigenvectors is called the eigenspace corresponding to λ. Let h_j be an eigenvector, $Ah_j = \lambda h_j$. If the equation

$Ah_j^{(1)} = \lambda h_j^{(1)} + h_j$ is solvable then the chain $\{h_j, h_j^{(1)}, \ldots, h_j^{(s_j)}\}$, $Ah_j^{(s_j)} = h_j^{(s_j)} + h_j^{(s_j-1)}$ is called the Jordan chain corresponding to the pair (λ, h_j). The number $s_j + 1$ is called the length of this chain if the equation $Ah - \lambda h = h_j^{(s_j)}$ has no solutions. If λ has finite algebraic multiplicity, then $s_j < \infty$. The vectors $h_j^{(m)}$ are called root vectors (or associated vectors). The collection of all eigenvectors and root vectors is called the root system of A. A system $\{g_j\}_{j=1}^{\infty}$ of vectors is called linearly independent if any finite subset of these vectors is linearly independent. Consider a system $\{g_j\}$ of linearly independent vectors in H. If for all j the vector g_j does not belong to the closure of the linear span of vectors $g_1, \ldots, g_{j-1}, g_{j+1}, \ldots$, then the system $\{g_j\}$ is called minimal. A minimal system $\{g_j\}$ forms a basis of H if any $g \in H$ can be uniquely represented as $g = \sum_{j=1}^{\infty} c_j g_j$. We shall write $A \in B(H)$ (or $A \in B$) if its root system forms a basis for H.

A minimal system $\{g_j\}$ forms a Riesz basis of H if there exists an isomorphism B (linear bijection of H onto H) which sends an orthonormal basis $\{f_j\}$ onto $\{g_j\}$, i.e., $Bf_j = g_j$ for all j. A minimal system $\{g_j\}$ forms a Riesz basis with brackets of H if there exists an isomorphism B which sends $\{F_j\}$ onto $\{G_j\}$, i.e., $BF_j = G_j$. Here $\{F_j\}$ is the collection of subspaces constructed as follows. Let $m_1 < m_2 < \ldots$ be an increasing sequence of integers; then F_1 is the span of vectors $f_1, \ldots, f_{m_1}$, F_j is the span of vectors $f_{m_{j-1}+1}, f_{m_{j-1}+2}, \ldots, f_{m_j}$, and G_j is defined similarly. Now we can give the basic definition in which a new word "basisness" is used.

Definition. A linear operator A with discrete spectrum possesses the basisness property if its root system forms a Riesz basis with brackets for H. In this case we write $A \in R_b(H)$ (or $A \in R_b$). If the root system of A forms a Riesz basis we write $A \in R(H)$ (or $A \in R$).

The purpose of this appendix is to give some sufficient conditions for $A \in R$ to be true. These conditions will essentially be conditions (1), (2). In the literature there are some results related to the question of basisness. In Kato [1, §V.4] a theorem on basisness for an operator $L + T$ is proved under the assumptions that the eigenvalues of L are simple, $\lambda_j - \lambda_{j-1} \to +\infty$ as $j \to \infty$, and T is bounded. In Gohberg-Krein [2] some conditions for the completeness of root systems of some nonselfadjoint operators are given. In Marcus [1], Kacnelson [1], Agranovich [1], Ramm [94], and Ramm [112] some conditions for $A \in R_b$ are given and in Ramm [94], [112] applications to diffraction and scattering theory are presented. One of the main results, Kacnelson [1] is that $A \in R_b$ if $p(1-a) \geq 1$. The assumption that L is selfadjoint can often be replaced by the assumption L is normal, provided that it is known a priori that the eigenvalues of L are concentrated near some rays in the complex plane.

In this appendix we give a simple method to prove that $A \in R$ if $p(1-a) \geq 2$. This method is based on some estimates of the resolvent of A.

The main result is the following:

<u>Theorem</u>. <u>Let</u> (1) <u>and</u> (2) <u>hold and</u> $p(1-a) \geq 2$. <u>Then</u> $A \in R$.

2. <u>Proof</u>: Let

$$P_j = -\frac{1}{2\pi i}\int_{C_j}(A - \lambda I)^{-1}d\lambda \tag{3}$$

denote the projection onto the root space L_j of the operator A corresponding to the eigenvalue $\lambda_j(A)$, where C_j is a circle with center $\lambda_j(A)$ sufficiently small that there are no other eigenvalues inside the circle. In order to prove that $A \in B$ it is sufficient to prove that

$$\sum_{j=1}^{N} P_j f \to f \quad \text{as} \quad N \to \infty, \quad f \in H, \tag{4}$$

where the arrow denotes convergence in H. In order to prove also that $A \in R$ it is necessary and sufficient to prove that Gohberg-Krein [2, p. 310, 334]

$$\sup\left|\sum_{j\in J} P_j\right| < \infty, \tag{5}$$

where the supremum is taken over all finite subsets of positive integers.

We start with the identity

$$\begin{gathered}(2\pi i\lambda)^{-1}f = -(2\pi i)^{-1}R_\lambda f + (2\pi i\lambda)^{-1}R_\lambda Af, \quad f \in D(A),\\ R_\lambda = (A - \lambda I)^{-1}\end{gathered} \tag{6}$$

and integrate this identity over the contour $\Gamma_m: |\lambda| = r_m = (\lambda_m+\lambda_{m+1})/2$. Note that the distance d_m between $\{\lambda_j\}$ and the circle $|\lambda| = r_m$ satisfies the inequality

$$d_m \geq (\lambda_{m+1} - \lambda_m)/2. \tag{7}$$

After integration we get

$$f = \sum_{j=1}^{N_m} P_j f + a_m + b_m, \tag{8}$$

where

$$a_m = (2\pi i)^{-1} \int_{\Gamma_m}^{-1} \lambda R_\lambda Lf d\lambda, \quad b_m = (2\pi i)^{-1} \int_{\Gamma_m}^{-1} \lambda R_\lambda Tf d\lambda. \tag{9}$$

It is easy to prove the following lemma.

Lemma 1. *Under the assumptions* (1), (2), *operator* $A = L + T$ *is closed, its spectrum is discrete and the eigenvalues of* A *lie in the set:*

$$K = \bigcup_{j=1}^{\infty} \{\lambda: \ |\lambda-\lambda_j| < |\lambda_j|^a Mq\}, \quad q > 1, \tag{10}$$

where M *and* a *are the constants from* (2).

While this statement can be found in the literature (Kato [1], Kacnelson [1], Ramm [112]) we present a proof for the convenience of the reader after the proof of the theorem.

To prove that $A \in B$ it is sufficient to prove that

$$a_m \to 0, \quad b_m \to 0 \quad \text{as} \quad m \to \infty. \tag{11}$$

Both terms can be considered similarly. Let us consider the first term. If $R_\lambda^0 = (L-\lambda)^{-1}$, then

$$R_\lambda = \{(L-\lambda)(I+R_\lambda^0 L^a L^{-a} T)\}^{-1} = (I+R_\lambda^0 L^a T_1)^{-1} R_\lambda^0, \tag{12}$$

$$T_1 = L^{-a}T, \quad \|T_1\| \le M,$$

$$|R_\lambda^0 L^a| = \sup_j \frac{|\lambda_j|^a}{|\lambda-\lambda_j|} \le \sup_j \frac{|\lambda_j|^a}{|r_m-\lambda_j|} \le M \frac{|\lambda_m|^a}{d_m} \le \frac{M}{m^{p(1-a)-1}}. \tag{13}$$

Here M denotes various constants and m is assumed to be large, so that it follows from (1) and (7) that $\lambda_m \sim cm^p$, $d_m \ge Mm^{p-1}$. It is clear now that $p(1-a) > 1$ implies the following estimate provided that $|\lambda|$ is sufficiently large and runs through the set $\{r_m\}$:

$$|R_\lambda^o L^a| \leq M|\lambda|^{-\gamma}, \quad \gamma = p^{-1}\{p(1-a)-1\} = 1-a-p^{-1} > 0. \tag{14}$$

Furthermore, we get

$$|R_\lambda^o| \leq \max_j \frac{1}{|\lambda-\lambda_j|} \leq \frac{M}{d_m} \leq \frac{M}{|\lambda|^{1-p^{-1}}}, \tag{15}$$

since for large m it follows from $\lambda_m \sim cm^p$ that $m \sim c_1\lambda_m^{1/p}$.

From (12), (14), (15) it follows that

$$|R_\lambda| \leq \frac{M}{|\lambda|^{1-p^{-1}}} \tag{16}$$

provided that $\gamma > 0$, i.e., $1-p^{-1} > a$. All estimates (13)-(16) are made under the assumptions that $|\lambda| = r_m$ and m is sufficiently large.

It is well known that the eigensystem of the selfadjoint operator L with discrete spectrum forms an orthogonal basis for H. For $A = L$, an identity of type (8) is

$$f = \sum_{j=0}^{N_m} P_j^0 f + a_m^0, \quad a_m^0 = (2\pi i)^{-1} \int_{\Gamma_m} \lambda^{-1} R_\lambda^0 Lf d\lambda, \tag{17}$$

where

$$P_j^0 = -(2\pi i)^{-1} \int_{C_j^0} R_\lambda^0 d\lambda \tag{18}$$

and C_j^0 is a small circle with center λ_j.

For the selfadjoint operator

$$f = \lim_{m\to\infty} \sum_{j=1}^{N_m} P_j f \quad \text{and} \quad a_m^0 \to 0 \quad \text{as} \quad m \to \infty. \tag{19}$$

Thus in order to prove that $a_m \to 0$ as $m \to \infty$ it is sufficient to prove that

$$a_m - a_m^0 \to 0 \quad \text{as} \quad m \to \infty. \tag{20}$$

To this end consider

$$|(R_\lambda - R_\lambda^0)Lf| = |R_\lambda T R_\lambda^0 Lf| \le M|R_\lambda||L^a R_\lambda^0||Lf|$$
$$\le M|\lambda|^{-2(1-p^{-1})+a}|Lf|,$$

$$|R_\lambda Tf| \le |(R_\lambda - R_\lambda^0)Tf| + M|R_\lambda^0 L^a||f| \le |R_\lambda^0 T R_\lambda Tf| + M|\lambda|^{-\gamma}|f| \tag{21}$$
$$\le M|R_\lambda^0 L^a||R_\lambda Tf| + M|\lambda|^{-\gamma}|f|$$
$$\le M|\lambda|^{-\gamma}|R_\lambda Tf| + M|\lambda|^{-\gamma}|f|.$$

If $\gamma > 0$ and $|\lambda|$ is sufficiently large we get

$$|R_\lambda Tf| \le M|\lambda|^{-\gamma}|f|. \tag{22}$$

If $\gamma > 0$ and $\gamma + 1-p^{-1} > 0$, i.e., $p(1-a) > 1$ and $p(2-a) > 2$, then (11) follows from (21), (22), and (9) for $f \in D(L)$. The idea of the following argument is to prove (11) for any $f \in H$ and thereby to prove that $A \in B$. To this end let us first give the proof for the simple case $A = L$. In this case the proof that $a_m^0 \to 0$ as $m \to \infty$ for any $f \in H$ can be given as follows: $a_m^0 = f - \sum_{j=1}^{N_m} P_j^0 f$ is a linear operator which is bounded since the P_j^0 are orthogonal projections. Thus if $a_m^0 = a_m^0(f) \to 0$ for all f in a dense set in H this is true for all $f \in H$. To apply this idea to a_m we must prove that $|\sum_{j=1}^{N_m} P_j| \le M$, where M does not depend on m. To prove this it is sufficient to prove that

$$I_m \equiv |\sum_{j=1}^{N_m}(P_j - P_j^0)| \le M. \tag{23}$$

We have

$$I_m \leq \frac{1}{2\pi}\left|\int_{\Gamma_m}(R_\lambda - R_\lambda^0)fd\lambda\right| \leq \frac{1}{2\pi}\left|\int_{\Gamma_m} R_\lambda T R_\lambda^0 f d\lambda\right|$$
$$\leq M\frac{|\lambda||f|}{|\lambda|^{1-p^{-1}+\gamma}} = \frac{M|f|}{|\lambda|^{1-a-2p^{-1}}} \tag{24}$$

Therefore if

$$p \geq \frac{2}{1-a}, \quad a < 1, \tag{25}$$

the above argument shows that $a_m(f) \to 0$ for all $f \in H$, so that $A \in B$. But actually inequality (24) shows more: if (25) holds, then $A \in R$ (i.e., the root system of A forms a Riesz basis without brackets of H). Indeed,

$$|\sum_J P_j| \leq |\sum_J P_j^0| + |\sum_J (P_j - P_j^0)| \leq M_1 + M_2 \leq M \tag{26}$$

for any set J of positive integers. This completes the proof of the theorem. □

Remark 1. Both the inequalities $p(1-a) > 1$ and $p(2-a) > 2$ follow from (25).

Proof of Lemma 1: From (12) it follows that $\lambda \notin \sigma(A)$ if $|R_\lambda^0 L^a| M < 1$. From (13) and (10) it follows that if $\lambda \notin K$, then

$$M|R_\lambda^0 L^a| \leq M \sup_j \frac{|\lambda_j|^a}{|\lambda - \lambda_j|} < \sup_j \frac{M|\lambda_j|^a}{Mq|\lambda_j|^a} \leq q^{-1} < 1,$$

so that $\lambda \notin \sigma(A)$. Thus $\sigma(A) \subset K$, where K is defined in (10). Discreteness of $\sigma(A)$ and the closure of A can be proved under weaker assumptions (§1.7).

3. Generalizations. Assumption (1) can be replaced by the following assumption:

$$\lambda_m^{a+1}(\lambda_{m+1}-\lambda_m)^{-2} \to 0 \quad \text{as} \quad m \to \infty, \tag{1'}$$

where a is defined by formula (2).

Proposition 1. *From* (1') *and* (2) *it follows that* $A \in R$.

Proof: Let $|\lambda| = (\lambda_{m+1}+\lambda_m)/2$, $d_m = \lambda_{m+1} - \lambda_m$, M be various positive constants which do not depend on m. We need to prove that: (i) $|\lambda|\ |R_\lambda - R_\lambda^0| \to 0$ as $|\lambda| \to \infty$, (ii) $|(R_\lambda - R_\lambda^0)L| \to 0$ as $|\lambda| \to \infty$, (iii) $|R_\lambda T| \to 0$ as $|\lambda| \to \infty$. We have: $R_\lambda - R_\lambda^0 = -R_\lambda T R_\lambda^0$, $|R_\lambda^0| \leq Md_m^{-1}$, $|R_\lambda^0 L^a| \leq M|\lambda_m|^a d_m^{-1}$, $|R_\lambda| \leq |R_\lambda^0||(I+R_\lambda^0 T)^{-1}| \leq Md_m^{-1}$, $|TR_\lambda^0| + |R_\lambda^0 T| \leq M|\lambda_m|^a d_m^{-1}$. Without loss of generality we can assume that L^{-1} exists (otherwise L can be replaced by $L + \varepsilon I$, where ε is a small number and $(L + \varepsilon I)^{-1}$ exists; in this case T would be replaced by $T - \varepsilon I$ and condition (2) holds for $T - \varepsilon I$ and $L + \varepsilon I$). From (1') it follows that $\lambda_m^a d_m^{-1} \to 0$ as $m \to \infty$, because $\lambda_m \to +\infty$ and $a < 1$. We have: (i) $|\lambda|\ |R_\lambda - R_\lambda^0| \leq |\lambda||R_\lambda T R^0| \leq M\lambda_m^{1+a} d_m^{-2} \to 0$, $m \to \infty$ (ii) $|(R_\lambda - R_\lambda^0)L| = |R_\lambda T R_\lambda^0 L| \leq M\lambda_m^{1+a} d_m^{-2} \to 0$, $m \to \infty$ (iii) $|R_\lambda T| \leq |(R_\lambda - R_\lambda^0)T| + |R^0 T| \leq M\lambda_m^{2a} d_m^{-2} + M\lambda_m^a d_m^{-1} \to 0$, $m \to \infty$.

Remark 2. If $\lambda_m \sim cm^p$ and $d_m \geq Mm^{p-1}$, then (1') implies that $p(1-a) > 2$. To get the condition $p(1-a) \geq 2$ as a sufficient condition for $A \in R$ we add the argument given following equation (22).

Remark 3. If a in (2) can be taken arbitrarily large negative and there exists some $b \in (-\infty,\infty)$ such that

$$d_m \geq M\lambda_m^b, \tag{1''}$$

then (1') holds.

Instead of (1) for a wide class of PDOs the following estimate is known:

$$\lambda_n = cn^p(1+O(n^{-\delta})), \quad c > 0, \quad p > 0, \quad \delta > 0. \tag{27}$$

In this case our arguments lead to

Proposition 2. *Let* $P(1-a) > 2$, $0 < \delta_1 < \delta$, *where* δ *is defined in* (27) *and* $c_1 > 0$ *is a constant. Then there exists a sequence of integers* $m_n \sim c_1 n^{1/\delta_1}$ *such that the system of subspaces* $\{P^{(n)}H\}_{n=1}^{\infty}$ *forms a Riesz basis of* H, *where* $P^{(n)} = \sum_{j=m_n}^{m_{n+1}} P_j$ *and* P_j *is defined by formula* (3). *This means that* $A \in R_b$ *and the sequence* m_n *defines the bracketing*.

The sequences $\{P^{(n)}H\}$ plays the role of the sequence $\{G_n\}$ of the subspaces defined in the Introduction. We need a few lemmas to prove this proposition.

Lemma 1. *If* $\lambda_n = cn^p(1 + O(n^{-\delta}))$, $\delta > 0$ *then* $N(\lambda) = \sum_{\lambda_n<\lambda} 1 = (\lambda c^{-1})^{1/p}(1 + O(\lambda^{-\delta/p}))$.

Proof: This statement follows from the fact that $\lambda = \lambda(n)$ and $N(\lambda)$ are reciprocal functions.

In what follows we assume that the assumption of Lemma 1 holds.

Lemma 2. *For sufficiently large* n *and* m, $n < m$, $0 < q_1 \leq nm^{-1} \leq q_2 < 1$ *there exist eigenvalues* $\lambda^{(1)}$ *and* $\lambda^{(2)}$, $\lambda_n \leq \lambda^{(1)} < \lambda^{(2)} \leq \lambda_m$ *such that* $\lambda^{(2)} - \lambda^{(1)} \geq c_1 m^{p-1}$

<u>and the interval</u> $(\lambda^{(1)}, \lambda^{(2)})$ <u>contains no eigenvalues</u>.

<u>Proof</u>: There are $m - n$ eigenvalues (counting multiplicity) in the segment $(\lambda_n, \lambda_m]$. Thus there exists at least a couple of eigenvalues $\lambda_n < \lambda^{(1)} < \lambda^{(2)} \leq \lambda_m$ such that there are no eigenvalues in the interval $(\lambda^{(1)}, \lambda^{(2)})$ and $\lambda^{(2)} - \lambda^{(1)} \geq (\lambda_m - \lambda_n)/(m-n) \geq c(m^p - n^p)/(m-n) - O(m^{p-\delta}/(m-n)$ $c_1 m^{p-1}$, where $c_1 = c_1(q_1, q_2)$. By c_1 we denote various positive constants.

<u>Lemma 3</u>. <u>Suppose that</u> $m = m(n)$, $1 - d(n) \leq m^{-1}(n)n \leq 1 - b(n)$ $b(n)/d(n) \geq c_1$, $b(n)n^{\delta} \to \infty$ <u>as</u> $n \to \infty$, $d(n) \to 0$, $n \to \infty$. <u>Then the conclusion of Lemma</u> 2 <u>holds</u>.

<u>Proof</u>: The proof is similar to that of Lemma 2. The last step is slightly different:

$$c\,\frac{m^p - n^p - O(m^{p-\delta})}{m - n} = cm^{p-1}\,\frac{1 - (\frac{n}{m})^p - O(m^{-\delta})}{1 - \frac{n}{m}}$$

$$\geq cm^{p-1}\,\frac{1 - (1 - b(n))^p - O(m^{-\delta})}{d(n)} \geq cm^{p-1}\,\frac{\frac{1}{2}pb(n)(1 - O(n^{-\delta}b^{-1}(n)}{d(n)}$$

$$\geq c_1 m^{p-1}.$$

Here we used the inequality $1 - (1 - x)^p \geq \frac{1}{2}px$ which holds for small x.

<u>Proof of Proposition 2</u>: We can take $b(n) = n^{-\delta_1}$, $0 < \delta_1 < \delta$, $d(n) = b(n)$. In this case $m_{n+1}/m_n =$ $1 + (b/m_n^{\delta_1})$ and $m_n \sim (\delta_1 b)^{1/\delta_1} n^{1/\delta_1}$. From this and Lemmas 3, 2 and the argument given in the proof in Section 2, Proposition 2 follows.

<u>Example 1</u>. Let $Qf = \int_\Gamma r_{st}^{-1} \exp(ikr_{st}) f(t)dt$, where Γ is a smooth closed surface in R^3, $k > 0$, $r_{st} = |s-t|$. Then

$Q = Q_o + Q_1$, where $Q_o = \text{Re } Q$, $Q_1 = i \text{ Im } Q$,

$$Q_o f = \int_\Gamma r_{st}^{-1} \cos(kr_{st}) f(t) dt, \quad Q_1 f = i \int_\Gamma r_{st}^{-1} \sin(kr_{st}) f(t) dt.$$

The operators Q_o, Q_1 are pseudo-differential of orders -1 and $-\infty$ respectively (Appendix 10) $\lambda_n(Q_o) \sim c_1 n^{-1/2}$, $c_1 = \text{const}$.

Let us assume that $L = Q_o^{-1}$ exists (without loss of generality, see Appendix 10). Then $\lambda_n(L) \sim cn^{1/2}$, $c = \text{const}$, so that $p = \frac{1}{2}$, where p is defined in (1). Since in the theorem the unperturbed operator is unbounded, setting $A = (Q_o + Q_1)^{-1} = (I + LQ_1)^{-1} L = L + T$, $T \equiv -(I+LQ_1)^{-1} LQ_1 L$, we assumed that $(Q_o+Q_1)^{-1}$ exists again without loss of generality; for $k > 0$ and k^2 is not an eigenvalue of the Laplace operator for the interior Dirichlet problem in the domain D with boundary Γ it is easy to prove that $(Q_o+Q_1)^{-1}$ exists. Since $\text{ord } LQ_1 L = -\infty$ we can take the number a in (2) to be negative and large, so that $p(1-a) > 2$. Thus $Q \in R_b$; if (1") holds, then $Q \in R$.

For complex k the order of $\text{Im } Q$ is equal to -3, $a = -1$ so that $p(1-a) = 1$ and $Q \in R_b$ but we cannot assert that $Q \in R$.

Example 2. Let $Qf = \int \exp(ikr_{xy}) r_{xy}^{-1} q(y) f(y) dy$, $k > 0$, $\int \equiv \int_{R^3}$. Operator Q plays the principal role in the potential scattering theory. Let us assume that $q \in C_o^\infty(\mathbb{R}^3)$, $q(x) \geq 0$. Then the operator $Q_1 f = \int \cos(kr_{xy}) r_{xy}^{-1} q(y) dy$ is a selfadjoint pseudo-differential operator of order -2 in $H = L^2(\mathbb{R}^3; q(x))$; the operator $Q_2 f = i \int \sin(kr_{xy}) r_{xy}^{-1} q(y) dy$ has order $-\infty$ because its kernel is infinitely smooth and

$q(y)$ is compactly supported; $\lambda_n(Q_1) \sim cn^{-2/3}$. Thus in this case, $p = 2/3$, a can be taken negative and as large as we want, inequality $p(1-a) \geq 2$ holds, and $Q \in R_b$. The root system of Q forms a Riesz basis of H if (1") holds. If q is not compactly supported, additional considerations are needed. It is easy to prove that $Qf \equiv 0$ implies $f = 0$, so that Q^{-1} exists.

Bibliographical Notes for Appendices

The result of Appendix 1 was announced in Ramm [74] and proved in Ramm [102]. It would be interesting to verify its efficiency, in particular for the analytic continuation problem by numerical experiments. If in equation (1), Appendix 1, $f(x)$ is known for $-1 \leq x \leq 1$ this equation can be solved explicitly (see Mushelishvili [1]). The results of Appendix 2 are closely connected with the results presented in Chapter 2. They are new, although an equation similar to equation (18) in Appendix 2 was used in Mushelishvili [1, p. 207] without a discussion of its characteristic values or of iterative processes. The results of Appendix 3 appeared in Ramm [20], although it seems no proof has been given of the fact that condition (30) allows the system (28)-(29) to be solved by an iterative process. The results of Appendix 4 (Ramm [40]) are closely connected with Section 4 of Chapter 2. The contents of Appendix 5 appeared in Ramm [110], while the contents of Appendix 6 has not previously been published. It would be interesting to solve numerically some practical problems using the results given in Appendices 5 and 6 and to study from the theoretical point of view numerical schemes for solving equation (13) of Appendix 6. Part of the results of Appendix 7 appeared in Ramm [110],[55]. The result of Appendix 8 was published in Ramm [116]. The result of Appendix 9 is new. Appendix 10 is essentially the paper by Ramm [110]. It is closely connected with Chapter IV. The result of Appendix 11 will appear in Ramm [112]. Appendix 10 is a self-contained introduction to the application of the theory of nonselfadjoint operators in diffraction. Some of the material of Chapter IV is included in this Appendix so that the reader can study Appendix 10 without going through the whole book. Appendix 11 is also written as a separate unit. It is closely connected with Appendix 10.

BIBLIOGRAPHY

Agmon, S. [1] Asymptotic formulas with remainder estimates for eigenvalues of elliptic operators, Arch. Rat. Mech. Anal., 28, (1968), 165-183.

[2] Lectures on Elliptic Boundary Value Problems, Van Nostrand, NY, 1965.

Agranovich, M. [1] Summability of series in root vectors of nonselfadjoint elliptic operators, Funct. Anal. and Appl., 10, (1976), 65-74.

Ahiezer, N. [1] Lectures on Approximation Theory, Ungar, N.Y., 1956.

Aizerman, M., Gantmaher, V. [1] Absolute Stability of Control Systems, Eds. Holden-Day, San Francisco, 1964.

Asakura, F. [1] On the Green function for $\Delta-\lambda^2$ with the third boundary condition in the exterior domain of a bounded obstacle, J. Math. Kyoto Univ., 18-3, (1978), 615-625.

Babich, V., Grigorjeva, N. [1] Asymptotic properties of solutions to some three dimensional wave problems, J. Sov. Math., 11, (1979), 372-412.

Baum, C. E. [1] Emerging technology for transient and broad-band analysis and synthesis of antennas and scatters, Proc. IEEE, 64, (1976), 1598-1616.

Baz, A., Zeldovich, Ja., Perelomov, A. [1] Scattering, Reactions, Decays in Nonrelativistic Quantum Mechanics, Nauka, Moscow, 1966 (in Russian).

Bethe, H. [1] Theory of diffraction by small holes. Phys. Rev., 66, (1944), 163-182.

Blaschke, W. [1] Disk and ball, Nauka, Moscow, 1967.

Blankenbecler, R., Sugar, R. [1] Phys. Rev., 136, (1964), 472-476.

Bogolubov, N., Mitropolsky, Ju. [1] Asymptotic Methods in Nonlinear Oscillation Theory, Gordon and Breach, N. Y., 1967.

Brézis, H. [1] Operateurs maximaux monotones et semi-groups de contractions dans les espaces de Hilbert, North-Holland, Amsterdam, 1973.

Brézis, H., Browder, F. [1] Existence theorems for nonlinear integral equations of Hammerstein type, Bull. Am. Math. Soc., 81, (1975), 73-78.

Browder, F. [1] Nonlinear operators and nonlinear equations of evolution in Banach spaces, Proc. Sympos. Pure. Math., vol. 18, part II, AMS, Providence, 1976.

Brown, W. [1] Dielectrics, Izd. Inostr. Lit., Moscow, 1961 (in Russian).

Buhgolz, G. [1] Calculation of Electrical and Magnetic Fields, Mir, Moscow, 1961.

Cesari, L. [1] Asymptotic Behavior and Stability Problems in Ordinary Differential Equations, Springer-Verlag, Heidelberg, 1971.

[2] Functional Analysis, Nonlinear Differential Equations and the Alternative Method, M. Dekker, N. Y., 1976.

Chadan, K., Sabatier, P. [1] Inverse Problems in Quantum Scattering Theory, Springer-Verlag, Berlin, 1977.

Debye, P. [1] Light scattering in solutions, J. Appl. Phys., 15, (1944), 338-349.

Decaurd, D., Foias, C., Pearcy, C. [1] Compact operators with root vectors that span, Proc. Am. Math. Soc., 76, (1979), 101-106.

Demidovich, B. [1] Lectures on Mathematical Theory of Stability, Nauka, Moscow, 1967 (in Russian).

Dunford, N., Schwartz, J. [1] Linear Operators, II, Wiley-Interscience, N. Y., 1963.

Dolezal, V. [1] Monotone Operators and Applications in Control and Network Theory, Elsevier, N.Y., (1979).

Dolph, C. [1] A saddle point characterization of the Schwinger stationary points in exterior scattering problems, SIAM Journ. Math. Anal., 5, (1957), 89-104.

[2] A current distribution for broadside arrays which optimizes the relation between beam width and side-lobe, Proc. IRE, 34, (1946), 335-348.

[3] The integral equation method in scattering theory, Problems in analysis, Princeton University Press, Princeton, 1970, 201-227.

Dolph, C., Ritt, R. [1] The Schwinger variational principles for one dimensional quantum scattering, Math Z., 65, (1956), 309-326.

Dolph, C., Scott, R. [1] Recent developments in the use of complex singularities in electromagnetic theory and elastic wave propagation. In Electromagnetic Scattering, Acad. Press, N. Y., (1978). 503-570.

Entch, R. [1] Uber Integralgleichungen mit positiven Kern, Journ. fur reine und angewandte Math., 141, (1912), 235-244.

Eskin, V. [1] Light Scattering in Polymer Solutions, Nauka, Moscow, 1973 (in Russian).

Fedorjuk, M. [1] Saddle Point Method, Nauka, Moscow, 1977.

Gahov, F., Cherskij, Ju. [1] Convolution Equations, Nauka, Moscow 1978 (in Russian).

Gajewski, H., Groger, K., Zacharias, K. [1] Nonlinear Operator Equations and Operator Differential Equations, Mir, Moscow, 1978.

Gerver, M. [1] An Inverse Problem for One-Dimensional Wave Equation, Nauka, Moscow, 1974 (in Russian).

Gohberg, I., Krein, M. [1] Systems of convolution integral equations on semiaxis, Uspehi Mat. Nauk, 13, (1958), 3-72.

[2] Introduction to the Theory of Linear Nonselfadjoint Operators, AMS, Providence, 1969.

[3] Theory and Applications of Volterra Operators in Hilbert Space, AMS, Providence, 1970.

Gohberg, I., Feldman, I. [1] Convolution Equation and Projective Methods of their Solutions, AMS, Providence, 1974.

Glaser, V., Martin, A., Grosse, H., Thirring, W. [1] A family of optimal conditions for the absence of bound states in a potential. In: Studies in Mathematical Physics: Essays in Honor of Valentine Bargmann, E. H. Lieb, B. Simon, and A. S. Wightman (Eds.), 169-194, Princeton Univ. Press, Princeton, 1976.

Glazman, I. [1] Direct Methods of Qualitative Spectral Analysis of Singular Differential Equations, Fizmatgiz, Moscow, 1963 (in Russian); Israel Scientific translation, 1965; Engl. transl., Davey, N. Y., 1965.

Grosse, H. [1] Bounds on scattering parameters, Acta Phys. Aust., 48, (1978), 215-228.

Gunter, N. [1] Potential Theory and its Applications in Mathematical Physics, GITTL, Moscow, 1953 (in Russian).

Hale, J. [1] Oscillations in Nonlinear Systems, McGraw Hill, N. Y., 1963.

Hellgren, G. [1] Questions of monoimpulse radiolocation theory, Foreign Radioelectronics, 12, (1962), 3-48.

Hönl, G., Maue, A., Westpfahl, K. [1] Theorie der Beugung, Springer-Verlag, Berlin, 1961.

Hörmander, L. [1] Linear Partial Differential Operators, Springer-Verlag, Berlin, 1963.

[2] The spectral function of an elliptic operator, Acta Math., 121, (1968), 193-218.

Howland, J. [1] Simple poles of operator-valued functions, J. Math. Anal. Appl., 6, (1971), 12-20.

Hsu, J., Meyer, A. [1] Modern Control Principles and Applications, McGraw-Hill, New York, 1969.

Hulst, Van de [1] Light Scattering by Small Particles, Mir, Moscow, 1961 (in Russian).

Il'in, V. [1] Existence of the reduced root system of a non-selfadjoint ordinary differential operator, Trudy Math. Inst. Steklova, 142, (1976), 148-155.

John, F. [1] Abhangigkeiten Zwischen den Flachenintegralen einer stetiger Function, Math. Ann., 111, (1935), 541-559.

Jossel, Ju., Kochanov, E., Strunskij, M. [1] Calculation of Electrical Capacity, Energija, Leningrad, 1969 (in Russian).

Kacnelson, V. [1] Conditions under which systems of eigen-vectors of some classes of operators form a basis, Funct Anal. and Applic., 1, (1967), 122-132.

Kantorovich, L., Akilov, G. [1] Functional Analysis, Macmillan, N. Y., 1964.

Kato, T. [1] Perturbation Theory for Linear Operators, Springer-Verlag, Berlin, 1966, MR 34# 3324.

Katznelson, J., Gould, L. [1] Construction of nonlinear filters and control systems, Information and Control, 5, (1962), 108-143.

Kiffe, T. [1] A Volterra equation with nonconvolution kernel, SIAM J. Math. Anal., 8, (1977), 138-149.

Kleinman, R. [1] The Rayleigh region, IEEE Proc., 53, (1965) 848-855.

Kolmogorov, A. [1] Interpolation and extrapolation of the stationary time series, Izvestija Acad. of Sci. USSR, ser. Math., 5, (1941),

Kontorovich, M. [1] Operational Calculus and Processes in Electrical Networks, Soviet Radio, Moscow, 1975.

Krasnoselskij, M. [1] Topological Methods in the Theory of Nonlinear Integral Equations, Macmillan, N. Y., 1964, (1956), MR 28# 2414.

[2] The Operator of Translation along the Trajectories of Differential Equations, AMS, Providence, 1968.

Krasnoselskij, M., Vainikko, G., Zabreiko, P., Rutickij, Ja., Stecenko, V. [1] Approximate Solution of Nonlinear Equations, Wolters-Noordhoff, Groningen, 1972 (1969), MR 27# 4271.

Krasnoselskij, M., Burd, V., Kolesov, Ju. [1] Nonlinear Almost-Periodic Oscillations, Nauka, Moscow, 1970 (in Russian).

Krasnoselskij, M., Zabreiko, P. [1] Geometrical Methods in Nonlinear Analysis, Nauka, Moscow, 1975 (in Russian).

Krein, M. G. [1] Integral equations on semiaxis with convolution kernel, Uspehi Mat. Nauk, 13, (1958), 3-120.

Krein, S. G. [1] Linear Differential Equations in Banach Space, AMS, Providence, 1972.

Krjanev, A. [1] Solution of unstable problems by iterative method, Dokl. Akad, Nauk USSR, 210, (1972), 20-22.

Kuhn, R. [1] Mikrowellenantennen, VEB Verlag, Berlin, 1964.

Landau, L., Lifschitz, E. [1] Electrodynamics of Continuous Media, Pergamon Press, N. Y., 1960.

Lax, P., Phillips, R. [1] Scattering Theory, Acad. Press, N. Y., 1967.

[2] A logarithmic bound on the location of the poles of the scattering matrix, Arch. Rat. Mech. Anal., 40, (1971), 268-280.

[3] Decaying modes for the wave equation in the exterior of an obstacle, Comm. Pure Appl. Math., 22, (1969), 737-787.

Lattes, R., Lions, J. [1] Methode de Quasi-Reversibilite et Applications, Dunod, Paris, 1969.

Lavrentjev, M., Vasiljev, V., Romanov, V. [1] Multidimensional Inverse Problems for Differential Equations, Nauka, Nobosibirsk, 1969 (in Russian).

Lefschetz, S. [1] Stability of Non-Linear Control Systems, Acad. Press, N. Y., 1965.

Levine, H., Schwinger, J. [1] On the theory of electromagnetic wave diffraction by an aperture in an infinite plane conducting screen, Comm. Pure Appl. Math., 3, (1950), 355.

Lifschitz, M. [1] Operators, Oscillations, Waves, AMS, Providence, 1973.

Lions, J. [1] Quelques methodes de resolution des problemes aux limites non lineares, Dunod, Gauthier-Villars, Paris, 1959, MR 41# 4326.

Londen, S. [1] On an integral equation in a Hilbert space, SIAM J. Math. Anal., 8, (1977), 950-969.

Majda, A., Taylor M. [1] The asymptotic behavior of the diffraction in classical scattering, Comm. Pure Appl. Math., 30, (1977), 639-669.

Malkin, I. [1] Theory of the Stability Motion, U.S. Atomic Energy Commission, 1958, 1959.

[2] Some Problems of Nonlinear Oscillation Theory, U.S. Atomic Energy Commission, 1958, 1959.

Marchenko, V., Hruslov, E. [1] Boundary Value Problems in Domains with Granular Boundary, Naukova Dumka, Kiev, 1974 (in Russian).

Marcus, A. [1] The root vector expansion of a weakly perturbed selfadjoint operator, Sov. Math. Doklady, 3, (1962), 104-108.

Middleton, D. [1] Introduction to Statistical Communication Theory, vol. 1,2, McGraw-Hill, N. Y., 1960.

Mihlin, S. [1] Variational Methods in Mathematical Physics, Macmillan, N. Y., 1964, MR 30# 2712.

[2] The Numerical Performance of Variational Methods, Wolters-Noordhoff, Groningen, 1971, MR 43# 4236.

Minkovich, B., Jakovlev, V. [1] Antenna Synthesis Theory, Soviet Radio, Moscow, 1969 (in Russian).

Morse, P., Feschback, M. [1] Methods of Theoretical Physics vols. 1, 2, McGraw-Hill, N. Y., 1953.

Mushelishvili, N. [1] Singular Integral Equations, Noordhof Int., Leyden, 1972.

Naimark, M. [1] Linear Differential Operators, Ungar, N. Y. 1967.

Newton, R. [1] Scattering of Waves and Particles, McGraw-Hill, N. Y., 1966.

Noble, B. [1] Wiener-Hopf Method for Solution of Partial Differential Equations, Pergamon Press, N. Y., 1958.

Odquist, K. [1] Uber die Randwertaufgaben der Hydrodynamik zaher Flussigkeiten, Math. Zeitschr., 32, (1930), 329-375.

Ortega, J., Rheinboldt, W. [1] Iterative Solution of Nonlinear Equations in Several Variables, Acad. Press, N. Y., 1970.

Parton, V., Perlin, P. [1] Integral Equations of Elasticity Theory, Nauka, Moscow, 1977 (in Russian).

Perov, A., Jurgelas Ju. [1] On convergence of an iterative process, Journ. Vycisl. Math. and Math. Phys., 17, (1977), 859-870.

Petkov, V. M. [1] High frequency asymptotics of the scattering amplitude for non-convex bodies, Comm. Part. Dif. Eq., 5, (1980), 293-329.

Petryshyn, W. [1] On the approximation-solvability of equations involving A-proper and pseudo-A-proper mappings, Bull. Am. Math. Soc., 81, (1975), 223-312.

Pisarenko, V., Rosanov, Ju. [1] On some problems for stationary processes which lead to integral equations of Wiener-Hopf type, Problemy Peredaci Informacii, 14, (1963), 113-135.

Pólya, G., Szegö, G. [1] Isopermetric Inequalities in Mathematical Physics, Princeton Univ. Press, Princeton, 1951.

Polak, E. [1] Computational Methods in Optimization, Acad. Press, N. Y., 1971.

Popov, M. M. [1] Diffraction losses of confocal resonator with mirrors of arbitrary shape, Dokl. Akad. Nauk USSR, 219, (1977), 63-65.

Popov, P. V. [1] Variational principles for spectrum of nonselfadjoint operators, Sov. Math. Doklady, 208, (1973).

Popov, V. M. [1] Hyperstability of Control Systems, Springer-Verlag, Berlin, 1973.

Ramm, A. G. [1] Filtering and extrapolation of some nonstationary random processes, Rad. Eng. Elect. Phys., 16 (1971), 68-75.

[2] Filtering of nonhomogeneous random fields, Optics and Spectroscopy, 27, (1969), 881-887; 26, (1969), 421-428, 808-812.

[3] A multidimensional integral equation with translation kernel, Diff. Eq., 7, (1971), 1683-1687, MR 44 #7235.

[4] On some class of integral equations, Diff. Eq., 9, (1973), 706-713, MR 49 #5749.

[5] Discrimation of the random fields in noises, Probl. of Inform. Transmission, 9, (1973), 192-205, MR 48 #13439.

[6] Investigation of a class of integral equations, Soviet Math. Dokl., 17, (1976), 1301-1305, MR 54 #3941, MR 55 #10106.

[7] Filtering and extrapolation of random fields and vectorial random processes in noises, Proc. of Intern. Confer., Prague, Sept. 1975, Acta Polytechn., Prace CVUT, III, (1975), 45-48.

[8] Filtering and extrapolation of random fields and vectorial random processes, Notices AMS, 22, (1975), A-708; 24, (1977), A-235.

[9] A new class of nonstationary random processes and fields and its applications, Proc. Tenth All-Union Sympos. on Methods of Representations and Analysis of Random Processes and Fields, Leningrad, vol. 3, (1978), 40-43.

[10] Eigenvalues of some integral operators, Diff. Eq., 14, (1978), 665-667.

[11] Investigation of a class of systems of integral equations, Proc. of Eighth Intern. Congress on Appl. Math., Weimar, (1978), 345-351.

[12] Approximate solution of some integral equations of the first kind, Diff. Eq., 11, (1975), 582-586; MR 51 #13613.

[13] Optimization of the resolving power of optical instruments, Optics and Spectroscopy, 42, (1977), 540-545 (with Rodionov).

[14] Apodization theory, Optics and Spectroscopy, 27, (1969), 508-514; 29, (1970), 290-394, 271.

[15] Resolution of optical systems, ibid., 29, (1970), 794-798; 422-424.

[16] Increase in the resolution of optical systems by means of apodization, ibid., 29, (1970), 594-599.

[17] On perturbations preserving asymptotic of spectrum, Atti. Acad. dei Lincei, (1978), 864, fasc 1, Jan. 1978.

[18] Solving an integral equation of potential theory by an iteration method, Sov. Phys. Dokl. 14, (1969), 425-527; MR 41 #9462.

[19] Approximate formulas for polarizability tensor and capacitance of bodies of an arbitrary shape, Sov. Phys. Doklady, 15, (1971), 1108-1111, MR 55 #1947.

[20] Calculation of the scattering amplitude for wave scattering from small bodies of an arbitrary shape, Radiofisika, 12, (1969), 1185-1197; MR 43 #7131.

[21] Approximate formulas for polarizability tensor and capacitance for bodies of an arbitrary shape, ibid., 14, (1971), 613-620; MR 47 #1386; PA 38695 (1974).

[22] Calculation of scattering amplitude of electromagnetic waves by small bodies of an arbitrary shape, ibid., 14, (1971), 1458-1460; EEA 4750 (1972).

[23] Calculation of the capacitance of a conductor placed in anisotropic in homogeneous dielectric, ibid., 15, (1972), 1268-1270; EEA 35522 (1972).

[24] Determination primary field from the characteristic of scattering by a small body, Radiotechnica i Electronica, 16, (1971), 679-681; EEA 27353 (1972).

[25] Iterative methods for solving some heat transfer problems, Engin. Phys. Journ., 20, (1971), 936-937.

[26] Electromagnetic wave scattering by small bodies of an arbitrary shape, Proc. of Fifth All-Union Sympos. on Wave Diffraction, Steklov Math. Inst. Akad. Nauk USSR, Leningrad, (1971), 176-186.

[27] Calculation of magnetization of thin films, Microelectronica, 6, (1971), 65-68 (with Frolov).

[28] Electromagnetic wave scattering by small bodies and related topics, Proc. Intern. Sympos. URSI, Nauka, Moscow, (1971), 536-540.

[29] Calculation of thermal fields by means of iterative processes, Proc. Fourth All-Union Conference on Heat and Mass Transfer, Mauka i Technika, Minsk, (1972), 133-137.

[30] Calculation of capacitance of a parallelepiped, Electricity, 5, (1972), 90-91 (with Golubkova and Usoskin), EEA 9403, (1973); PA 13670 (1973).

[31] On skin-effect theory, Journ. Techn. Phys., 42, (1972), 1316-1317, PA 15218 (1973).

[32] A remark on the theory of integral equations, Diff. Eq., 8, (1972), 1177-1180; MR 47 #2284.

[33] Scattering matrix for light scattered by a small particle of arbitrary shape, Optics and Spectroscopy, 37, (1974), 68-70, PA 27831 (1975).

[34] Scattering of waves by small particles, ibid., 43, (1977), 307-312.

[35] Scalar scattering by the set of small bodies of an arbitrary shape, Radiofisika, 17, (1974), 1062-1068.

[36] New methods of calculation of static and quasistatic electromagnetic waves, Proc. Fifth Intern. Sympos. "Radioelectronics 74," Sofia, vol. 3, (1974), 1-8 (report 12).

[37] Plane wave scattering by a system of small bodies of an arbitrary shape, Abstracts of Eighth Intern. Acoust. Congr., London, (1974), p. 573.

[38] Estimates of some functionals in quasistatic electrodynamics, Ukrain. Phys. Journal, 5, (1975), 534-543, PA 78 75406.

[39] On integral equation of the first kind, Notices AMS, 22, (1975), A-457; A class of integral equations, Math. Nachrichten, 92, (1979), 21-23.

[40] Two-sided estimates of the scattering amplitude at low energies, J. Math. Phys., 21, (1980), 308-310.

[41] A boundary value problem with a discontinuous boundary value, Diff. Eq., 13, (1976), 656-658, MR 54 #10830.

[42] An iterative method of solving integral equation for third boundary-value problem, ibid., 9, (1973), 1593-1596, MR 48 #6861.

[43] On estimates of thermal resistance ofor bodies of complex shape, Engin. Phys. Journ., 13, (1967), 914-920, PA 28398 (1968).

[44] On estimates of temperature field for bodies of complex shape. In Investigation of Nonstationary Heat and Mass Transfer, 64-70, Nauka; Technika, Minsk, (1966).

[45] Estimates of thermoresistances, Proc. Third All-Union Conference on Heat and Mass Transfer, Minsk, 1968, 12-17.

[46] Exsitence uniqueness and stability of the periodic regimes in some nonlinear networks, Proc. Third Intern. Sympos. on Network Theory, Split, Yugoslavia, Sept. 1975 623-628.

[47] An iterative process for the calculation of periodic and almost periodic oscillations in some nonlinear systems, Rad. Eng. Elect. Physics, 21, (1976), 137-190.

[48] A new method of calculation of the stationary regimes in some nonlinear networks, Proc. of Intern. Conference on Computer-Aided Design of Electron. and Microwave Systems, England, Hull, July 1977.

[49] Existence of periodic solutions of certain nonlinear problems, Diff. Eq., 14, (1977), 1186-1197.

[50] Stability of equation systems, ibid., 15, (1978), 1188-1193.

[51] Existence uniqueness and stability of solution to some nonlinear problems, Proc. Intern. Congress on Appl. Math., Weimar, (1978), 352-356.

[52] Existence, uniqueness and stability of solutions to some nonlinear problems, Applic. Analysis, (to appear).

[53] Stationary regimes in passive nonlinear networks, In Nonlinear Electromagnetics, Ed. P. L. E. Uslenghi, Acad. Press, N. Y. 1980.

[54] Calculation of the quasistationary states in quantum mechanics, Sov. Phys. Dokl., 17, (1972), 522-524, PA 10359 (1973).

[55] On exterior diffraction problems, Rad. Eng. Elect. Physics, 7, (1972), 1064-1067, MR 51 #4864; PA 42958 (1973); EEA 26877 (1973).

[56] Diffraction losses in open resonators, Optics and Spectroscopy, 40, (1976), 89-90, PA 71124 (1976).

[57] On simultaneous approximation of a function and its derivative by interpolation polynomial, Bull. Lond. Math. Soc., 9, (1977), 283-288.

[58] Approximation by entire functions, Notices AMS, 23, (1976), A-9; Izvestiya VUZ, Mathem. 22, (1978), 54-58.

[59] Finding of the shape of a domain from the scattering amplitude, Proc. Third All-Union Sympos. on Wave Diffraction, Moscow, Nauka, (1964), 143-144.

[60] Reconstruction of the shape of a domain from the scattering amplitude, Radiotechnika i Electronica, 11, (1965), 2068-2070.

[61] Signal restoration from its values in a discrete sequence of points, Rad. Eng. Elect. Physics, 11, (1965), 1664-1666.

[62] Antenna synthesis with prescribed pattern, 22nd All-Union Conference Dedicated the Day of Radio, Moscow, 1966, Antenna Section, 9-13.

[63] Statement and numerical solution of inverse ionospheric problem, ibid., Section on Wave Propagation, 3-6.

[64] Reconstruction of potential and domain from scattering amplitude, ibid., 7-10.

[65] On numerical differentiation, Mathematics, Izvestija Vusov, 11, (1968), 131-135.

[66] On equations of the first kind, Diff. Eq., 4, (1968), 1062-1064.

[67] On antenna synthesis theory, In Antennas, N5, (1969), 35-46, Ed. "Svjaz" Moscow.

[68] Nonlinear antenna synthesis problems, Sov. Phys. Dokl. 14, (1969), 532-535, MR 41 #4904.

[69] Optimal solution of antenna synthesis problem, ibid., 180, (1968), 1071-1074.

[70] Optimal solution of linear antenna synthesis problem, Radiofisika, 12, (1969), 1842-1848, MR 43 #8223.

[71] On nonlinear problems of antenna synthesis, Radiotechnika i Electronika, 15, (1970), 15-22.

[72] Nonlinear problem of plane antenna synthesis, ibid., 15, 500-503.

[73] Reconstruction of the shape of a reflecting body from the scattering amplitude, Radiofisika, 13, (1970), 727-732.

[74] Stable solution of an integral equation arising in applications, Notices AMS, 24, (1977), A-295; Comp. Rend. de l'Acad. Bulg. des Sci., 32, N6, (1979), 715-717.

[75] Existence in the large of solutions to systems of nonlinear differential equations, ibid., 24, (1977), A-17.

[76] Investigation of the scattering problem in some infinite domains with infinite boundaries, Vestnik Leningradskogo Univers., N7, (1963), 45-66; 19, (1963), 67-76; 13, (1964), 153-156; N1, (1966), 176.

[77] Spectral properties of the Schrödinger operator in domains with infinite boundaries, Sov. Math. Dokl., 4, (1963), 1303-1306, MR 31 #3297; Matem. Sborn., 66, (1965), 321-343, MR 30 #3297; ibid., 72, (1967), 638, MR 34 #7994.

[78] Analytic continuation of solutions of the Schrödinger equation in spectral parameter and behavior of solution to nonstationary problem as $t \to \infty$, Uspehi Mat. Nauk, 19, (1964), 192-194.

[79] Statement of diffraction problems in domains with infinite boundaries, Proc. Third All-Union Sympos. on Wave Diffraction, Nauka, Moscow, (1964), 28-31.

[80] Conditions under which the scattering matrix is analytic, Sov. Phys. Dokl., 9, (1965), 645-647, MR 32 #2049, PA 2397 (1965).

[81] Analytic continuation of the resolvent kernel of the Schrödinger operator in spectral parameter and limit amplitude principle in infinite domains, Dokl. Ak. Nauk. Azerb. SSR, 21, (1965), 3-7.

[82] On wave diffusion, Mathematics, Izvestija Vusov, N2, (1965), 136-138, MR 32 #1451.

[83] On a method of solving the Dirichlet problem in infinite domains, ibid., 5, (1965), 124-127, MR 32 #7933.

[84] Behavior of solution to a nonstationary problem as $t \to +\infty$, ibid., 1, (1966), 124-138, MR 33 #7674.

[85] Necessary and sufficient conditions for limit amplitude principle, ibid., N5, (1978), 96-102.

[86] Necessary and sufficient condition for limit amplitude principle, Sov. Math. Dokl., 163, (1965), 981-983, MR 33 #7673.

[87] Domains where resonances are absent in the three dimensional scattering problem, Sov. Phys. Dokl., 11, (1966), 114-116, MR 34 #3902, PA 25824 (1966).

[88] Spectrum of the Schrödinger operator with spin-orbit potential, Sov. Phys. Dokl., 11, (1967), 673-675, MR 34 #7993, PA 6289 (1967).

[89] Some theorems about equations with parameter in Banach space, Doklady Ak. Nauk Azerb. SSR, 22, (1966), 3-6, MR 33 #7963.

[90] Some theorems on analytic continuation of the Schrödinger operator resolvent kernel in spectral parameter, Izvestija Akad. Nauk Armjan. SSR, ser. Math., 3, (1968), 443-468, MR 42 #5563.

[91] Asymptotic behavior of eigenvalues and eigenfunction expansions for the Schrödinger operator with increasing potential in domains with infinite boundaries, ibid., 4, (1969), 393-399, MR 42 #3451.

[92] Green function study for differential equations of the second order in domains with infinite boundaries, Diff. Eq., 5, (1969), 1111-1116.

[93] Eigenfunction expansions for exterior boundary problem, ibid., 7, (1971), 565-569, MR 44 #2094.

[94] Eigenfunction expansion corresponding the discrete spectrum, Rad. Eng. Elect. Physics, 18, (1973), 364-369, MR 50 #1641, PA 23255 (1974), EGA 15274 (1974).

[95] Perturbation of network small vibration frequencies upon introduction of damping, Prikl. Math. and Mech., 33, (1969), 315-317.

[96] The exponential decay of a solution of a hyperbolic equation, Diff. Eq., 6, (1970), 1598-1599, MR 44 #631.

[97] On the limiting amplitude principle, ibid., 4, (1968), 370-373, MR 37 #1755.

[98] Some integral operators, ibid., 6, (1970), 1096-1106, MR 42 #8339.

[99] Eigenfunction expansion for nonselfadjoint Schrödinger operator, Sov. Physics, Doklady, 15, (1970), 231-233, MR 42 #703.

[100] Optimal harmonic synthesis of generalized Fourier series and integrals with randomly perturbed coefficients, Radiotechnika, 28, (1973), 44-49, EEA 13686 (1974).

[101] Investigation of some classes of integral equations and their applications. In Abel Inversion and Its Generalizations, N. G. Preobrazhensky, Ed., pp. 120-179, Acad. Sci. USSR, Siberian Dept., Inst. Theor. and Appl. Mech., Novosibirsk, 1978.

[102] Variational principles for spectrum of compact nonselfadjoint operators, J. Math. Anal. Appl., (to appear).

[103] Theoretical and practical aspects of singularity and eigenmode expansion methods, Tech. Rep., #2051, Math. Res. Center, University of Wisconsin, Madison, Wisc., March (1980). Also IEEE Trans. on Antennas and Propag., (1980).

[104] Linear filtering of some vector nonstationary random processes, Math. Nachrichten, 91, (1979), 269-280.

[105] On nonlinear equations with an unbounded operator, Ibid., 92, (1979), 13-20.

[106] Electromagnetic wave scattering by small bodies of an arbitrary shape, Proc. Intern. Symposium - Workshop, Columbus, Ohio, June 1979, Pergamon Press.

[107] Investigation of a class of nonlinear integral equations and calculation of passive nonlinear networks, Nonlinear Vibr. Problems, (to appear).

[108] Investigation of a class of systems of integral equations, Journ. of Math. Anal. and Appl., (to appear).

[109] Perturbations preserving asymptotic of spectrum of linear operators, Journ. of Math. Anal. and Appl., (to appear).

[110] Nonselfadjoint operators in diffraction and scattering, Math. Meth. in the Appl. Sci., 2, (1980).

[111] Analytical results in random fields filtering theory, Zeitschr. Angew. Math. Mech., (1980).

[112] On the basis property for the root vectors of some nonselfadjoint operators, J. Math. Anal. Appl., (to appear).

[113] Nonlinear equations without restrictions on growth of nonmonotone nonlinearity, Lett. Math. Phys.

[114] A variational principle for resonances, J. Math. Phys., (1980).

[115] Stable solutions of some ill-posed problems, Nav. Res. Lab. Report N4269, June (1980).

[116] A uniqueness theorem for a Dirichlet problem, Siberian Math. Journ., 10, (1978), 1421-1423.

Rayleigh, J. [1] Scientific Papers, Cambridge, 1922 (in particular papers from Phil. Mag., vols. 35, 41, 44).

[2] The Theory of Sound, Dover, N. Y., 1945.

Remes, E. [1] General Numerical Methods of Chebyshev Approximation, Izd. Ak. Nauk Ukr. SSR, Kiev, 1957.

Ronkin, L. [1] Introduction to the Theory of Entire Functions in Several Variables, Nauka, Moscow, 1971, (in Russian).

Rosenwasser, E. [1] Oscillations of Nonlinear Systems, Nauka, Moscow, 1969 (in Russian).

Schechter, M. [1] Modern Methods in Partial Differential Equations, McGraw-Hill, N. Y., 1977.

Seeley, R. [1] Refinement of the functional calculus of Calderon and Zygmund, Proc. Konik. Neder. Acad., Ser. A, 68, (1965), 521-531.

Shubin, M. [1] Pseudo-differential operators and the Spectral Theory, Moscow, Nauka, 1978 (in Russian).

Slepian, D. [1] Estimation of signal parameters in the presence of noise, Trans. IRE, v. PGII-3, (1954), 68.

Stevenson, A. [1] Solution of electromagnetic scattering problems as power series in the ratio (dimension of scatterer/wave length), J. Appl. Phys., 24, (1953), 1134-1142.

Tikhonov, A., Areenin, V. [1] Solutions of Ill-Posed Problems, Halsted Press, N. Y., 1977.

Tsyrlin, L. [1] On a method of solving of integral equations of the first kind in potential theory problems, J. Vycisl. Math. and Math. Phys., 9, (1969), 235-238.

Vainberg, B. [1] On eigenfunctions of an operator corresponding to the poles of the analytic continuation of the resolvent through the continuous spectrum, Math. USSR Sborn., 16, (1972), 307-322.

[2] On the analytic properties of the resolvent for a class of operator pencils, Math. USSR Sborn., 6, (1968), 241-272.

Vainberg, M. [1] Variational Methods and Monotone Operators Method, Wiley, N. Y., 1974.

Vainberg, M., Trenogin, V. [1] Bifurcation of Solution of Nonlinear Equations, Noordhoff Int., Leiden, 1974.

Van Trees, H. [1] Detection, Estimation and Modulation Theory, Wiley, N. Y., 1967.

Voitovich, N., Kacenelenbaum, B., Sivov, A. [1] Generalizated Method of Eigenoscillations in Diffraction Theory, Nauka, Moscow, 1977 (in Russian).

Wainstein, L. [1] Static problems for circular hollow cylinder of finite length, Journ. Tech. Phys., 32, (1962), 1165-1173; 37, (1967), 1181-1188.

[2] Open Resonators and Open Waveguides, Soviet Radio, Moscow 1966 (in Russian).

Wiener, N. [1] The Extrapolation, Interpolation, and Smoothing of Stationary Time Series, Wiley, N. Y., 1949.

Wiener, N., Hopf, E. [1] Uber eine Klasse singularer Integra Gleichungen, Sitzber der Berliner Akad. Wissensch., (1931), 696-706.

Wiener, N., Masani, P. [1] The prediction theory of multivariable stochastic processes, Acta Math., 98, (1957), 111-150; 99, (1958), 93-137.

Yaglom, A. [1] Introduction to the theory of stationary random functions, Uspehi Mat. Nauk, 7, (1952), 3-168; Trudy Mosc. Mat. Obšč., 4, (1955), 333-374.

Youla, D. [1] The solution of a homogeneous Wiener-Hopf int ral equation occurring in the expansion of second-order stationary random functions, Trans. IRE, v. IT-3, (1957), 187.

Zabreiko, P., Koshelev, A., Krasnoselskij, M., Mihlin, S., Rakovscik, L., Stecenko, V. [1] Integral Equations, Noordhoff Int., Leiden, 1975.

Zadeh, L., Ragazzini, J. [1] Optimum filters for the detection of signals in noise, Proc. IRE, 40, (1952), 1223.

Zuhovickij, S., Avdeeva, L. [1] Linear and Convex Programming, Nauka, Moscow, 1967 (in Russian).

LIST OF SYMBOLS

finite function = compactly supported function

iff = if and only if

$\equiv$ is equal by definition or identically

$\mathbb{R}^r$ - r-dimensional Euclidean space

D - domain in $\mathbb{R}^r$ with boundary $\partial D = \Gamma$

$\overline{D} = D + \Gamma$, $\Omega = \mathbb{R}^r \setminus D$

$H_t = H_t(D) = W_2^t(D)$ Sobolev space, $H_{-t}(D) = H_{-t}$ its dual space, $\overset{\circ}{W}{}_2^t$ - the closure of $C_0^\infty(D)$ in the norm of $W_2^t(D)$, p. 17

$L^p(D)$ - Lebesgue space

$C^k(D)$ - the space of k-times continuously differentiable functions with the usual norm

$C_0^\infty(D)$ - the set of finite in D functions from $C^\infty(D)$

$P(\lambda), Q(\lambda)$ - polynomials

$p = \deg P(\lambda)$, $q = \deg Q(\lambda)$

$J, I = (-\infty, \infty)$; $I_+ = (0, \infty)$

$\sigma(h)$ - order of singularity of a distribution h, p. 17

L - elliptic self-adjoint operator in $L^2(\mathbb{R}^r)$

$s = \operatorname{ord} L$

Λ - spectrum of L

$\Phi(x,y,\lambda)$, $d\rho(\lambda)$ - spectral kernel and spectral measure of L, p. 14

$1(t) = \begin{cases} 0, & t < 0 \\ 1, & t \geq 0 \end{cases}$

$\delta(t)$ - delta-function

$\delta_{ij} = \begin{cases} 1, & i = j \\ 0, & i \neq j \end{cases}$

$\mathscr{R}$ - class of kernels $R(x,y) = \int_\Lambda P(\lambda)Q^{-1}(\lambda)\Phi(x,y,\lambda)d\rho(\lambda)$

$\lambda_j(D)$ - eigenvalues of operator $R\phi = \int_D R(x,y)\phi(y)dy$

H - Hilbert space

X - Banach space

$\emptyset$ - empty set

$\{0\}$ set consisting of only zero

A - linear operator

$N(A) = \text{Ker } A$

$D(A) = \text{dom } A$

$R(A) = \text{im } A = \text{range } A$

$\lambda_n(A)$ - eigenvalues of A

$s_n(A)$ - singular values of A

L_n - n-dimensional linear subspace in H, $L_n^\perp$ its orthogonal complementation

H_A - Hilbert space with inner product

$[u,v] = (Au,v), \quad u \in D(A), \quad (Au,u) \geq m(u,u), \quad m > 0$

$\to$ strong convergence in H

$\rightharpoonup$ weak convergence in H

$\rho(f,L)$ - the distance between element f and subspace L

∂_N^j - derivative along normal N to the surface Γ

$[\partial_N^j F]|_\Gamma$ - jump of the derivative when crossing Γ along norma

$\sigma(A)$ - spectrum of linear operator A, p. 66

$\sigma_d(A)$ - discrete spectrum of A, p. 66

$\sigma_c(A)$ - continuous spectrum of A, p. 66

$\dotplus$ - direct sum, Π - direct complementation

Δ - Laplacian

∇ - gradient

Im, Re - imaginary, real

supp - support

ort - unit vector

emf - electromotive force

tr - trace

$A|_M$ - restriction of operator A to a subspace M

$\sigma_r(A)$ - rest spectrum of A, p. 66

$A[f,f]$ - quadratic form, p. 66

$\alpha_{ij}(\gamma)$ - polarizability tensor, p. 94

β_{ij} - magnetic polarizability tensor, p. 94

P_i - dipole electrical moment, p. 94

M_i - magnetic dipole moment, p. 95

$\langle u \rangle$ - statistical average, p. 110

f_E - scattering amplitude, p. 96

$\psi(t,s) = (\partial/\partial N_t)(1/r_{st})$, $r_{st} = |s-t|$

$A\sigma = \int_\Gamma (2\pi)^{-1}\psi(t,s)\sigma(s)ds$

V - volume of D

S - area of Γ

$Z(p)$ - impedance of a linear two-port, p. 160

$Y(p) = Z^{-1}(p)$ - admittance of the two-port, p. 160

B_2 - Besicovich space of almost periodic functions, p. 159

$u^{[j]}$ - p. 32

H_j - p. 32

S^2 - unit sphere in $\mathbb{R}^3$

$A_J = (A-A^*)/2i$ - imaginary part of linear bounded operator

$A_R = (A+A^*)/2$ - real part of linear bounded operator

W_D - p. 210

ℓ^2 - the Hilbert space of sequences $x = (x_1, x_2, \ldots)$ with $\|x\| = (\sum_{j=1}^{\infty} |x_j|^2)^{1/2}$, $(x,y) = \sum_{j=1}^{\infty} x_j y_j^*$

B(I) - the space of bounded measurable on I functions with th norm: $|u| = \sup_{x \in I} |u(x)|$

$\partial/\partial N_i$, $\partial/\partial N_e$ - interior and exterior normal derivative on Γ, p. 89

[,] - vector product

cl = closure, p. 66

For some of the symbols we do not give the page number becaus they are standard.

AUTHOR INDEX

ISBN 0-387-**90540**-5
ISBN 3-540-**90540**-5